T0406762

Society of Earth Scientists Series

Series editor

Satish C. Tripathi, Lucknow, India

For further volumes:
http://www.springer.com/series/8785

The Society of Earth Scientists Series aims to publish selected conference proceedings, monographs, edited topical books/text books by leading scientists and experts in the field of geophysics, geology, atmospheric and environmental science, meteorology and oceanography as Special Publications of The Society of Earth Scientists. The objective is to highlight recent multidisciplinary scientific research and to strengthen the scientific literature related to Earth Sciences. Quality scientific contributions from all across the Globe are invited for publication under this series.

Santosh Kumar · Rishi Narain Singh
Editors

Modelling of Magmatic and Allied Processes

Editors
Santosh Kumar
Department of Geology
Centre of Advanced Study
Kumaun University
Nainital
India

Rishi Narain Singh
CSIR-National Geophysical Research Institute
Hyderabad
India

ISSN 2194-9204 ISSN 2194-9212 (electronic)
ISBN 978-3-319-06470-3 ISBN 978-3-319-06471-0 (eBook)
DOI 10.1007/978-3-319-06471-0
Springer Cham Heidelberg New York Dordrecht London

Library of Congress Control Number: 2014940708

Printed on acid-free paper

Springer is part of Springer Science+Business Media (www.springer.com)

Preface

The igneous rocks present on the surface of the Earth are the imprints of the complex physico-chemical processes in the interior of the Earth. Formation of subsurface magma chambers and the evolutionary processes that govern the changes in the composition of magmas during its ascent and emplacement are important to understand the origin of various types of magmatic rocks. Magma mixing and mingling have now been recognized as major magmatic processes both in plutonic and volcanic environments. On the basis of study of textural, mineralogical, and chemical criteria, viable models of magma production and evolution can be proposed. Hydrothermal fluids generated by diverse crustal and mantle-related geological processes are found as significant ore-depositing agents. Oxidized granitoids are mostly responsible for the origin of metallic (copper and gold) sulfide deposits particularly of porphyry type. Thus, these processes are also important for assessing the mineral deposits of magmatic origin.

In order to understand these geological and geophysical processes, we invited Dr. Santosh Kumar, Professor and Head, Department of Geology, Kumaon University, Nainital, India, a noted researcher of igneous petrology and Dr. Rishi Narain Singh, INSA Scientist, CSIR-NGRI, Hyderabad, India who is known for his valuable contributions in the field of mathematical modelling of various geological and geophysical processes to edit a book on the *Modelling of Magmatic and Allied Processes*. Valuable contributions on the subject have been made by various active scientists. I sincerely thank the editors and the contributors for this important edition in The Society of Earth Scientists Series.

Satish C. Tripathi

Contents

Introduction

Modelling of magmatic and allied processes is essentially an integration of all information (field relation, petrography, mineralogy, geochemistry, and geophysics) of an igneous rock suit or province in order to simulate physical and chemical attributes of magmatic processes, which commonly operate in volcanic, subvolcanic, and plutonic environments. Factors responsible for magma differentiation (partial melting, magma generation, accumulation, transport, emplacement, crystallization, immiscibility, mingling, mixing, diffusion, etc.) can be parameterized and used further to test and validate several hypotheses, which are originally framed on the basis of intelligent field and petrographic practices. Matching (or reproduction) of calculated magma differentiation trends with the observed values is thus an essential component of petrogenetic (or geochemical) modelling of magmatic processes. During the last few decades, many improved methods and new techniques have been envisaged, which need to be learned in order to keep pace with the fast changing scenario in the science of modelling of the magmatic and allied processes. The chapters contributed by competent subject experts in this book will surely be useful to understand the physical and chemical processes involved in the magma generation and its evolution.

In “Magmatic Processes: Review of Some Concepts and Models”, **Santosh Kumar** has reviewed and described the methods and conceptual models of magmatic differentiation caused by crystallization and magma mixing processes. Apart from describing these two major processes, some subsidiary magmatic processes are also discussed. The validity of models has been explained in the light of field and textural observations. **R. N. Singh and A. Manglik** discuss the various physical and chemical parameters (P, T, fluid mass, and Xi), which primarily control the petrological properties of deep-seated rocks in space and time. They present derivations of widely used equations used to construct geotherms in continental and oceanic lithosphere degree and depth distribution of partial melt with depth, distribution of trace and radioactive elements with melting models and crustal evolution models. Inversion model of petrological data using these forward models will help to understand the nature of mantle source region and quantify the processes that operated at deeper (mantle) levels. Equations describing the mantle melting processes have also been derived. **K. Vijaya Kumar and K. Rathna** demonstrate lucidly the quantitative models of two most fundamental processes of magma differentiation, viz., mantle melting and fractional crystallization that essentially control elemental variations during the evolution of mantle-derived magmas, instead of a result of mantle heterogeneity. It has been further argued that fractional crystallization accompanied with assimilation (AFC) is capable of generating large trace element variation. Indian examples have been cited in support of these models. **R. N. Singh and A. Manglik**, in another contribution, explain the role of mantle cooling by thermal convection, driven by decaying radioactive elements, as a mechanism of large-scale geological and tectonic processes. Parameterized model of thermal convection is used to understand the evolution of mantle temperatures and geotherms of convecting mantle throughout the geological history. **R. N. Singh and J. Ganguly** review the different types of models of paleogeotherms for steady-state and transient conditions of continental lithosphere. Effect on style of mantle convection in the thermal structure of the lower part of the lithosphere is also reviewed. Based on heat-flow data and P-T stability of mantle xenolith hosted in kimberlites, steady-state paleogeotherm of Dharwar craton (Early Earth) is discussed. Several other transient thermal models, involving CO_2 advection and emplacement of thrust sheets, are also described. Calculations show that metamorphic field profile, defined by temperature maxima experienced by rocks after exhumation by erosion, is not affected by different style of thrusting phenomenon. **Igor Borska and Igor Petrík** offer a detailed insight into the significance and use of principal accessory phases while modelling typology and evolutionary processes of granitic magma. Principles and methods of studying the P-T evolution and nature of host magma type using morphometric analysis and compositions of accessory minerals are presented citing some examples from igneous terrains of the Western Carpathians. **Ewa Słaby and others** have envisaged new tools for 3D Concentration and Distribution Models (DC-DMs) and fractal statistics, which can be potentially used to describe the complex patterns of elemental distribution during crystal growth particularly in hybridizing magmatic system. In situ LA-ICP-MS analyzed elemental concentration of various domains of crystal can be used to simulate both DC-DMs and fractal dimension of evolving magma system. **N. C. Pant** describes principles, methods, and applications of two conventional microanalytical techniques: scanning electron microscopy (SEM) and electron probe micro analysis (EMPA), which are invariably used to characterize and to infer the operative processes in the evolution of magmatic rocks.

Some hydrothermal fluids are genetically linked and evolved from magmatic system at extreme condition of magma differentiation. The evolutionary aspects of hydrothermal system and its economic potential originated from felsic magma system are also covered in this book. **R. Sharma and P. Srivastava** thoroughly review and discuss the nature and evolution of hydrothermal fluids of magmatic origin which are trapped as fluid inclusions. Hydrothermal fluid processes responsible for the formation of economic deposits are also explained quoting some type examples of gold, porphyry type, and tungsten deposits in India. **S. Ishihara and A. Imai** describe the oxidized nature of granitic magma and its potential to generate porphyry type gold–copper deposits using chemistry of hydrous mafic silicates and apatite citing some examples from well-known deposits of the world. The key role of fluorine, chlorine, and sulfate as chief carrier of ore elements in hydrothermal system and evolving oxygenated environment of felsic melts during Precambrian-Phanerozoic are also briefly discussed.

In the last "Mass Balance Modelling of Magmatic Processes in *GCDKit*", **V. Janoušek and J.-F. Moyen** provide a brief introduction about the utility of a open-source software package Geochemical Data Toolkit (*GCDkit*; http://www.gcdkit.org) written in *R* language for handling, calculating, and plotting the geochemical data of igneous rock suite. They demonstrate *plugin* modules for numerical modelling of magmatic processes with special reference to forward and reverse mass-balance calculation of fractional crystallization. The recent version (2013) of *GCDkit* can be downloaded freely from the above website.

Although an attempt has been made to synthesize the scope and conceptual qualitative and quantitative methods and models of magmatic and allied processes in this book, there may still be a lot remaining to be covered entire aspects of modelling the magmatic processes.

Santosh Kumar and Rishi Narain Singh

Magmatic Processes: Review of Some Concepts and Models

Santosh Kumar

Abstract Magmas are commonly high-temperature, high-entropy silicate solutions of wide compositional range, and may crystallize to form a variety of igneous rocks viz. ultramafic, mafic, felsic to intermediate igneous rock types. Most igneous rocks in space and time may or may not be part of a co-magmatic suite but may have evolved by a number of major and subsidiary magmatic processes operating from source to sink regions. Two essentially important major processes reviewed and discussed herein, fractional differentiation and mixing of magmas, which may operate either separately or concurrently and are commonly responsible for the textural and chemical evolution of most igneous rocks. The methods and validity of qualitative and quantitative geochemical models of these processes are also described and evaluated in the light of field and textural observations.

1 Introduction

Natural magmas are polycomponent silicate melts from which mineral phases separate during crystallization according to their solubilities, primarily governed by chemical potentials of components as function of temperature, pressure and composition (e.g. Dickson 2000). The diversity in mineralogical and chemical compositions of igneous rocks suggests their origin and evolution from primary magmas by the process of *magmatic differentiation.* Commonly, homogeneous and heterogeneous modes of differentiation may be recognized by which magma may undergo compositional changes. In *homogeneous* differentiation, only magma as primary melt is itself involved whereas *heterogeneous* differentiation causes separation of either melts or crystals from magma (Muller and Saxena 1977).

S. Kumar (✉)
Department of Geology, Centre of Advanced Study,
Kumaun University, Nainital 263002, India
e-mail: skyadavan@yahoo.com

S. Kumar and R. N. Singh (eds.), *Modelling of Magmatic and Allied Processes*,
Society of Earth Scientists Series, DOI: 10.1007/978-3-319-06471-0_1,

Most geochemical models of magmatic differentiation assume formation of crystals homogeneously throughout the cooling magma chamber. However, crystallization may proceed along the walls and roof of a magma chamber causing boundary layer fractionation, and the products of such processes would be different from those formed by homogeneous crystallization that can be modeled numerically (Nielsen and DeLong 1992).

It is very difficult to recognize the nature of primary magmas because of uncertainty in the composition of the source region and the later processes acted upon them modifying their compositions significantly. Magmas may contain some residues derived from source regions and/or early fractionated crystals before they erupt on surface or emplaced in magma chamber at crustal depth. A number of theories and models on the accumulation and style of magma ascent and emplacement has been proposed (e.g. Clemens and Mawer 1992; Petfort et al. 2000; Dietl and Koyi 2011; Ferré et al. 2012), which are not discussed herein. Magmas as mixtures of melt plus crystals ascend to higher levels along fractures or as diapirs, and because of heat loss or buoyancy the outermost magma begins to crystallize and cease its upward movement (Paterson and Vernon 1995 and references therein). Several models have been proposed to explain the mechanism of magmatic differentiation which may occur at any scale (mm–kms) in a variety of tectono-magmatic environments. Differentiation of magmas may be accompanied by a number of subsidiary but significant processes, such as liquid immiscibility, thermo-gravitational diffusion, melt-melt interaction, crystal-liquid fractionation, crystal-charged magma mixing, and partial assimilation of wall rocks (Wyllie 1971; Wilson 1993). Such processes of magmatic differentiation, operating either separately or in combination, have commonly contributed in the evolution of magmas (Fig. 1). It is, therefore, almost impossible to postulate a unique genetic model of magma evolution due to the intricacies and complexities of the involved processes. Nonetheless, viable models of magma production and evolution can be proposed based on textural, mineralogical and chemical criteria. In most cases, two essentially important processes, fractional crystallization (separation of crystals from parental melts) and mixing (hybridization) of magmas, separately or concurrently, have been suggested responsible for the textural and chemical evolution of magma.

2 Liquid State Differentiation

An initially homogeneous magma may separate into two or more compositionally distinct magmas by the processes of *liquid immiscibility*. For example, many tholeiitic basalts contain two co-existing glass phases. There is credible laboratory evidence of liquid immiscibility between alkaline silicate liquid and carbonate-rich fluids, which forms a strong genetic link to understand the evolution of the carbonatite-ijolite–nephelinite rock association. Such unmixing or exsolution or magma splitting is restricted to magmas of evolved composition and is not a

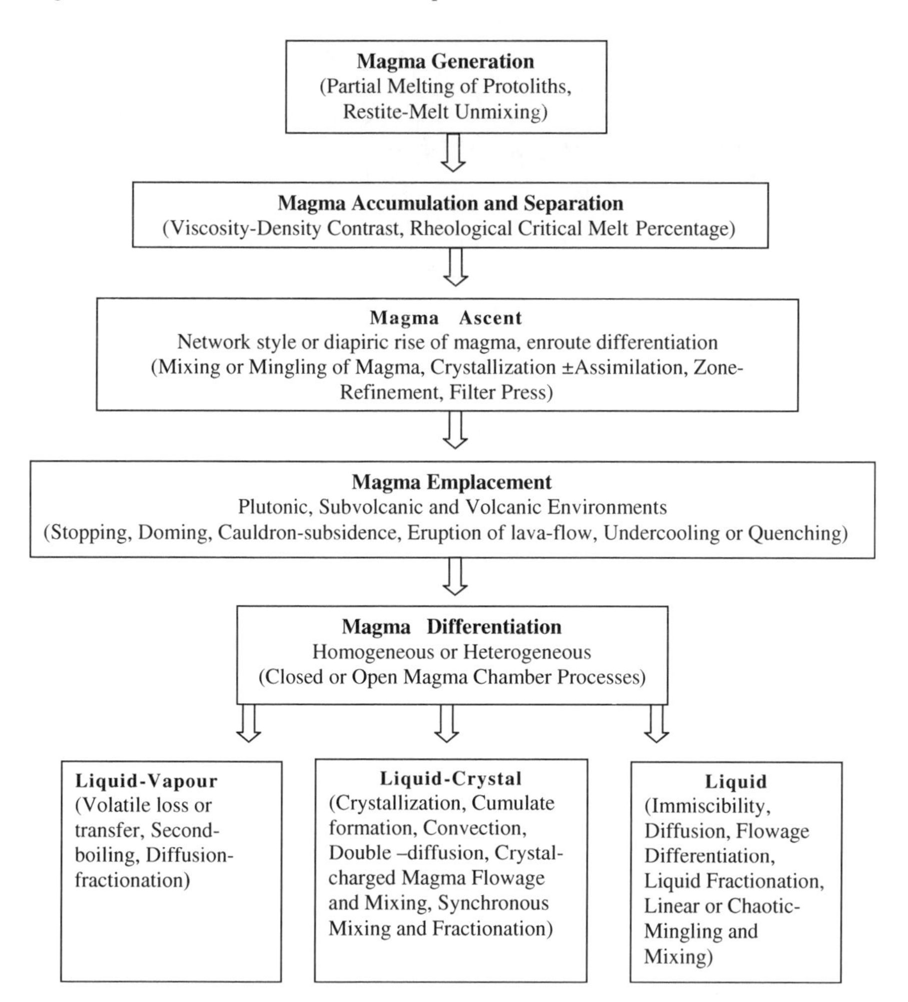

Fig. 1 Summary of major and subsidiary processes responsible of magma generation, emplacement and differentiation that occur in liquid or crystal-liquid or liquid-vapour conditions (based on Wyllie 1971; Muller and Saxena 1977; McBirney et al. 1985; Clemens and Mawar 1992; Wilson 1993; Kumar 2010)

significant process during differentiation of more primitive magmas (Wilson 1993). Compositional gradients in magmas may cause diffusion and redistribution of elements that are result of *Soret effect*. It may also occur in homogeneous, non-convecting magmas that are subjected to thermal gradients. Unlike the normal trend in fractional crystallization, the hotter parts of a magma chamber may be silica-rich whereas the colder regions may be of iron-rich mafic compositions. Thus, care must be taken when dealing with chilled-margin materials because gradational compositions of an igneous body might have been produced by the Soret effect. The diffusion of chemical species in silicate melts governs the kinetics of most magmatic processes including partial melting, fractional crystallization, magma mixing and crystal growth. Different components of a silicate melt might diffuse in different directions, depending upon diffusion coefficients, in the same temperature gradient. Perugini et al. (2006) observed that even at a micrometric length-scale, small volumes of magma can be strongly influenced by the coupled action of chemical diffusion and chaotic flow fields because of *diffusion fractionation*. Some regions of the co-existing magmas with contrasting temperature and compositions may partially to completely equilibrate chemically by the process of chemical diffusion depending upon resident time of liquid condition and diffusion coefficient of chemical species (e.g. Kumar and Rino 2006 and references therein).

It has been experimentally shown that, upon crystallization or melting at the walls of the shallow chamber, the liquid fraction may segregate to form compositionally distinct magmas that are still largely in liquid state (McBirney et al. 1985). This explains the occurrence of common types of volcanism and differentiation patterns observed in many shallow-level plutons.

3 Crystal-Liquid Separation and Associated Processes

Crystal fractionation is considered as the dominant process of magmatic differentiation, where an effective physical separation of phases, normally one liquid and the others crystalline, takes place. Conceptually crystallization of magma in a chamber forms a mixture of solids (phases) and residual liquids. Magma differentiation is dominantly driven by residual melt extraction from a partially crystalline magma chamber. Initially gravity settling of crystals was considered the most plausible mechanism of crystal accumulation on floor or wall of the magma chamber but on closer examination it is clear that other mechanisms such as in situ crystallization, flowage differentiation, diffusive exchange, compaction (filter-press) and convective fractionation in a crystallizing boundary layer may also be equally or partly effective to explain the process of crystal-melt separation at varying degrees. This is because many crystallizing magmas behave as *Bhingham* liquids, and thus even the dense ferromagnesian minerals may not be able to sink if they are unable to overcome the field strength of the magma. *Flowage differentiation* calls upon shear stresses in magma to help moving the crystals. *Convecting* magma can transport crystals in suspension to proximal or distal depositional sites forming typical cumulates such as schliers, mineral aggregates, clots or layers.

Fractional differentiation of magma may form a framework of touching crystals, commonly termed as cumulate, and need not essentially imply a process of crystal settling as discussed earlier. Cumulate terminology was indeed developed for describing the textural relations of the Skaergaard layered intrusion, which was formed by fractional crystallization of a single batch of tholeiitic parent (e.g. Wager et al. 1960; Wager and Brown 1968; Jackson 1967; Irvine 1982, 1987; Wadsworth 1985). *Primocryst* refers to unzoned, early-formed crystals in contact with each other in a magma which may be referred to as *cumulus crystal*. The most interior calcic part of cumulus plagioclase can also be termed the primocrystic part of the crystal according to definition given by Maaløe (1985). Cumulate nomenclature can even be used for all fractionated minerals which do not touch each other because of the presence of postcumulus or intercumulus materials (Fig. 2a, b). Postcumulus refers to events occurring after the development of initial cumulus fabric whereas intercumulus refers to the products of such postcumulus processes. After the formation of cumulus crystals, the intercumulus liquid may solidify in situ to produce postcumulus crystals, which are defined by the analogous primary porosity (e.g. Irvine 1982), residual porosity (e.g. Morse 1979) representing the amount of trapped liquid at complete solidification.

Mineral layering (rhythmic or cryptic) may be developed in cumulate rocks, primarily controlled by differentiation mechanisms such as crystal nucleation, resorption and coarsening phenomena, crystal sorting, in situ crystallization and current transport (Irvine 1982; McBirney 1995), other than density current, size of crystals and gravitational segregations. Layering may result from the differences in the rates of chemical and thermal diffusion, and one simplest process is double-diffusion which resembles the mechanism proposed for oscillatory crystallization and zoning of plagioclase (McBirney 1984).

Cumulates are more common in mafic magmas than the felsic magmas because of the viscosity contrast of residual liquids. In high viscosity felsic magma, convection in the magma chamber may inhibit cumulate-forming process (e.g. Ewart et al. 1975) or may result in a texture that is not a typically cumulative (e.g. Bachmann et al. 2007). However, crystal fractionation in felsic magma may result in the accumulation of crystalline materials at the margin of the magma chamber by the process of centripetal accretion similar to as observed in Tuolumne Intrusive Series (Fourcade and Allégre 1981) and Modra Massif (Cambel and Vilinovič 1987).

In *filter pressing,* a mat of crystals compacts under its own weight and expels less (or more) dense interstitial or residual melt. Another mechanism is *gas-driven* filter press in which gas-saturated residual melts from nearly solidified magma having a large amount of crystal-mush are driven out in a propagating fracture to form dykes or veins. Melt can also flow through a dense mat of groundmass crystals and can be driven by the differential pressures between small, recently-nucleated vesicles (higher pressure) and larger, early-formed vesicles (lower pressure). The small vesicles are formed because of crystallization of anhydrous groundmass minerals, resulting in the exsolution of gases, a process termed as *second boiling* (Sisson and Bacon 1999). Latent heat of crystallization may induce the effect of second boiling. Eichelberger et al. (2006) have demonstrated a mechanism of aplite dyke formation in a mostly crystallized felsic magma chamber which has gained sufficient strength to support the external anisotropic stress field. Melt pressure exceeds least principal stress ($P_m > \sigma_3$) when the strength of the crystal framework is also exceeded, resulting in propagation of a fracture as a melt dyke in plane perpendicular to the least principal stress (Fig. 3). Deering and Bachmann (2010) have recently suggested an upper limit of

Fig. 2 **a** Plagioclase cumulus suspended in high proportion of intercumulus clinopyroxene forming gabbro cumulate. *Crossed Polars*. Base of photo equals 4 mm. **b** Olivine and plagioclace cumulus phases with minor amount of intercumulus clinopyroxene forming cumulate. *Crossed Polars*. Base of photo equals 4 mm. Locality: Phenai Mata Igneous Complex, near River Heran, Chalamali village, Baroda, India

extraction at 50 % melt because removing more than 50 % of melt will exceed 75 % crystals in the residue, and consequently permeability will drop which will hinder melt extraction severely.

In situ crystallization is commonly evident in mineral assemblages that include zoned crystals as an extended sequence of crystallization, particularly true in the case of a small and relatively rapidly cooled igneous body like Skaergaard intrusion. In *fractional crystallization* (Rayleigh distillation law) equilibrium is assumed only between the surface of the crystallizing phases and the melt, and crystallized minerals are assumed to become isolated from the residual melt and accumulate on the floor or walls of the magma chamber. The liquid path (LLD: liquid line of descent) for fractional crystallization is almost identical to that for equilibrium crystallization but the crystal compositional path is quite different (e.g. Ragland 1989). Magma generated in diverse tectonic settings, such as dry-reduced magmas in hot-spot divergent margins or in wetter and more oxidizing in arc environments will evolve along different LLD, thus leading to different trace element evolution (Deering and Bachmann 2010). Geochemical modelling of cumulus and associated processes is discussed in "Geochemical Modelling of Melting and Cumulus Processes: A Theoretical Approach".

3.1 Assimilation of Solid Rocks

Assimilation of crustal rocks (deeper lithology and/or country-rocks) could be an important process in determining the compositional diversification of magmas during its ascent and emplacement, particularly for deep-crustal magma reservoirs.

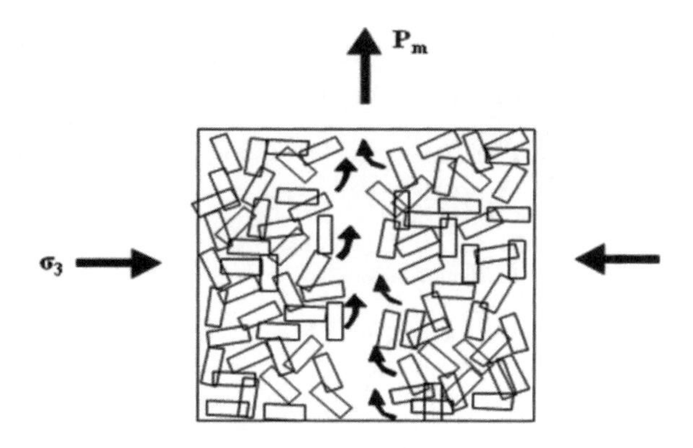

Fig. 3 Mechanism of aplite (melt) dyke formation at a late stage of felsic magma chamber evolution when pressure of residual pore fluid (P_m) exceeds least principal stress (σ_3) (after Eichelberger et al. 2006). See text for explanation

Assimilation coupled with fractional crystallization (AFC) can be an important process in the evolution of much continental magmas. Assimilation of low-density crust and synchronous fractionation increases buoyancy required for ascent, and passages for ascent are created by removal or stopping of overlying crustal rocks into the magma chamber. Bulk assimilation of magma thus represents summation of interaction with components from all levels of crust traversed by magma (e.g. Beard et al. 2005). Conventional FC models assume that crystals are removed instantaneously from the magma as soon as they are produced. However, recent studies suggested that the crystals are suspended within the magma body for a certain period, affecting the whole-rock composition in response to intra-grain isotopic zoning, which enabled to develop a mass-balance model for *assimilation and imperfect fractional crystallization* (AIFC) responsible for magma evolution (Nishimura 2012).

3.2 Replenishment of Magma Chamber

Most magma chambers are episodically replenished by new pulses of magma, periodically tapped and continuously fractionated (Wilson 1993). In a magmatic system undergoing paired recharge and fractionation, the LLD for the major elements is similar to that produced by fractional crystallization. For example, in a simple ternary system crystallizing ol + cpx + pl (Fig. 4), adding a pulse of more primitive magma will push it back into the olivine phase field from where it will evolve back towards the ol + cpx cotectic. In such a delayed long-term situation the amount of ol + cpx fractionated from the system will be higher than those in a closed system.

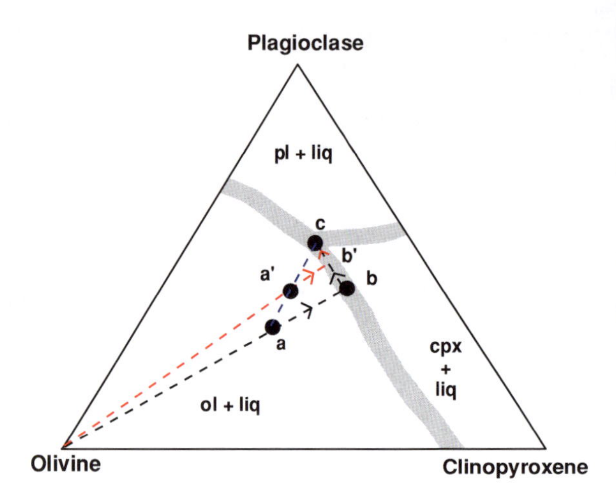

Fig. 4 Magma replenishment viewed in the context of simplified ternary system (after Wilson 1993). Magma '*a*' in a chamber crystallizing the assemblage ol + pl + cpx would lie at the ternary eutectic '*c*'. Recharging the chamber with new pulse of magma '*a*', followed by complete mixing, would generate a new magma composition a' which would evolve back towards the eutectic '*c*' along the LLD a'-b'-c

4 Magma Mingling and Mixing

Magma mingling and mixing have been recognized as major magmatic processes both in plutonic and volcanic environments. Interaction between coeval mafic and felsic magmas has been viewed as a prime cause of thermal rejuvenation as a result of mafic magma underplating and volatile supply to crustal-derived felsic melts. Water content, crystal size, degree of initial crystallinity, mass fraction, composition and temperature are important parameters determining the rheology of interacting mafic and felsic magmas (Frost and Mahood 1987). Low viscosity, minimal rheological differences, and thermal equilibrium between coeval mafic and felsic magma will produce convective overturn forming a hybrid magma zone (Huppert et al. 1984). After thermal equilibrium the interacting magma system is open for chemical and mechanical (already crystallized mineral) exchanges. Crystals may reveal a complex record of open system processes during magma ascent from deeper to shallower levels in the conduit, in response to decompression crystallization and H_2O degassing (Humphreys et al. 2006).

Enclaves represent all kind of *lithic materials* enclosed within granitoids (Didier 1973; Vernon 1983). Enclaves can be broadly classified into several types depending upon their relationships with the enclosing granitoids. (1) *Xenoliths* are partly digested or undigested fragments of country rock or of an enroute deeper-derived lithology. Country rock xenoliths should be confined to the margin of the pluton; (2) *Cognate* (autolith) enclaves must have cogenetic affiliation with felsic host, as early-crystallized phases forming cumulates *or* segregation of mafic phases *or* early-formed mafic border facies of felsic magma itself; (3) *Restite* represents the refractory residue left after partial melting; (4) Mafic or mafic–felsic hybridized magma globules, commonly referred to as *microgranular enclaves*, undercooled and mingled into relatively cooler, partly crystalline felsic melt;

(5) Synplutonic mafic dyke may intrude and disrupt at waning stage of felsic magma evolution, and take the form of the angular to subangular enclaves.

Enclaves must bear some relevant field, petrographic, mineralogical, geochemical and isotopic signatures relevant to operative magmatic processes. One of the most common indicators of magma mixing and mingling is the occurrence of *micro-granular enclaves* (ME) of contrasting compositions with respect to felsic host, mostly forming the calc-alkaline igneous complexes. Therefore, the ME in granitoids serve as a potential tool to understand the processes of coeval mafic and felsic magma interaction in the plutonic environment. Likely processes of mingling and hybridization resulting from mafic magma injection into a felsic magma at various stages of its crystallization are shown in Fig. 5. The ME generally form rounded to elongated shapes on two-dimensional surfaces when mafic magma interacts with felsic magma at its initial or intermediate (partly crystalline) stages of evolution. Synplutonic mafic dykes may inject into most crystallized felsic magma and may disrupt to form angular to subangular (brecciated) enclaves. Thus the mafic–felsic magma interacting system may truly represent a MASLI (composite mafic–silicic intrusive) system (Wiebe 1994). The ME may be aligned in the direction of magma mingling and flow and the degree of elongation correlates with the intensity of flow foliation in the host granitoids, suggesting that both ME (semi-solidified) and granitoids were deformed during semi-crystalline magmatic flow conditions (Vernon et al. 1988).

It is important to distinguish between magma mingling (or co-mingling) and magma mixing processes. Magma mixing causes homogenization of interacting melt phases and the conversion of early crystals to partly dissolved (corroded) forms in a new hybrid magma, whereas mingling or co-mingling involves partial mixing or interpenetration of felsic-mafic magmas without pervasive changes (Kumar et al. 2004). The occurrence of mafic enclaves inside the ME (i.e. composite enclaves, Kumar 2010), abundance of features like magmatic flowage, mafic/felsic xenocrysts, acicular apatites, and pillow-like shapes of the ME etc. indicate co-existence and mixing of magmas with contrasting compositions and temperatures. Other features supporting magma mixing are (1) a more-or-less rounded shape of ME with occasional crenulated chilled margin (2) an intermediate composition between the composition of the felsic host and mafic end-members (3) rounded crystals through partial dissolution and coated by another mineral in equilibrium with host granitoids giving rise to rapakivi-like texture, and (4) presence of quartz ocelli (Fig. 6a). Quenched enclaves may represent the composition of pristine mafic magma or hybrid (intermediate) magma. In some large undercooled ME ($d > 12$ cm) residual (rhyolitic) melt may have been driven out of enclaves into the partly crystalline host magma (Kumar et al. 2004).

The ME range in size from 1 cm to several meters across and their contacts are commonly sharp with the host granitoids but diffused contacts are also noted because of quenching of mafic (enclave) magma against cooler and semi-crystalline granitoid melt (Fig. 6b), which implies that felsic and mafic magmas coexisted. The ME with serrate or cuspate margins, with lobes convex towards the host granitoids, may also be observed (e.g. Vernon 1983). Near the contact of host granitoids the ME generally show mineral alignment along the contact outline.

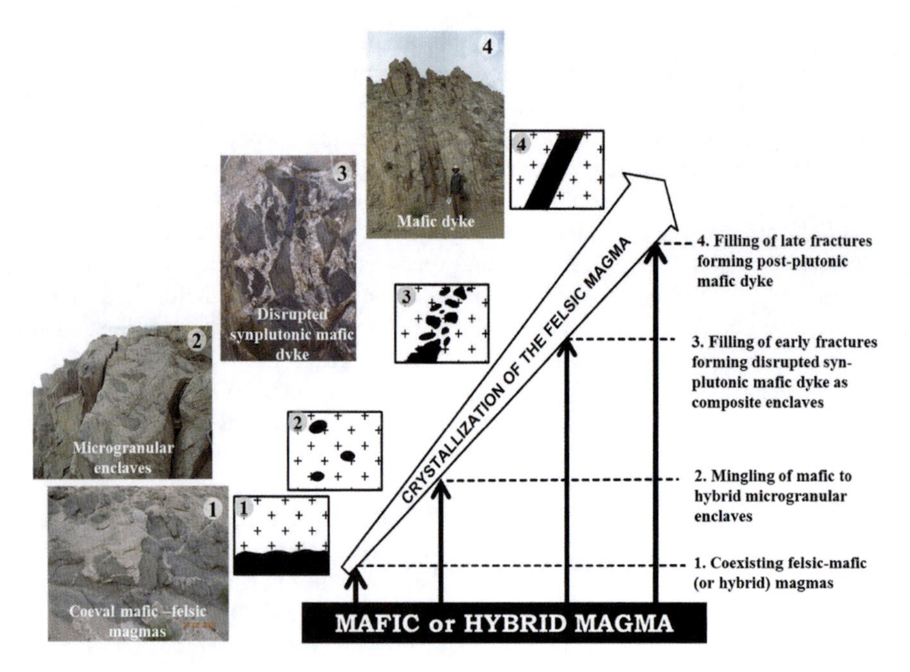

Fig. 5 Schematic presentation of various types of magma interactions resulting from injection of mafic or hybrid magma into felsic magma at different stages of its crystallization (slightly modified after Barbarin 1989, Barbarin and Didier 1992). Field features shown against various stages (*1–4*) of mafic–felsic interacting magma system can be observed in the central and eastern parts of the Ladakh batholith (Kumar 2010). See text for explanation

Felsic and mafic grains from the granitoids may penetrate partly into the ME particularly where semi-solidified ME interacts with partly crystalline felsic host magma, suggesting crystal-charged magma interactions. On the contrary, country rock xenoliths may show a reaction signature with the host granitoids suggesting solid-liquid interaction.

At any stage of mafic–felsic magma mingling and mixing, the magma system may be frozen (solidified). Magma interaction is indeed a chaotic process during which a portion of mafic magma can survive mixing process (Perugini et al. 2003). In the same system, mixing process may be characterized by chaotic regions in which intense hybridized regions represent *active mixing region* (AMR) whereas less efficient mixing dynamic regions are called *isolated mixing regions* (IMR), which may contain blobs of magmatic enclaves. More recently, Perugini and Poli (2012) reviewed the intricacies of magma interaction processes both in plutonic and volcanic environments and suggested that time spent by the magmatic system in the molten or partial molten state is a crucial factor for the preservation of magma mixing fingerprints. They further argued that the new conceptual models of *chaotic mixing* and *diffusion fractionation* may pose serious problems for the

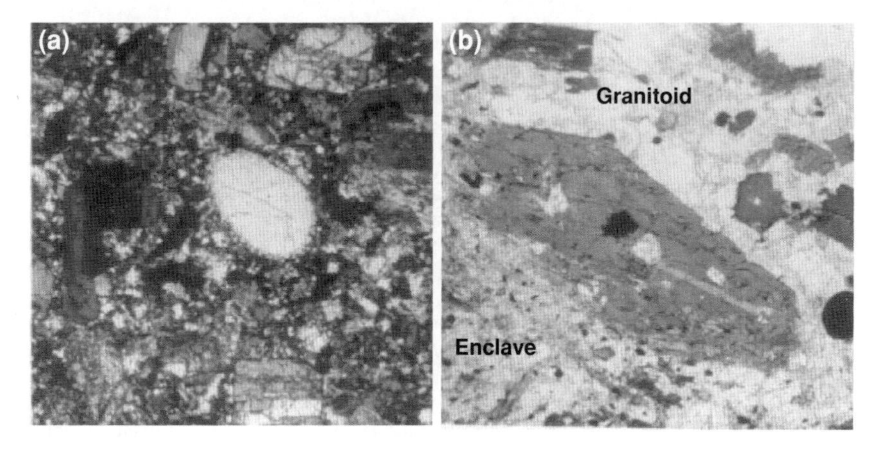

Fig. 6 **a** Ocellar quartz in quartz diorite, which formed in a magma mixing environment. *Crossed Polars*. Base of photo equals 4 mm. **b** Enclave and granitoid contact showing distinct chilled margins imprinted on lower crystal faces of amphibole which suggests quenching of enclave magma against relatively cooler and partly crystalline granitoid melt. *Crossed Polars*. Base of photo equals 4 mm. Locality: Hodruša Štiavnica Intrusive Complex, Central Slovakia (after Kumar 1995)

interpretation of compositional variability of igneous rocks if rely only on pre-existing conventional models of magma mixing. Magmas affected by coupled action of chemical diffusion and chaotic field flow, and trapped as melt inclusions will provide misleading information about melt composition (Perugini et al. 2006).

5 Qualitative Assessment of Magmatic Processes

5.1 Geochemical Variations

Magmas are in a viscous state due to entrained restite commonly derived from the source region and the presence of early-formed crystals (McBirney 1993). Crystal-charged magmatic processes are therefore considered most important in the evolution of igneous bodies whereas processes occurring in the liquid state are not so significant in the bulk evolution of igneous magmas. In the classical work of Harker (1909) it was realized that the great diversity and compositional variations within many igneous rock bodies can be attributed to differentiation processes. Advancement in theories and ideas of magmatic differentiation postulated that the geochemical variations of any igneous rock suite alone cannot point to the operative processes. For example, near-linear variations on Harker plots (Fig. 7) can be caused by several possible processes such as fractional crystallization (or crystal fractionation), mixing of magma end-members and melt-restite separation during progressive partial melting (Wall et al. 1987; Clemens 1989). It has been further

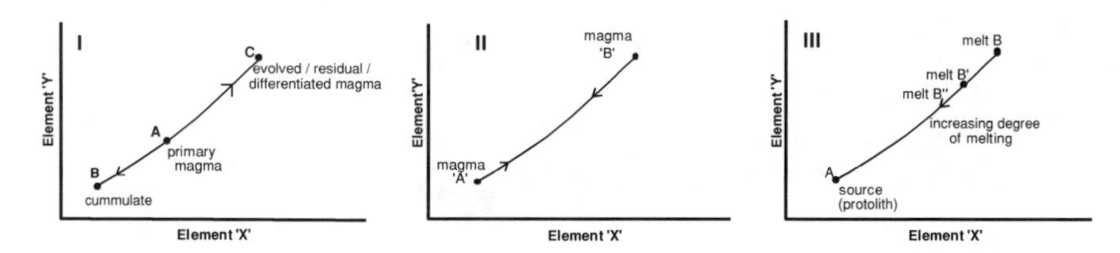

Fig. 7 Linear geochemical variation of igneous rocks can be viewed in three major ways. I: Crystal fractionation of a primary magma (*A*), forming cumulates (*B*) and subsequently evolved residual or differentiated magma (*C*). Element '*X*' (mostly SiO_2) or any other parameter can be chosen as "*index of differentiation*". II: The same geochemical variation can be generated by the mixing of two magma end-members '*A*' and '*B*' in various proportions, forming members of a hybridized (mixed) igneous rock suite. III: The same geochemical variation can be formed by the processes of melt-restite unmixing or separation during progressive melting of a source region (protolith '*A*'), which will form a small melt fraction initially at '*B*' and then gradually follows the compositional path of B-B'-B'' with increasing degree of melting

argued that typical *Rayleigh fractionation* (i.e. disequilibrium crystallization) will result in a curved linear trend on Harker plots whereas *phenocryst unmixing* (term used to the explain the process in which degree of fractional or equilibrium crystallization occurs in a cooling magma followed by separation of compositionally fixed *phenocryst* assemblage from residual melt before final crystallization) would generate a linear trend on Harker plots (Clemens and Stevens 2012). It has also been pointed out that how the data are plotted, e.g. expansion of the SiO_2 axis on Harker plots, can significantly linearise a curvilinear data-set.

Many igneous rocks preserve physical, textural and chemical characteristics that also point to differentiation processes involved in their formation and evolution. Chemical features of crustally-derived granitic rocks, in particular, are highly influenced by the presence of restitic materials (melted or unmelted parts of protoliths), accessory minerals (Fe-Ti oxides, titanite, zircon, monazite, apatite etc. as poikilitic inclusions in early crystallized phases) and peritectic phases (newly-formed crystalline products of the melting reaction). The elements with low solubilities in granitic melts with varying proportions of *peritectic assemblage entrainment* (PAE) accompanied by co-entrainment of accessory phases, are considered mainly responsible for the chemical variations of granitic rocks whereas concentration of elements with high solubilities in felsic melts may simply reflect protolith compositions (Clemens and Stevens 2012). For the identification of magmatic processes, chemical variations of igneous rocks should therefore be combined with field, textural, mineral and other physical observations.

Conventionally, chemical elements in magma are present in an ionic state or have polymer structures that form silicate chains during crystallization (Masson 1965). Recently, Vigneresse et al. (2011) introduced a new and advanced concept of hard-soft acid-base (HSAB) interaction to characterize a magma that consists either of solid, melt or an exsolved gaseous phase. This concept of HSAB has offered a new insight into the way magma differentiates. For example, felsic magma commonly follows a trend towards higher hardness driven by increasing silica content whereas almost constant hardness of mafic/ultramafic magmas are driven by ferro-magnesian minerals with nearly equal hardness values, and hence these evolve towards minimum electrophilicity. This bimodal chemical evolution of magmas satisfactorily explains the natural occurrence of most common igneous differentiation trends; one tholeiitic trend effectively determined by iron enrichment and another evolving towards silica enrichment with little or no iron enrichment, as commonly observed in calc-alkaline series, corresponding to Fenners' and Bowens' trends respectively (Keelmen 1990 and references therein).

5.2 Pearce Element Ratios

Pearce element ratios (PER) are eminently suited to test the internal petrological hypothesis concerning the mechanism of magmatic differentiation for basaltic magmas (Pearce 1968, 1970, 1990; Russell and Nicholls 1990). A conserved (or excluded) element during magmatic differentiation can be defined as one element that is *neither* added to the system by assimilation *nor* removed by fractionation (bulk distribution coefficient, $D = 0$). The *conserved* element is common to both axes (orthogonal axes) as denominator. Commonly a single conserved element is used as denominator but functions of more than one conserved element can also be used. Molar mineral components taking part in melting or crystallization can be calculated for any set of composition (Russell and Nicholls 1988; Stanley and Russell 1989; Pearce 1990). The use of PER diagrams in petrology has received strong criticism due to difficulties disentangling any geological effect from the spurious correlation effect (e.g. Rollinson and Roberts 1986). Pearce (1987) convincingly explained that (i) the divisor must be constant and the trend of variation should not pass through the origin, and (ii) the slope and intercept of an observed or simulated trend are much more important than the correlation between the ratios. Nicholls (1988) further argued that PER diagrams provide an unambiguous test of petrologic hypothesis because they are based on the stoichiometry of rock-forming minerals. The PER have been successfully applied to identifying magmatic processes of volcano-plutonic mafic igneous complexes (e.g. Trupia and Nicholls 1996; Kumar 2003).

5.3 Synchronous Mixing-Fractionation Trends

Mixing of two coeval liquids may define a straight line on variation diagrams as long as the liquids are not concurrently fractionating. However concurrent mixing-fractionation processes of coexisting magmas may provide complex chemical evolutionary trends. Mafic to hybrid enclaves in granitoids may represent various stages of interactions between mafic and felsic magmas (e.g. Barbarin and Didier 1992; Janoušek et al. 2004; Barbarin 2005; Słaby and Martin 2008; Bora et al. 2013).

The ME in granitoids causes physical and chemical complexities in magma chambers and as a whole will largely affect the evolutionary history of the granitic pluton. Chemical variations of compatible elements against silica play significant role in recognizing the operative processes involved in the temporal and chemical evolution of mafic, hybridized and granitoid rocks as schematically presented in Fig. 8. Magma with the lowest silica content and enriched in compatible elements may represent a mafic end-member, which can evolve along a trend of decreasing compatible elements and concurrently mixed with fractionating felsic melt forming a series of hybrid rocks. These combined processes are regarded as mixing-fractionation of coeval mafic and felsic magmas during syn-crystallization with increasing polymerization and crystal loads (e.g. Barbarin 2005; Słaby and Martin 2008; Bora et al. 2013). However, non-colinearity or high data-scatter of elements would have been caused by non-linear (chaotic) mixing and diffusive fractionation processes (e.g. Perugini et al. 2008; Słaby et al. 2011).

6 Semi-Quantitative Assessment of Crystal-Fractionation

Geochemical variation diagrams are not considered to have much potential for examining the petrological hypothesis precisely, unless combined with other geological and petrographical evidence. Compositional variations in many igneous rock sequences may show good coherence, suggesting fractional crystallization has played a dominant role but this must be tested against petrographic or field criteria. Often a hypothesis emerges first from a consideration of field and petrographic observations, and is then tested against chemical dataset. In the formulation of a crystal fractionation hypothesis, the constructed chemical variations should be capable of showing both the liquid and fractionating minerals, which is possible in a two-element variation diagram, commonly referred to as a *mixing* calculation (Cox et al. 1979), similar to as Harker plot with additional consideration of solid compositions. Diagrams constructed based on the principles of mixing calculation are capable of explaining *addition* or *subtraction* (or 'extract') of phases but do not imply a specific mechanism. A basic principle of mixing calculation lies in the *lever rule* as commonly used in the phase diagrams. Two chemical parameters X and Y may represent percentages of oxides or parts per million of trace elements or any other weight expression of analytical data (Fig. 9). Addition of 'B' composition to 'A', the resulted mixture M will evolve in a straight line A-B, depending upon the relative proportions of 'A' and 'B' in the mixture 'M' (Fig. 9a). Similarly extraction of 'A' from a parent 'M' will evolve residual liquid towards 'B', where A-M-B is a lever with point 'M' at the fulcrum. At any specific point of mixture such as 'M' the proportion of two end-members 'A' and 'B' can be calculated. Bulk extract E formed by crystallization of two phases A and B from parent P will evolve the residual liquid towards D (Fig. 9b).

Given a known liquid path (LLD from parent P to daughter D liquid represented by whole rock compositions) caused by fractionation of multiple phases, it may be

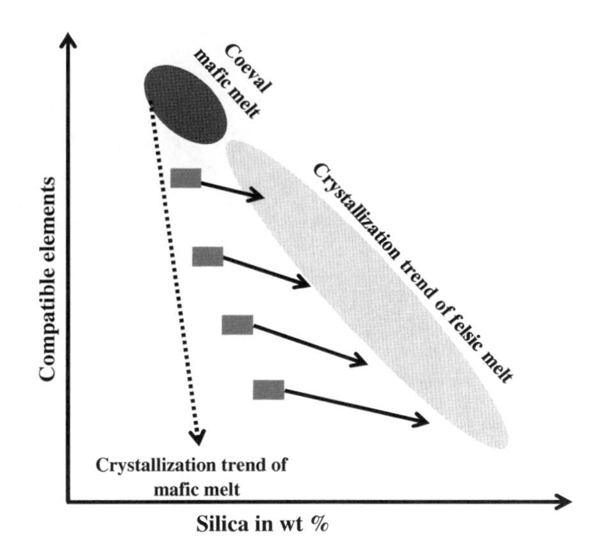

Fig. 8 Schematic presentation of mixing-fractionation trends of coeval mafic (*black small ellipse*) and felsic (*grey large ellipse*) magmas. Broken arrow represents fractionation path of mafic magma whereas *solid arrows* with *solid rectangles* represent various stages of mixing trends synchronous with progressive fractionations of both magmas forming series of hybrid rocks (Based on Słaby and Martin 2008; Słaby et al. 2011; Bora et al. 2013, Ewa Słaby, pers. comm.)

possible to construct an extract polygon (triangle A-B-C) geometrically using compositions of each fractionating mineral. No unique solution for bulk extract E can be determined alone from this diagram (Fig. 9c) because of the fact that the intersection of a line and a triangle in the same plane is a line. However, with the help of several pairs of such two-element extract diagrams, the proportion of extract phases can be determined by transformation of values obtained by intersection of non-equilateral chemical polygons with LLD vectors onto a equilateral triangle formed by fractionating three-phases, which will provide proportions of fractionating phases from a parental magma, and subsequently the bulk cumulate composition. The presence of inter-cumulus trapped liquid may, however, cause a mismatch of a typical cumulate composition.

All LLD may not essentially be a single straight line. There may be a break or point of inflection in the slope of the LLD because of increasing number of phases in the crystallizing assemblages somewhat analogous to crystallization of magma in primary phase field A, at cotectic A-B, and at eutectic A-B-C in a ternary phase diagram (e.g. Ragland 1989; Wilson 1989,1993).

7 Quantitative Assessment of Crystal Fractionation

As the member of phases participating in the fractionation increases, graphical methods of fractionation become more difficult to apply. Petrographic mixing of solids (crystallization of phases forming a mixture of solids i.e. cumulate ± trapped melt) and residual liquids have therefore provided a conceptual basis to envisage the mathematical method for modelling the multiple phase fractionation of magma (e.g. Bryan et al. 1969; Wright and Doherty 1974; Albarède 1995). In this method a

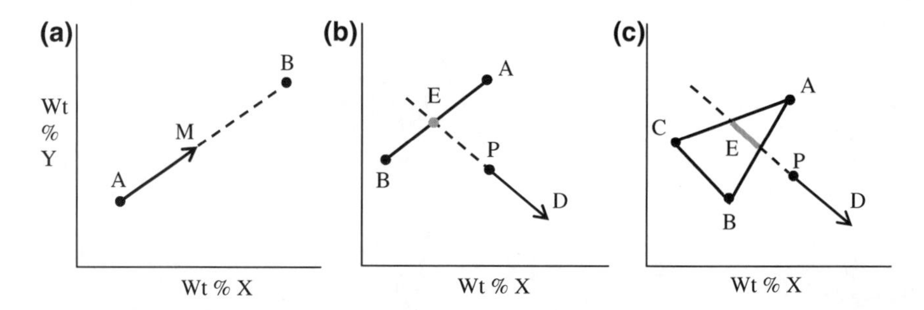

Fig. 9 **a–c** *X–Y* element variations (**a**) showing the evolution of mixture '*M*' in a straight line as a result of addition of '*A*' to '*B*' (**b**) Bulk extraction '*E*' of two phases *A–B* from parent '*P*' evolving daughter '*D*' (**c**) Bulk extraction '*E*' (*thick grey line*) of three phases *A-B-C* from a parent '*P*' evolving daughter '*D*' (after Cox et al. 1979; Ragland 1989; Wilson 1989). See text for explanation

mixture of given minerals fractionates from a parental magma to produce daughter (residual) liquid in a best-fit to the regression line of compositions. This method has the advantage of using more phase assemblages but has certain inherent limitations when it applies more complex phase assemblages. In "Mass Balance Modelling of Magmatic Processes in *GCDkit*" method and application of *R*-language based software *Geochemical Data Toolkit (GCDkit 3.0)* are described. The programme can compute the proportion of the fractionating minerals and calculated daughter magma composition from a parental melt using minerals and whole rock data-set based on the basic principle of chemical mass-balance. From the obtained solution, the bulk composition of cumulate can also be calculated using the composition of the fractionating minerals and their fractions constituting the bulk cumulate. The obtained sum of the residual squares of the model may be slightly more than the mathematically acceptable value ($\Sigma r^2 \ll 1.00$) but can be considered reasonably valid provided the model is petrographically and geologically consistent (e.g. Naslund 1989; Kumar and Kmeť 1995; Kumar 2002). Hunter and Sparks (1987) carried out an exhaustive mass-balance calculation for the evolution of Skaegaard parental magma producing a substantial amount of silica-rich residual melt that received criticism (e. g. McBirney and Naslund 1990) on the firm ground that one can achieve the desired solution in a pure mathematical sense by adjusting and manipulating some compositions and weighing factors of elements without field and petrological constraints.

Fractional crystallization (Rayleigh fractionation) is widely used to constrain trace element evolution of crystallizing melt as described by the equation (Gast 1968; Neumann et al. 1954):

$$\mathbf{C}_{\mathbf{L}}^{i} = \mathbf{Co^{i}F^{(Di-1)}}$$

where, C_L^i = concentration of element i in the residual melt, C_O^i = concentration of element i in the initial melt, F = fraction of melt left, and D^i = bulk partition coefficient of the crystallizing mineral assemblage for element i. The distribution coefficients (Kd^s) for various phases in magma are experimentally determined from the glass and phenocrysts.

There may be a large difference between calculated and observed residual magmas for certain incompatible elements by a factor of two to three, which can be explained by assimilation of crustal melt with the residual magma during the fractional crystallization process (AFC) or concomitant fractionation-mixing of coeval magmas. Alternatively, use of inappropriate Kd^s for some chemically very sensitive phases to their respective melts may result in erroneous values. Quantitative modelling of fractional differentiation thus requires precise Kd^s, which are generally functions of temperature, pressure, oxygen fugacity and bulk composition of magma.

In quantitative modelling of fractional crystallization the damping effect of Mg/Fe during the progress of fractionation may lead to erroneous results that can be minimized by the use of multiphase Rayleigh fractionation (Morse 2006). The modified Rayleigh equation for multiphase fractionation is:

$$\mathbf{C = CoF_L^{f\alpha(Di-1)}}$$

where C and Co are compositional terms stated in terms of X_1^L, and refer to the mole fractions of end-members within the binary solution as projected from other components or phases, F_L = fraction of liquid remaining, f_α = fraction of the active crystal phase relative to total crystals, and D = partition coefficient X_1^S/X_1^L. By active crystal phase (here α) is meant that causes C to evolve, with all other phases being passive.

More recently, the concept of *concurrent fractional and equilibrium crystallization* (CFEC) in a multi-phase magmatic system in the light of experimental results on diffusivities of elements and other species between minerals and melts has been proposed, and applied successfully to demonstrate coherent and scattering of elemental trends of melts of Bishop Tuff by the CFEC (Sha 2012). Langmuir (1989) proposed a model of in situ crystallization at the cooling interface of magma chamber, which was further developed by McBirney (1995). O'Hara and Fry (1996) provided an explicit numerical solution for observed lateral variations in the mass fraction crystallized with position in the magma chamber by the integration of residual liquids from crystallization in addition to other factors such as imperfect fractional crystallization, refilling of magma chambers during fractionation, and in situ crystallization.

8 Mixing Test of Two-Magma End-Members

A magma system may form by mixing (hybridization) of two magma end-members. This hypothesis can be tested using elemental concentrations of participating magma end-members and a mixing equation (Fourcade and Allègre 1981), which will estimate the proportion of mafic to felsic components needed to produce a hybridized magma system. If the concentration of an element (i) in hybrid or

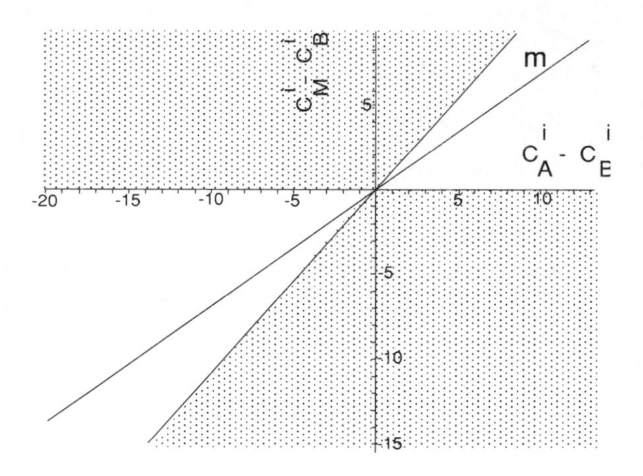

Fig. 10 $C^i_A - C^i_B$ versus $C^i_M - C^i_B$ mixing-test variation diagram, where C^i_A = content of '*i*' element in felsic magma *A*; C^i_B = content of '*i*' element in mafic magma *B*; C^i_M = content of '*i*' element in hybrid magma *M*; and *m* = fraction of felsic magma in the mixture (hypothetical regression line) represented by slope. *Grey* region represents domain of invalid mixing between *A* and *B* magma end-members (after Fourcade and Allègre 1981)

mixture (C^i_m) is formed by two-component i.e. felsic (C^i_A) and mafic magma (C^i_B) end-members in various weight proportions (x), then the mixing equation can be written as:

$$
\begin{aligned}
C^i_m &= x\,C^i_A + (1 - x)\,C^i_B \\
C^i_m &= x\,C^i_A + C^i_B - xC^i_B \\
C^i_m - C^i_B &= x\left(C^i_A - C^i_B\right) \qquad (1)
\end{aligned}
$$

Equation (1) can be used to calculate the proportion (x) of felsic to mafic magma in the mixture for each element (i). If we plot (C^i_m–C^i_B) *versus* (C^i_A–C^i_B) for a chemical database, a straight line would result whose slope will provide the mass proportion (x) of felsic magma in the mixture (Fig. 10). In this plot, mixing domain will be delineated by the slopes between *zero* and *one* because the sum of proportions of felsic and mafic magmas that formed the hybrid magma will never exceed one.

Simple two-component linear mixing can be used to geochemically model the hybridized magma system but for non-linear and complex chaotic mixing system fractal (scale-invarient) geometry can be more useful and demonstrative (e.g. Perugini and Poli 2012, and references therein).

Acknowledgments I extend sincere thanks to R. N. Singh, Senior INSA Scientist, CSIR-NGRI, Hyderabad, who helped me to understand the numerical basis of magmatic processes. Fruitful discussion with Igor Petrík, Vojtech Vilinovič, Igor Broska, H. R. Naslund, Ewa Słaby has enhanced knowledge of modelling the magma chamber processes. Sita Bora and Hansa Joshi are thanked for assisting me while finalizing this chapter. Science and Engineering Research Board (SERB), New Delhi is highly acknowledged for financial support. David R. Nelson is especially thanked for fruitful and generous comments that highly improved the earlier version. Permissions from Elsevier and Veda Publishing House for using some materials in this chapter are gratefully acknowledged.

References

Albarède F (1995) Introduction to geochemical modeling. Cambridge University Press, Cambridge, p 543

Bachmann O, Miller CF, de Silva S (2007) The volcanic-plutonic connection as a stage for understanding crustal magmatism. J Volcanol Geotherm Res 167:1–23

Barbarin B (1989) Importance des différents processus d'hybridation dans les plutons granitiques du batholite de la Sierra Nevada, Californie. Schweiz Mineral Petrogr Mitt 69:303–315

Barbarin B (2005) Mafic magmatic enclaves and mafic rocks associated with some granitoids of the central Sierra Nevada batholiths, California: nature, origin, and relations with the host. Lithos 80:155–177

Barbarin B, Didier J (1992) Genesis and evolution of mafic microgranular enclaves through various types of interaction between coexisting felsic and mafic magmas. Trans Royal Soc Edinb Earth Sci 83:145–153

Beard JS, Ragland PC, Crawford ML (2005) Reactive bulk assimilation: a model for crust-mantle mixing in silicic magmas. Geology 33:681–684

Bora S, Kumar S, Yi K, Kim N, Lee TH (2013) Geochemistry and U-Pb SHRIMP zircon chronology of granitoids and microgranular enclaves from Jhirgadandi Pluton of Mahakoshal belt, Central India Tectonic Zone India. J Asian Earth Sci 70–71:99–114

Bryan WB, Finger LW, Chayes F (1969) Estimating proportions in petrographic mixing equations by least squares approximation. Science 163:926–927

Cambel B, Vilinovič (1987) Geochémia and petrológia granitoidných hornín Malých Karpát. Veda, Slovak Academy of Science, Bratislava, p 247

Clemens JD (1989) The importance of residual source material (restite) in granite petrogenesis: a comment. J Petrol 30:1313–1316

Clemens JD, Mawer CK (1992) Granitic magma transport by fracture propagation. Tectonophysics 204:339–360

Clemens JD, Stevens G (2012) What controls chemical variation in granitic magmas? Lithos 134:317–329

Cox KG, Bell JD, Pankhurst RJ (1979) The interpretation of igneous rocks. George Allen and Unwin, London, p 450

Deering CD, Bachmann O (2010) Trace element indicators of crystal accumulation in silicic igneous rocks. Earth Planet Sci Lett 297:324–331

Didier J (1973) Granites and their enclaves: the bearing of enclaves on the origin of granites. Development in petrology, vol 3. Elsevier, Amsterdam, p 393

Dietl C, Koyi H (2011) Sheets within diapirs—results of a centrifuge experiment. J Struct Geol 33:32–37

Dickson FW (2000) Chemical evolution of magma. J Geodyn 30:475–487

Eichelberger JC, Izbekov PE, Browne BL (2006) Bulk chemical trends at arc volcanoes are not liquid line of descent. Lithos 87:135–154

Ewart A, Hildreth W, Carmichael ISE (1975) Quaternary acid magma in New Zealand. Contrib Mineral Petrol 51:1–27

Ferré EC, Olivier G, Montanari D, Kalakay TJ (2012) Granite magma migration and emplacement along thrusts. Int J Earth Sci 101:1673–1688.

Fourcade S, Allègre CJ (1981) Trace elements behaviour in granite genesis; a case study: the calc-alkaline plutonic association from the Quérigut complex (Pyrénées, France). Contrib Mineral Petrol 76:177–195

Frost TP, Mahood GA (1987) Field, chemical and physical constraints on mafic–felsic interaction in the Lamarck granodiorite, Sierra Nevada, California. Geol Soc Am Bull 99:272–291

Gast PW (1968) Trace element fractionation and the origin of tholeiitic and alkaline magma types. Geochim Cosmochim Acta 32:1057–1086

Harker A (1909) The natural history of igneous rocks. Methuen, London, p 384

Humphreys MCS, Blundy JD, Sparks RJS (2006) Magma evolution and open-system processes at Shiveluch volcano: insight from phenocrysts zoning. J Petrol 47:2303–2334
Hunter RH, Sparks RSJ (1987) The differentiation of the Skaergaard Intrusion. Contrib Mineral Petrol 95:451–461
Huppert HE, Sparks RJS, Turner JS (1984) Some effects of viscosity on the dynamics of replenished magma chamber. J Geophys Res 89:6857–6877
Irvine TN (1982) Terminology for layered intrusions. J Petrol 23:127–162
Irvine TN (1987) Glossary of terms for layered intrusions. In: Parson I (ed) Origin of igneous layering. D. Reidel Publishing Company, Dordrecht, pp 641–647
Jackson ED (1967) Ultramafic cumulates in the Stillwater, Great Dyke and Bushweld intrusion. In: Wyllie PJ (ed) Ultramafic and related rocks. Wiley, New York, pp 20–38
Janoušek V, Braithwaite CJR, Bowes DR, Gerdes A (2004) Magma-mixing in the genesis of Hercynian calc-alkaline granitoids: an integrated petrographic and geochemical study of the Sazava intrusion, Central Bohemian Pluton, Czech Republic. Trans Roy Soc Edin Earth Sci 91:15–26
Keelmen PB (1990) Reaction between ultramafic rock and fractionating basaltic magma I. Phase relations, the origin of calc-alkaline magma series, and the formation of discordant dunite. J Petrol 31:51–98
Kumar S (1995) Microstructural evidence of magma quenching inferred from microgranular enclaves hosted in Hodruša granodiorites, Western Carpathians. Geol Carp 46:379–382
Kumar S (2002) Calculated magma differentiation trends of discrete tholeiitic and alkaline magma series of Phenai Mata Igneous Complex, Baroda district, Gujarat, western India. J Appl Geochem 4:93–102
Kumar S (2003) Pearce Element ratios applied to model basic rock members of Phenai Mata Igneous Complex, Baroda district, Gujarat. J Geol Soc India 61:565–572
Kumar S (2010) Mafic to hybrid microgranular enclaves in the Ladakh batholith, northwestern Himalaya: Implications on calc-alkaline magma chamber processes. J Geol Soc India 76:5–25
Kumar S, Kmeť J (1995) A calculated magma differentiation trend of Hodruša Štiavnica Intrusive Complex, western Carpathians. Bull Czech Geol Surv 70:15–18
Kumar S, Rino V (2006) Field evidence of magma mixing from microgranular enclaves in Palaeoproterozoic Malanjkhand granitoids, central India: evidence of magma mixing, mingling, and chemical equilibration. Contrib Mineral Petrol 152:591–609
Kumar S, Rino V, Pal AB (2004) Field evidence of magma mixing from microgranular enclaves hosted in Palaeoproterozoic Malanjkhand granitoids, central India. Gond Res 7:539–548
Langmuir CH (1989) Geochemical consequences of *in situ* crystallization. Nature 340:199–205
Maaløe S (1985) Principles of igneous petrology. Springer, New York, p 373
Masson CR (1965) An approach to the problem of ionic distribution in liquid silicates. Proc Roy Soc London A287:201–221
McBirney AR (1984) Igneous petrology. Freeman Cooper and Company, California, p 504
McBirney AR (1993) Igneous Petrology. Jones and Bartlett, Boston, p 572
McBirney AR (1995) Mechanism of differentiation in the Skaergaard intrusion. J Geol Soc London 152:421–435
McBirney AR, Naslund HR (1990) The differentiation of the Skaergaard intrusion: a discussion on Hunter and Sparks (Contrib Mineral Petrol 95:451–461). Contrib Mineral Petrol 104:235–247
McBirney AR, Baker BH, Nilson RH (1985) Liquid fractionation. Part I: basic principles and experimental simulations. J Volcanol Geotherm Res 24:1–24
Morse SA (1979) Kiglapait geochemistry. II: Petrography. J Petrol 20:591–624
Morse SA (2006) Multiphase Rayleigh fractionation. Chem Geol 226:212–231
Mueller RF, Saxena SK (1977) Chemical petrology (with application to the terrestrial planets and meteorites). Springer, New York, p 394
Naslund HR (1989) Petrology of the Basistoppen Sill, East Greenland: a calculated magma differentiation trend. J Petrol 30:299–319

Nielsen RL, DeLong SE (1992) A numerical approach to boundary layer fractionation: application to differentiation in natural magma systems. Contrib Mineral Petrol 110:355–369
Neumann H, Mead J, Vitaliano CJ (1954) Trace element variation during fractional crystallization as calculated from the distribution law. Geochim Cosmochim Acta 6:90–99
Nicholls J (1988) The statistics of Pearce element diagrams and the Chayes closure problem. Contrib Mineral Petrol 99:11–24
Nishimura K (2012) A mathematical model of trace element and isotopic behaviour during simultaneous assimilation and imperfect fractional crystallization. Contrib Mineral Petrol 164:427–440
O'Hara MJ, Fry N (1996) The highly compatible trace element paradox—fractional crystallization revisited. J Petrol 37:859–890
Paterson SR, Vernon RH (1995) Bursting the bubble of ballooning plutons: a return to nested diapirs emplaced by multiple processes. Geol Soc Am Bull 107:1356–1380
Pearce TH (1968) A contribution to the theory of variation diagrams. Contrib Mineral Petrol 19:142–157
Pearce TH (1970) Chemical variations in the Palisades sill. J Petrol 11:15–32
Pearce TH (1987) The identification and assessment of spurious trends in Pearce-type ratio variation diagrams: a discussion of some statistical arguments. Contrib Mineral Petrol 97:529–534
Pearce TH (1990) Getting the more from your data: applications of Pearce Element Ratio analysis. In: Russel JK, Stanley CR (eds) Theory and application of Pearce element ratios to geochemical data analysis. Short Course No 8, Geological Association of Canada, Vancouver, pp 99–129
Perugini D, Poli G (2012) The mixing of magmas in plutonic and volcanic environments: analogies and differences. Lithos 153:261–277
Perugini D, De Campos CP, Dingwell DB, Petrelli M, Poli D (2008) Trace element mobility during magma mixing: preliminary experimental results. Chem Geol 256:146–157
Perugini D, Petrelli M, Poli G (2006) Diffusive fractionation of trace elements by chaotic mixing of magmas. Earth Planet Sci Lett 243:669–680
Perugini D, Poli G, Christofides G, Eleftheriadis G (2003) Magma mixing in the Sithonia pluonic complex, Greece: evidence from mafic microgranular enclaves. Mineral Petrol 78:173–200
Petfort N, Cruden AR, McCaffrey KJW, Vigneresse J-L (2000) Granite magma formation, transport and emplacement in the earth's crust. Nature 408:669–673
Ragland PC (1989) Basic analytical petrology. Oxford University Press, New York, p 369
Rollinson HR, Robert CR (1986) Ratio correlations and major element mobility in altered basalts and komatiites. Contrib Mineral Petrol 93:89–97
Russell JK, Nicholls J (1988) Analysis of petrologic hypothesis with Pearce element ratios. Contrib Mineral Petrol 99:25–35
Russell JK, Nicholls J (1990) Formulation and testing of scientific hypothesis. In: Russell JK, Stanley CR (eds) Theory and application of Pearce Element Ratio to geochemical data analysis. Geological Association of Canada, Short Course No 8, Vancouver, pp 1–10
Sha LK (2012) Concurrent fractional and equilibrium crystallization. Geochim Cosmochim Acta 86:52–75
Sisson TW, Bacon CR (1999) Gas-driven filter pressing in magmas. Geology 27:613–616
Słaby E, Martin H (2008) Mafic and felsic magma interactions in granites: the Hercynian Karkonosze pluton (Sudetes, Bohemian Massif). J Petrol 49:353–391
Słaby E, Smigielski M, Smigielski T, Domonik A, Simon K, Kronz A (2011) Chaotic three-dimensional distribution of Ba, Rb, and Sr in feldspar megacrysts grown in an open magmatic system. Contrib Mineral Petrol 162:909–927
Stanley CR, Russell JK (1989) Petrological hypothesis testing with Pearce Element Ratio diagrams: derivation of diagram axis. Contrib Mineral Petrol 103:78–89
Trupia S, Nicholls J (1996) Petrology of recent lava flows, Volcano Mountain, Yukon Territory, Canada. Lithos 37:61–78

Vigneresse JL, Duley S, Chattaraj PK (2011) Describing the chemical character of a magma. Chem Geol 287:102–113

Vernon RH (1983) Restite, xenolith, and micrograniotoid enclaves in granites. Clarke Memorial Lecture. Proc Roy Soc N S W 116:77–103

Vernon RH, Etheridge MA, Wall VJ (1988) Shape and microstructure of microgranitoid enclaves: indicators of magma mingling and flow. Lithos 22:1–11

Wall VJ, Clemens JD, Clarke DB (1987) Models for granitoid evolution and source compositions. J Geol 95:731–749

Wilson M (1989) Igneous petrogenesis. Unwin Hyman Ltd, London, p 466

Wilson M (1993) Magmatic differentiation. J Geol Soc London 150:611–624

Wright TL, Doherty PC (1974) A linear programming and least squares computer method for solving petrologic mixing problems. Geol Soc Am Bull 81:1995–2008

Wadsworth WJ (1985) Terminology of postcumulus processes and products in the Rhum layered intrusion. Geol Mag 122:549–554

Wager LR, Brown GM (1968) Layered igneous rocks. Oliver and Boyd Ltd, Edinburg, p 588

Wager LR, Brown GM, Wardsworth WJ (1960) Types of igneous cumulates. J Petrol 1:73–85

Wiebe RA (1994) Silicic magma chambers a traps for basaltic magmas: the Cadillac mountain intrusive complex, Mount Desert Island. J Geol 102:423–427

Wyllie PJ (1971) The dynamic earth. Wiley, New York, p 416

Models for Quantifying Mantle Melting Processes

R. N. Singh and A. Manglik

Claude Allegre stubbornly passed on to his students the habit of turning his perception of any geological process into equations that could eventually be tested against measurements....

Albarede (1995)

Abstract Partial melting of mantle and crustal rocks is an important process for the genesis of a suite of igneous rocks seen at the surface of the earth. These rocks preserve the imprints of the complex physico-chemical processes in the earth's interior in the form of their distinct end-member geochemical and isotopic compositions. Spatial and temporal variations in temperature, pressure, fluid mass and concentration of chemical species basically control the petrological property of rocks. This chapter describes basic framework of petrological modelling approach to quantify the deeper processes. Most frequently used equations for geotherm construction for continental and oceanic lithosphere, degree of partial melting and its distribution with depth due to perturbation in geotherms, partition of trace elements and radioactive elements in various partial melting models, crustal evolution and chemical geodynamics models are presented with their derivations.

1 Introduction

The present structure and composition of the earth has been arrived at by physico-chemical processes which are involved in the cooling of the earth. Within the earth, thermal conduction is not an efficient mechanism for bringing heat to the surface from the deep interior. Heat convection, supported by convective instability arguments, is a more favored mechanism. Convection brings materials from the deep interior to the base of the lithosphere leading to their partial melting as a result of decompression. Melts having lesser density rise towards the surface and bring deeper heat to the surface and near-surface rocks, from where the heat is

R. N. Singh (✉) · A. Manglik
CSIR-National Geophysical Research Institute, Uppal Road, Hyderabad, India
e-mail: rishiprema@gmail.com

S. Kumar and R. N. Singh (eds.), *Modelling of Magmatic and Allied Processes*,
Society of Earth Scientists Series, DOI: 10.1007/978-3-319-06471-0_2,

transported by conduction process. Thus, partial melting is a vital process for earth's evolution. Why and how such a process takes place needs quantitative answers. It should be appreciated that reliable knowledge is gained by confronting data with models. Enough knowledge exists to use principles and their mathematical forms to understand what has undergone with the earth and also fathom what lies ahead. As geology is a historical science, reductionism, philosophically speaking, will not be able to trace computationally all geological trajectories, connecting all space-time events, holistically. What can then be done reasonably? We should reduce the whole problem into subsets of characteristic processes and model each one separately to arrive at understanding of these underlying processes. Based on such knowledge, one can then construct geological history of an event or a location or the whole earth. In an interesting book titled "*Melting the Earth: The history of ideas on volcanic eruptions*" (Sigurdsson 1999), historical evolution of the idea of the melting in the earth has been discussed. Starting from Greeks to current understanding of the pressure release melting has been narrated in this book. Ideally, melting can occur in the earth due to the following processes:

(a) Temperature changes.
(b) Pressure changes.
(c) Composition changes.
(d) Melting temperature depression by the presence of volatiles, especially water.

All the above are possible, but pressure release melting is the most dominant process.

Melting can provide good estimate of the thermal condition at its source regions. We thus need to know the distribution of the temperature, pressure and composition with depth, and the relationship of melt generation with changes in these distributions. To further constrain the physico-chemical processes, we need to model how chemical species partition amongst competing phases and components. Albarede (1995) and Shaw (2006) have given an excellent theoretical treatment of the subject.

2 Geotherm

Calculation of partial melting requires the knowledge of the geotherm in the lithosphere. In the lithosphere heat transport takes place via heat conduction and the sources of the heat are radiogenic heat sources and heat flow from the mantle. Here the relevant balance law is energy conservation and constitutive law is the Fourier law. We shall derive thermal structure for both continental and oceanic regions. McKenzie et al. (2005) have formulated comprehensive thermal model of both continental and oceanic lithospheres including the effects of temperature dependence of physical properties.

2.1 Continental Lithosphere

Thermal structure is determined using heat flow and heat generation data. Worldwide analysis of these data sets has yielded following general inferences:

(1) Cratons show low surface heat flow values whereas orogens show high values. There has been a decrease in the heat flow values over the geological history. Radiogenic heat too decreases with time.
(2) There is scatter in the data over mean values. This scatter can be due to a variety of causes such as hydrothermal circulation.
(3) Radiogenic heat sources in general contribute about 40 % to the surface heat flow and the remaining is attributed to the deeper sources.
(4) Surface heat flow and heat generation are linearly related. The slope of such a curve gives the depth scale of the radiogenic heat and intercept with heat flow axis gives heat flow from the interior.

The last mentioned relationship is used to derive depth dependence of the radiogenic heat in the crust. The steady state temperature is given by the solution of the following heat conduction equation:

$$\frac{d}{dz}\left(K(T)\frac{dT}{dz}\right) + A(z) = 0, \tag{1}$$

with the following boundary conditions:

$$T = T_S, \quad \text{at} \quad z = 0; \qquad K(T)\frac{dT}{dz} = Q_S, \quad \text{at} \quad z = 0. \tag{2}$$

Here T denotes temperature, K the thermal conductivity, $A(z)$ the radiogenic heat distribution, and z the depth. The variation of the radiogenic heat with depth is given by:

$$A(z) = A_0 \exp(-z/d). \tag{3}$$

For a homogeneous layer with constant K and A, the solution of Eq. (1) yields the following expression for temperature distribution:

$$T(z) = T_S + Q_S\frac{z}{K} - A\frac{z^2}{2K}. \tag{4}$$

For a stratified crustal model, thermal structure can be obtained by using the following relationship between temperature T_n and heat flux Q_n at the base of the nth layer having uniform properties (K_n and A_n) and thickness $h_n(z_n - z_{n-1})$ with temperature T_{n-1} and heat flux and Q_{n-1} at the surface of the layer:

$$T_n = T_{n-1} + Q_{n-1}\frac{z_n - z_{n-1}}{K_n} - A_n\frac{(z_n - z_{n-1})^2}{2K_n}, \tag{5}$$

$$Q_n = Q_{n-1} + A_n(z_n - z_{n-1}). \tag{6}$$

A_n can be chosen to represent any general form of radiogenic heat in the crust. Given values of the concentrations (c_i in ppm) of isotopes of U, Th and K and their decay energies (e_i in J/kg s), the radiogenic heat is given by:

$$A = \rho \sum_i e_i c_i \tag{7}$$

The decay energies of U, Th and K are respectively 9.66×10^{-2}, 2.65×10^{-2} and 3.58×10^{-6} mW/kg. For most rocks, A varies within the range of 0.008–8.0 μW/m^3. The base of the lithosphere is assumed to have temperature of about 1,330 °C. The above formalism can be used to obtain the steady state thermal regime of the lithosphere.

Sometimes the normal heat production in the crust cannot yield partial melt such as in granite formation which is a major constituent of the continental crust. In this case we need to resort to generating transient thermal regimes in the lithosphere. Tectonic deformation and erosion can lead to redistribution of heat sources in the crust which can produce higher temperature transiently. One can also assume horizontal flows of material which can transport heat to longer distances and lead to partial melting. The mathematical model to construct such regimes will require solutions of the following advection-diffusion equation:

$$\rho C_p^{'}\left(\frac{\partial T}{\partial t} + V.\nabla T\right) = \nabla.(K.\nabla T) + A, \tag{8}$$

with suitable initial and boundary conditions. Here, V is advection velocity in the medium, ρ is the density and $C_p^{'}$ is the isobaric specific heat capacity.

2.2 Oceanic Lithosphere

Thermal structure of the oceanic lithosphere in its simplest form is constructed by using a cooling half space model. Mathematically, we need to solve the following problem:

$$\rho C_p^{'}\frac{\partial T}{\partial t} = K\frac{\partial^2 T}{\partial z^2}; \qquad T(0,t) = T_S, \qquad T(z \to \infty, t) = T_m, \qquad T(z,0) = T_m. \tag{9}$$

Here K, T_m *and* T_S represent thermal conductivity, mantle temperature, and surface temperature (other variables are already defined). Solution of this equation with the above initial and boundary conditions is expressed in terms of error function as (Turcotte and Schubert 2002):

$$\frac{T(z,t) - T_S}{T_m - T_S} = \mathrm{erf}\left(\frac{z}{\sqrt{4Kt/\rho C_p'}}\right). \tag{10}$$

If the oceanic lithosphere is spreading at the rate v, then time can be related to horizontal distance from the ridge as $t = x/v$. Thus, the horizontal and vertical variation in the temperature is given by:

$$\frac{T(z,x) - T_S}{T_m - T_S} = \mathrm{erf}\left(\frac{z}{\sqrt{4Kx/\rho C_p' v}}\right). \tag{11}$$

For the error function the following approximate formula can be used (Abramowitz and Stegun 1964):

$$\mathrm{erf}(x) \approx 1 - \left(a_1 t + a_2 t^2 + a_3 t^3\right)\exp\left(-x^2\right), \tag{12}$$

where

$$t = \frac{1}{(1 + 0.47047x)}, \quad a_1 = 0.34802, \quad a_2 = -0.09587, \quad a_3 = 0.74785.$$

Given the values of parameters appearing in the above equation, thermal structure can be easily determined. Variation of heat flow and subsidence with age of the oceanic lithosphere is given by (Turcotte and Schubert 2002):

$$Q = K\frac{T_m - T_S}{\sqrt{\pi Kt/\rho C_p'}}, \qquad w = \frac{2\rho_m \alpha (T_m - T_S)}{\rho_m - \rho_S}\sqrt{Kt/\pi\rho C_p'} \tag{13}$$

3 Pressure Distribution Within Lithosphere

At depths, stresses can be decomposed into isotropic pressure and deviatoric stresses. Except near the surface region, stresses can be taken largely as pressure P which is governed by the following equation:

$$\frac{dP}{dz} = \rho g. \tag{14}$$

Here P, ρ and g are pressure, density of rocks and acceleration due to gravity. Integrating this equation gives $P(z)$ as:

$$P(z) = \int_{0}^{z} \rho g dz. \tag{15}$$

If the material properties of a given crustal column are constant, then we have:

$$P(z) = \rho g z. \tag{16}$$

Taking value of crustal thickness as 40 km, density as 2,800 kg m^{-3} and acceleration due to gravity as 9.80 ms^{-2}, the pressure at the base of the crust is $P \approx 1.1$ GPa ≈ 11 kbar. For some geological problems it would be desirable to know both hydrostatic (given above) and deviatoric stresses, required for both compressive and extensional regimes. We would then need to know all components of stress regime which requires much more involved process of calculations.

4 Degree of Partial Melting

As a parcel of rock rises upward, it experiences a decrease in pressure but its temperature does not get equilibrated with the surrounding material. This adiabatic decompression leads to partial melting as the rock parcel crosses the solidus curve. The slope of the solidus curve or of any equilibrium boundary for a reaction is defined by the Claussius-Clayperon relation (Ganguly and Saxena 1987). This relationship is derived for one component system in two phases by equating Gibbs free energy of both phases, G_1 and G_2 at pressures and temperatures (P, T) and $(P + dP, T + dT)$. We, thus, have along the phase separation boundary:

$$G_1 = G_2 \quad \text{and} \quad G_1 + dG_1 = G_2 + dG_2, \tag{17}$$

which gives:

$$dG_1 = dG_2. \tag{18}$$

Change in the Gibbs free energy (defined as $G = E + PV - TS$) is obtained as:

$$dG = dE + PdV + VdP - TdS - SdT. \tag{19}$$

Here, E is internal energy, V is volume and S is entropy. Using first and second laws of thermodynamics ($dE = TdS - PdV$), we get

$$dG = VdP - SdT. \tag{20}$$

From Eqs. (18) and (20), we get:

$$V_1 dP - S_1 dT = V_2 dP - S_2 dT. \tag{21}$$

This gives the Claussius-Clayperon relation as:

$$\left(\frac{dP}{dT}\right) = \frac{\mathrm{S}_2 - \mathrm{S}_1}{\mathrm{V}_2 - \mathrm{V}_1} \equiv \frac{\Delta S}{\Delta V}. \tag{22}$$

where $\Delta \mathrm{S}$ and ΔV represent change in the entropy and volume, respectively, due to reaction which in this case is the transformation of a solid assemblage to melt (Ganguly and Saxena 1987). For olivine taking the values of ΔV and ΔS as 0.434 J/MPa/g and 0.362 J/K/g, respectively, we get dT/dP as $\sim$120 K/GPa. This needs to be compared with the adiabatic gradient followed by a parcel of mantle material ascending to the surface. Under equilibrium condition in which the entropy of a system is conserved:

$$dS = \left(\frac{\partial S}{\partial T}\right)_P dT + \left(\frac{\partial S}{\partial P}\right)_T dP. \tag{23}$$

Partial derivatives in the above equation can be written in terms of heat capacity C_P and coefficient of thermal expansion α as:

$$dS = \left(\frac{C_P}{T}\right) dT - \alpha V dP = \left(\frac{C_P}{T}\right) dT - \frac{\alpha}{\rho} dP. \tag{24}$$

Equating dS as zero, the adiabatic gradient can be obtained as:

$$\frac{dT}{dP} = \frac{\alpha T}{\rho C_P}, \tag{25}$$

where α *is* the coefficient of thermal expansion. This is sometimes referred as isentropic gradient to distinguish it from adiabatic gradient of system that is subjected to internal entropy production (Ganguly 2008). Using Eq. (16), this equation can be reduced to:

$$\frac{dT}{dz} = \frac{g \alpha T}{C_P}. \tag{26}$$

The above equation is used to find the adiabatic gradient for a given rock type and, thus, the depth at which partial melting starts once the solidus curve is intersected. We can estimate the adiabatic gradient for olivine using the values of α, C_p, V and T as $2.7 \times 10^{-5}\,\mathrm{K}^{-1}$, 193J/(K-mol), 43.8 J/(MPa-mol), and 1650 K, respectively. This gives the adiabatic gradient as $\sim$10 K GPa^{-1}, which is much smaller than inverse of the Claussius-Clayperon slope obtained earlier. Thus, a parcel of the mantle material rising along the adiabatic gradient will experience partial melting.

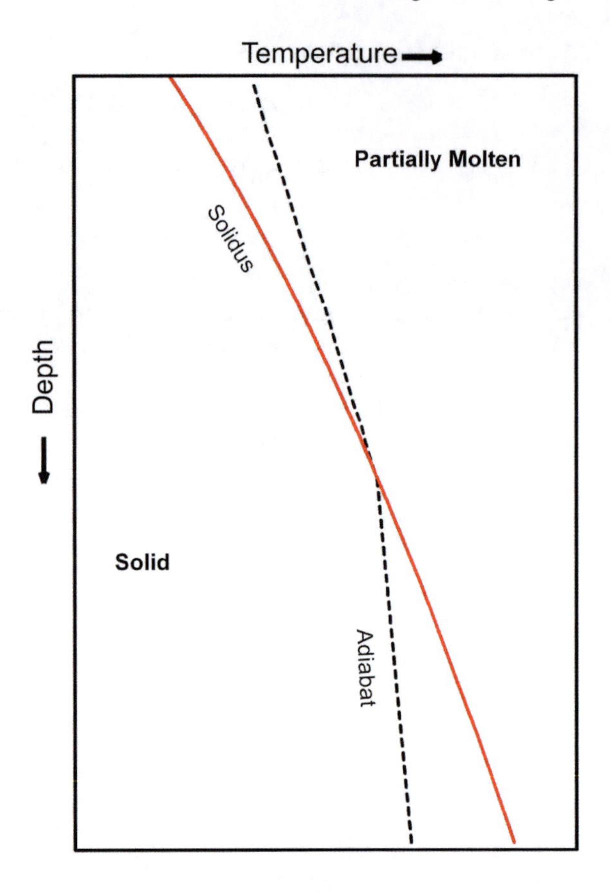

Fig. 1 A sketch of *solidus* and *adiabatic curves*

Partial melting of the rock adds another effect, the release of latent heat due to which further rise of the rock parcel deviates from the adiabatic path given by Eq. (26). Some parameterized models for partial melting taking into consideration these effects have been proposed (Fig. 1).

Recently, Ganguly (2005) has derived the following expression for temperature gradient for the irreversible adiabatic decompression of isoviscous materials rising from great depths in the mantle:

$$\left(\frac{dT}{dz}\right)_{Q(ir)} = \frac{\rho_r}{\rho}\left(\frac{dT}{dz}\right)_S + \frac{g}{C_p}\left(1 - \frac{\rho_r}{\rho}\right) - \frac{2}{C_p}\left(\frac{du}{dz}\right). \tag{27}$$

Here, ρ_r is density of rising mantle rock and u is vertical velocity (positive downward). Ganguly (2005) showed that the temperature of rising mantle rocks will be higher than given by adiabatic gradient depending on the value of the ratio ρ_r/ρ. When $\rho_r/\rho < 0.94$ the ascending mantle rocks would be heated up instead of cooling under adiabatic condition leading to increased melt productivity.

Equations (22) and (25) set the conditions for the onset of partial melting. A quantity of interest is the amount of melt generated when the mantle material rises after it intersects the solidus. For this, we follow the changes in the entropy as the material moves along the solidus. It is given by:

$$\frac{dS^s}{dP} = \frac{C_p^s}{T}\left(\frac{dT}{dP}\right)_{2\varnothing} - \alpha^s V^s. \tag{28}$$

Here, subscript $2\varnothing$ indicates that the gradient is along the solidus and superscripts s and l refer to solid and liquid phase, respectively. If solid and melt together have entropy S_0, then we have:

$$\mathrm{S_0} = \mathrm{fS^l} + (1-f)S^s, \tag{29}$$

where f is melt fraction. From the above equation, we get:

$$f = \frac{S_0 - S^s}{S^l - S^s} = \frac{S_0 - S^s}{\Delta S}. \tag{30}$$

Differentiating Eq. (30) with respect to P and substituting Eq. (28) we get:

$$\left(\frac{\partial f}{\partial P}\right)_S = -\frac{1}{\Delta S}\left[\frac{C_p^s}{T}\left(\frac{\partial T}{\partial P}\right)_{2\varnothing} - \alpha^s V^s\right]. \tag{31}$$

This equation can further be evaluated by using solidus gradient equation as:

$$\left(\frac{\partial f}{\partial P}\right)_S = -\frac{1}{\Delta S}\left[\frac{C_p^s}{T}\frac{\Delta V}{\Delta S} - \alpha^s V^s\right]. \tag{32}$$

We can evaluate the changes in melt fraction with pressure using data for heat capacity, change in volume and entropy at fusion, coefficient of thermal expansion and temperature. Calculations show that melt fraction increases with decrease in pressure.

Besides above modelling, following relationship has been used in the literature (Ahern and Turcotte 1979) to calculate the degree of partial melting, f:

$$f = \mathrm{a}\{\exp(b(T - cz - d)) - 1\}, \tag{33}$$

where $\mathrm{a} = 0.4$, $\mathrm{b} = 3.65 \times 10^{-3}\mathrm{K}^{-1}$, $\mathrm{c} = 3.0 \times 10^{-3}\mathrm{K\,m}^{-1}$, and $\mathrm{d} = 1{,}100\ °\mathrm{C}$. This relationship does not hold for very low degree (<5 %) of partial melting. McKenzie and Bickle (1988) and Ellam (1992) have given the following expression for calculating f:

$$f - 0.5 = T' + \left(T'^2 - 0.25\right)\left(0.4256 + 2.988T'\right), \tag{34a}$$

$$f = 2.7157 \times 10^{-7}z^3 - 5.0715 \times 10^{-5}z^2 - 3.7816 \times 10^{-4}z + 0.30929, \tag{34b}$$

where $T' = (T - (T_s + T_l)/2)/(T_l - T_S) \cdot T_S$ and T_l denote solidus and liquidus temperatures. The second relationship has been used to quantify partial melting due to rise of hot mantle plume (Ellam 1992).

There has been discussion in the literature on the variation of melting with depth. McKenzie (1984) and McKenzie and Onions (1991) have argued for decrease in degree of melting with decrease in pressure whereas Ahern and Turcotte (1979) argue for increase in melting as pressure decreases. Asimov et al. (1997) have argued for increase in melt productivity with decrease in pressure both for batch and fractional melting under isentropic condition.

5 Changes in the Temperature After Melt Extraction

We shall take a simple two-layered model of lithosphere with no radiogenic heat and thermal structure as discussed in Foucher et al. (1982):

$$T(z) = \begin{cases} \frac{T_m z}{H}, & 0 \le z \le H \\ T_m + G(z - H), & z \ge H \end{cases}. \tag{35}$$

In this thermal model, the lower region of the lithosphere has a constant thermal gradient, denoted by G, connecting to adiabatic gradient of mantle instead of having a constant temperature.

After stretching of the lithosphere by a factor β the thermal structure changes to:

$$T = \begin{cases} T'_m(\beta/H)z, & 0 \le z \le H/\beta \\ T'_m + G(z - H/\beta), & z \ge H/\beta \end{cases}, \tag{36}$$

where $T'_m = T_m - G(H - H/\beta)$. With partial melting, the above expression modifies to:

$$T = \begin{cases} T'_m(\beta/H)z - fL/C, & 0 \le z \le H/\beta \\ T'_m + G(z - H/\beta) - fL/C, & z \ge H/\beta \end{cases}. \tag{37}$$

Here, L and C are latent heat of fusion and specific heat, respectively. f is function of temperature. After its substitution, we can obtain the thermal structure of the stretched lithosphere.

6 Depth Distribution of Partial Melt

The distribution of degree of partial melt with depth can be calculated by substituting the expression for temperature T given by Eq. (36) into equation for degree of partial melting given by Eqs. (33) and (34a) (Foucher et al. 1982). This yields an equation for f as the function of depth. This calculation can be done numerically. It is seen that the top of the melt zone is at the depth H/β and bottom is at the depth where degree of partial melting is zero ($z=H_f$). This is given by:

$$H_f = \frac{T_m - GH - d}{c - G} \quad \text{for} \quad \beta > H/H_f. \tag{38}$$

The degree of partial melting increases from zero at H_f to maximum at the depth of H/β as:

$$f = f_0(1 - \frac{z}{H_f}). \tag{39}$$

We can also calculate the thickness of the layer (H_b) formed by emplacement of melts as liquid igneous body (density ρ_{al}) as:

$$H_b = \frac{\rho_a}{\rho_{al}} \int_{H/\beta}^{H_f} f_0 \left(1 - \frac{z}{H_f}\right) dz. \tag{40}$$

Here, ρ_a is the density of asthenosphere. This expression is evaluated as:

$$H_b = \frac{\rho_a}{2\rho_{al}} \left(1 - \frac{\beta_c}{\beta}\right)^2 H_f f_0 z, \tag{41}$$

where $\beta_c = H/H_f$.

7 Partition Coefficients for Trace Elements

Trace elements have been found to trace the melting processes. These elements follow principles of dilute solutions. Their distributions amongst phases are easily characterized as these elements do not interact with each other and follow Henry's law. The distribution in any two phases is described in terms of partition function. The total mass of elements is conserved within all phases. Elements remaining in solid or released in melt are characterized by their partition coefficient values. Shaw (2006), Albarede (1995), and Ganguly (2008) can be referred for original references.

If mass of component i in a phase α is m_i, with phase mass being M_i, the concentration of component i in this phase is defined as $c_i^\alpha = m_i/M_i$. For total number of components being p, we have:

$$\sum_{i=1}^{p} c_i^\alpha = 1. \tag{42}$$

Let concentration of a component i in phases α and β be denoted by c_i^α and c_i^β, respectively. The partition coefficient is defined as:

$$D_i^{\beta-\alpha} = c_i^\beta / c_i^\alpha. \tag{43}$$

A bulk partition coefficient can be defined for use in petrology as rocks contain a large number of minerals which are involved in partial melting.

Let mineral proportions of a rock before partial melting be denoted by X_i^0 and after melting by X_i. The concentration of element in the rock is defined as:

$$c_S = \sum_i c_i X_i. \tag{44}$$

The concentration in the melt can be defined using partition coefficient ($D^{i\text{-}m}$)when mineral and melt are in equilibrium as:

$$c_l = c_i / D^{i-m}. \tag{45}$$

Combining the above two equations, we get:

$$c_S = c_l \sum_i D^{i-m} X_i. \tag{46}$$

Considering total mass of rock before melting as M_0, after melting the rock has mass as M and melt as L. If f is the melt fraction, we have:

$$L = f\, M_0 \quad \text{and} \quad M = (1-f)M_0. \tag{47}$$

We now write for each mineral:

$$MX_i = M_0 X_i^0 - Lp_i, \tag{48}$$

where p_i is the proportion of mineral i in the melt. Thus, total concentration of an element in the solid is given by:

$$c_S = c_l \sum_i \frac{\left(D^{i-m} X_i^0 - p_i D^{i-m} f\right)}{(1-f)}. \tag{49}$$

Thus, the bulk partition coefficient D is given by:

$$D = (D_0 - Pf)/(1-f), \tag{50}$$

where $D_0 = \sum_i D^{i-m} X_i^0, P = \sum_i p_i D^{i-m}$.

The above formalism is for non-modal melting. For modal melting, we have $X_i^0 = p_i$. In this case, we have $D_0 = P = D$. Elements having small values of $D(<1)$, are called incompatible and others compatible elements. Incompatible and compatible elements partition preferentially into melt and solids, respectively.

8 Batch Melting

Consider mass of solid M_0 with concentration of some element as c_S^0. After partial melting, let the volume of the melt be L with concentration of the element as c_l and in the residual solid c_S with mass as M. Total melt fraction attains equilibrium with the solid and only then it is extracted fully. Thus, we have the following relationship (Shaw 2006; Ganguly 2008):

$$M_0 = L + M, \quad c_S^0 M_0 = c_l L + c_S M. \tag{51}$$

Let us now define melt fraction and partition coefficient as:

$$f = \frac{L}{M_0}, \qquad D = \frac{c_S}{c_l}. \tag{52}$$

Using these relationships, we find the concentrations in liquid and residual solid as (Fig. 2):

$$c_l = \frac{c_S^0}{f + D(1-f)}, \qquad c_S = \frac{Dc_S^0}{f + D(1-f)}. \tag{53}$$

This is also called equilibrium melting. Figure 2 shows graphically the relationship (Eq.53(i)) for various values of D. Here total melt remains in contact with matrix. For non-modal melting, above equations are changed as:

$$c_l = \frac{c_S^0}{D_0 + f(1-P)}, \qquad c_S = \frac{c_S^0(D_0 - Pf)}{(1-f)(D_0 + f(1-P))}. \tag{54}$$

9 Fractional Melting

In this case, melt fraction is extracted from the solid as soon as it is formed or exceeds a critical limit. Here we treat the first case. Let the mass of solid converted to melt be given by dM with concentration c_l. We can then write (Shaw 2006; Ganguly 2008):

$$c_l dM = d(c_S M), \quad \frac{c_S dM}{D} = c_S dM + M dc_S, \quad \frac{dc_S}{c_S} = \left(\frac{1}{D} - 1\right) dM/M. \tag{55}$$

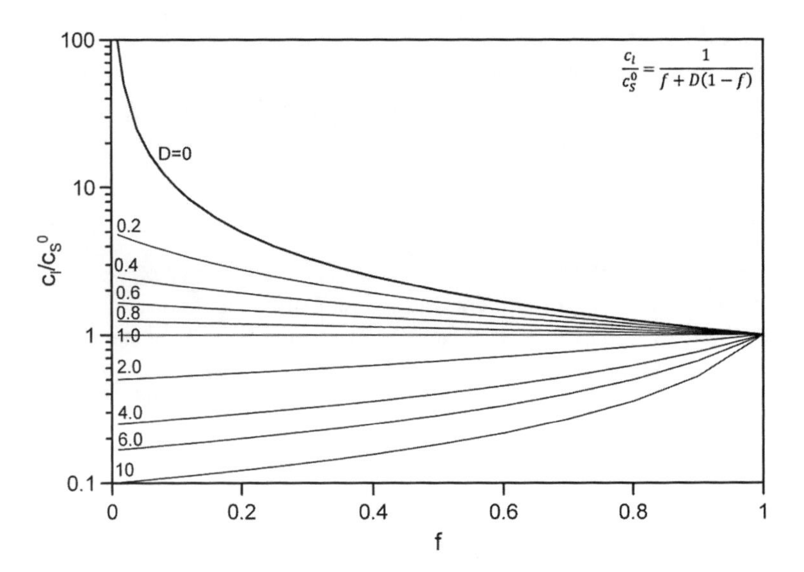

Fig. 2 Relative concentration of trace elements in liquid for partial melting by batch melting model described by Eq. (53)

Integrating the above equation, we get:

$$c_l = \left(\frac{c_S^0}{D}\right)(1-X)^{\frac{(1-D)}{D}}, \quad c_S = \left(c_S^0\right)(1-X)^{(1-D)/D}. \tag{56}$$

Here $X = L/M_0$. This is the amount of melt fraction extracted from the sources. Average concentration in melt is obtained by averaging over 0–X (Fig. 3):

$$\bar{c}_l = (1/X)\int_0^X c_l dX = c_S^0\left(1-(1-X)^{1/D}\right)/X. \tag{57}$$

These two melting models, batch and fractional, are the end member models. For non-modal melting, we get:

$$c_l = \left(c_S^0/D_0\right)(1-PX/D_0)^{(1-P)/P}, \quad c_S = c_S^0(1-PX/D_0)^{1/P}/(1-X). \tag{58}$$

The average concentration over all X is given by:

$$\bar{c}_l = c_S^0\left(1-(1-PX/D_0)^{1/P}\right)/X. \tag{59}$$

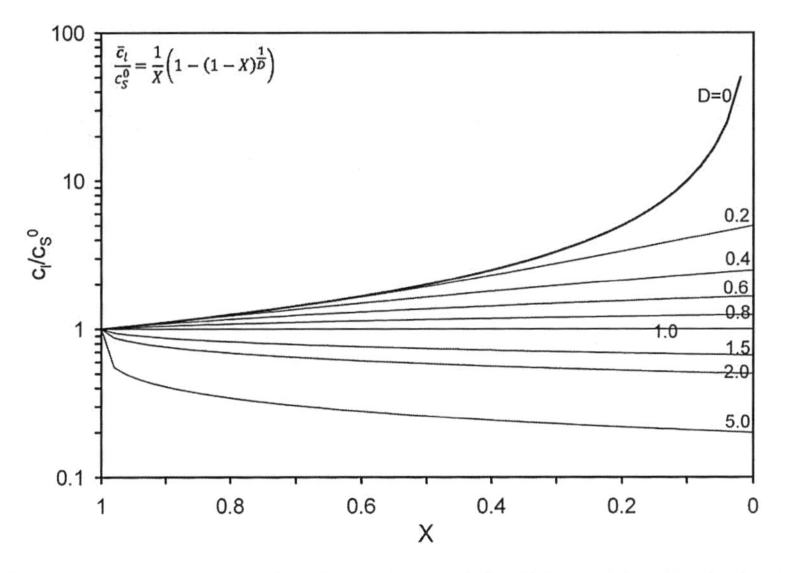

Fig. 3 Relative average concentration of trace elements in liquid for partial melting by fractional melting model described by Eq. (57)

10 Continuous Melting Model

McKenzie (1985) and Williams and Gill (1989) discussed a model wherein a fraction of partial melt remains always within the matrix. Let m be the melting rate, i.e., the amount of the matrix transformed into melt per unit time. We then have;

$$\frac{d}{dt}(\rho_l \phi c_l) = c_s m, \qquad \frac{d}{dt}(\rho_S (1-\phi) c_S) = -c_l m. \tag{60}$$

Combining these two equations, we get:

$$\frac{dc_l}{dt} = -\alpha c_l. \tag{61}$$

where

$$\alpha = \frac{(1-D)mF}{(\phi \rho_l)}, \qquad F = \left[1 + \frac{D\rho_S(1-\phi)}{(\rho_l \phi)}\right]^{-1}.$$

We then have:

$$c_l = c_l^0 e^{-\alpha t}. \tag{62}$$

Here, c_l^0 is the initial composition of the melt. McKenzie (1985) has shown that the extracted melt, denoted by X, is related to the melt production rate as:

$$X = 1 - \exp(-rt), \qquad r = m/(\rho_l\phi + \rho_S(1-\phi)). \tag{63}$$

From here, one can get expression for t as:

$$t = -\log(1-X)/r. \tag{64}$$

Thus, the concentration c_l is written as:

$$c_l = c_l^0(1-X)^{G(1-D)}, \qquad G = \frac{\rho_l\phi + \rho_S(1-\phi)}{\rho_l\phi + \rho_S(1-\phi)D}. \tag{65}$$

It can also be shown that:

$$c_l^0 = Gc_0, \tag{66}$$

where c_0 is the initial concentration in the source. We then get:

$$c_l = c_0D(1-X)^{G(1-D)}. \tag{67}$$

This is the expression for the continuous melt. This can be integrated over $[0, X]$ to get an expression for aggregated continuous melt as:

$$\bar{c}_l = c_0G\frac{1-(1-X)^{G(1-D)+1}}{X(1+G(1-D))}. \tag{68}$$

From above two equations expressions for batch melting ($X = 0$ and $\phi = f$) and fractional melting ($\phi = 0$) can be derived.

11 Fractionation of Radioactive Elements

We shall now present expressions for the radioactive element distributions in melts and matrix following McKenzie (1985) and Williams and Gill (1989). Let the concentration and half-life of elements be denoted by c and λ respectively. Let $c_S^P\left(c_l^P\right)$ represent concentration of parent element in solid (melt). We have the following equations for distribution of parent element in solid and melt as:

$$\frac{d}{dt}\left((1-\phi)\rho_S c_S^P\right) = -c_l^P m, \qquad \frac{d}{dt}\left(\phi\rho_l c_l^P\right) = -c_S^P m. \tag{69}$$

Combining these two equations, we get:

$$\left((1-\phi)\rho_S D^P + \phi\rho_l\right)\frac{dc_l^P}{dt} = \left(D^P - 1\right)mc_l^P, \tag{70}$$

where $D^P = c_S^P / c_l^P$. This equation can be further simplified to:

$$\frac{dc_l^P}{dt} = -\alpha_P c_l^P, \tag{71}$$

where

$$\alpha_P = \frac{(1-D^P)mF_P}{(\rho_l\phi)}, \qquad F_P = \left[1 + \frac{(1-\phi)\rho_S}{\phi\rho_l}D^P\right]^{-1}. \tag{72}$$

For daughter element, $c_S^D\left(c_l^D\right)$, it is necessary to include the decay as its life is small compared to the half-life of the parents. We have:

$$\begin{aligned}\frac{d}{dt}\left\{(1-\phi)\rho_S c_S^D + \phi\rho_l c_l^D\right\} = & c_S^D m - c_l^D m + \lambda_P\left\{(1-\phi)\rho_S c_S^P + \phi\rho_l c_l^P\right\} \\ & - \lambda_D\left\{(1-\phi)\rho_S c_S^D + \phi\rho_l c_l^D\right\}.\end{aligned} \tag{73}$$

This equation can further be reduced to:

$$\frac{d}{dt}c_l^D = -\lambda_D c_l^D + \lambda_P a_P - a_D, \tag{74}$$

where

$$\lambda_D = \frac{(1-D^D)mF_D}{\phi\rho_l}, \qquad a_P = \frac{\lambda_P c_l^D F_P}{F_D}, \qquad a_D = \lambda_D c_l^D. \tag{75}$$

We have now two equations for both parent and daughter elements distribution in the melt. We can derive from them equations for batch and continuous melting.

11.1 Batch Melting

Here, $m = 0$. Thus, we have in terms of activity a_p $(= \lambda_p c^P)$:

$$\frac{da_P}{dt} = 0, \quad \frac{da_D}{dt} = \lambda_D\left(\frac{F_D}{F_P}a_P - a_D\right). \tag{76}$$

The solutions of the above equations are:

$$a_P = A, \quad a_D = Be^{-\lambda_D t} + \frac{F_D}{F_P} a_P. \tag{77}$$

For $r = a_D/a_P$, we have the following expression from above equations:

$$r = Be^{-\lambda_D t} + F_D/F_P. \tag{78}$$

At $t = 0$ we have:

$$r(t=0) = r_0 = \left(\frac{A}{a_P} + r_e\right), \qquad r_e = F_D/F_P \tag{79}$$

We thus get:

$$r = r_0 e^{-\lambda_D t} + r_e\left(1 - e^{-\lambda_D t}\right). \tag{80}$$

Thus, constant of decay of r depends upon λ_D. When t tends to infinity, $r = r_e$. This value will be 1 only if $D^D = D^P$.

11.2 Continuous Melting

Here m and ϕ are constant. We, therefore, have:

$$\frac{da_P}{dt} = -\alpha_P a_P, \qquad \frac{da_D}{dt} = -\alpha_D a_D + \lambda_D\left(\frac{F_D}{F_P} a_P - a_D\right). \tag{81}$$

The solutions of these equations are:

$$\begin{aligned} a_P &= a_P(0) + e^{-\alpha_P t} \\ a_D &= a_D(0)e^{-(\lambda_D+\alpha_D)t} + \frac{\lambda_D F_D}{(\lambda_D + \alpha_D - \alpha_P)F_P} a_P(0)\left(e^{-\alpha_P t} - e^{-(\lambda_D+\alpha_D)t}\right). \end{aligned} \tag{82}$$

Numerical applications of the above formalism have been given in Williams and Gill (1989).

12 Chemical Geodynamics

Chemical fractionations in the earth have involved major geodynamical phenomena. Thus, trace element evolutions, which can characterize fractionation phenomena, can also yield the nature of the geodynamical processes. Issues such as mass, duration, and time of the crust extracted from the mantle can be estimated using the evolution of trace elements in these reservoirs. Rb-Sr and Sm-Nd data in the crust, MORB and OIBs can be used for this purpose. MORB and OIBs can give estimates of depleted and undepleted reservoirs. Observations of isotopic systems define normalized isotope ratio (ε) and fractionation factor (f) as (Turcotte and Schubert 2002):

$$\varepsilon = \left[\left(\frac{c^{D*}/c^{D}}{c_S^{D*}/c_S^{D}}\right) - 1\right] \times 10^4, \quad f = \left(\frac{c^{P}/c^{D}}{c_S^{P}/c_S^{D}}\right) - 1. \tag{83}$$

Here, c^P and C^{D^*} refer to mole densities of parent and daughter isotope and c^D refers to reference isotope of daughter element. Normalization is done with respect to bulk silicate earth (denoted by subscript S). The expressions for spontaneous separation of crustal reservoir (τ_c) and ratio of mass of crust and mantle reservoirs (M_c/M_m) are given (Turcotte and Schubert 2002) as:

$$\begin{aligned} \tau_c &= \frac{\varepsilon_{mp}}{Qf_{mp}}, \frac{M_c}{M_m} = \left[D_{si}\left(1 - \frac{f_{cp}}{f_{mp}}\right) - 1\right]^{-1}, \\ Q &= 10^4 \frac{c_{s0}^{P}}{c_{s0}^{D*}} \lambda, D_{si} = \frac{<c_c^{D*}>}{<c_s^{D*}>} = \frac{<c_c^{D}>}{<c_{s0}^{D}>}. \end{aligned} \tag{84}$$

Here, p refers to present values and $\langle \ldots \rangle$ refers to enrichment factor at the time of crust separation from mantle. Subscripts c and m refer to crust and mantle, respectively. λ is the decay constant. This formalism has been applied using Sm-Nd and Rb-Sr isotope systems.

13 Crustal Evolution

Crust has been generated by mantle melting. Further layered structure of the crust has been formed due to intra-crustal melting. Attempts have been made to carry out population dynamics of crustal elements which have been generated over the earth's history and survived in spite of ongoing erosion and subduction. We shall give below a highly simplified treatment of the evolution of crust. Present volume of the crust is 7×10^9 km^3, having grown over time period of 3.8 Ga. Denoting volume of crust as, $V(t)$ at time t, the conservation equation for V can be written as (Gurnis and Davis 1985):

$$\frac{dV}{dt} = \dot{V}_g - \dot{V}_r, \tag{85}$$

where $\dot{V}_g$ and $\dot{V}_r$ denote rates of growth and removal. We need to relate rates of growth and removal with the existing volume of continental crust and other crust modifying processes such as subduction and erosion. Following forms for these two rates have been considered:

$$\dot{V}_g = C_g \dot{A}^{\alpha}, \qquad \dot{V}_r = C_r \dot{A}^{\beta} V^{\nu}. \tag{86}$$

Here C_g, C_r, α, β, and ν are constants and $\dot{A}$ equals the average plate velocity times ridge length. At present $\dot{A} = 3$ km^3/yr. Heat flow (q)is proportional to square root of plate velocity. So we can take $q \propto \sqrt{\dot{A}}$. Thus, the equation for V is:

$$\frac{dV}{dt} = C_g q^{2\alpha} - C_r q^{2\beta} V^{\nu}. \tag{87}$$

Given the time dependence of q as:

$$q(t) = \exp(-\lambda t), \quad \lambda = \ln 2(\tau_{1/2}), \quad \tau_{1/2} = 2 - 4\text{Ga}, \tag{88}$$

the form of $V(t)$, the volume of crust, can be determined. Following special cases have been solved in Gurnis and Davis (1985):

(i) No recycling ($C_r = 0$, $\alpha = 1$):

$$V(t) = (C_g/2\lambda)\left(1 - e^{-2\lambda t}\right). \tag{89}$$

(ii) Constant recycling rate ($\alpha = 1$, $\beta = 0 = \nu$):

$$V(t) = (C_g/2\lambda)\left(1 - e^{-2\lambda t}\right) - C_r t. \tag{90}$$

(iii) Rate dependent recycling ($\nu = 0$, $\beta \neq 0$):

$$V(t) = \frac{C_g - C_r}{2\lambda}\left(1 - e^{-2\lambda t}\right), \quad \text{for } \beta = 1. \tag{91}$$

$$V(t) = (C_g/2\lambda)\left(1 - e^{-2\lambda t}\right) - (C_r/4\lambda)\left(1 - e^{-4\lambda t}\right), \quad \text{for } \beta = 2 \tag{92}$$

(iv) Volume dependent removal rate ($\alpha = 1, \nu = 1$):

$$V(t) = \frac{C_g}{2\lambda - C_r}\left(e^{-C_r t} - e^{-2\lambda t}\right), \qquad \text{for } \beta = 0. \tag{93}$$

$$V(t) = \frac{C_g}{C_r}\left(1 - e^{-C_r t}\exp\left(\frac{C_r}{2\lambda}e^{-2\lambda t}\right)\right), \quad \text{for } \beta = 1. \tag{94}$$

These equations can be used to find evolution of the volume of the crust.

14 Melt Extraction Velocity

As pressure release melting is the main process for partial melt generation, it is necessary to know how mantle upwells (passive, active or both). Melt distribution and flow can be envisioned as porous flow or channel flow. In case of porous flow models, the velocities are found to be of the order of few meters per year whereas faster velocities are possible in the case of channel flows. Which one of these is more reasonable for MORB or other types of melt sources can be determined by looking into signatures of short time processes. Stracke et al. (2006) have analyzed U-series decay isotopes in the Icelandic lavas and found evidences of short times for melt transport to the surface. Thus, channel flow model fits better with available data.

15 Mantle Heterogeneities

When we characterize mantle and crust as reservoirs, we take them to be homogeneous, ignoring small-scale heterogeneities. It has been shown that mantle convection assimilates the subducted materials with chaotic mixing. Evidences of such stirring have also been seen in the rocks brought at the surface. Quantification of such processes will require constructing suitable mathematical models and confronting with data. There are two ways to make progress. One can make fully physics based models. But problem here is that there are too many physical and chemical parameters to be known for making progress, the uncertainty of result being decided by least known parameter. With current level of development this is not profitable approach. Another approach is to use statistical methods. One can here theoretically obtain the probability distribution of isotopic ratios and confront them with observed data to know the nature of controlling parameters, such as fraction of melt, average duration between melting and nature of mantle stirring. Significant progress has been made in this direction. Rudge et al. (2005) have derived a comprehensive statistical framework, extending formulations and ideas of several previous workers (Allegre and Lewin 1995a, b; Slater et al. 2001; Kelllog et al. 2002; Meinborn and Anderson 2003), and applied it to model the isotopic variability of MORB. Melt fraction is found to be 0.5 % and average time since parcel last melted as 1.4–2.4 Ga.

16 Concluding Remarks

Modelling of the partial melting processes in general requires the study of multiphase media including matrix deformation. The detailed petrological modelling will need to consider space- and time- variations of temperature, stress, fluid mass and concentration of chemical species which are coupled (Sleep 1974; McKenzie 1984; Ribe 1987; Dobran 2001; Steefel et al. 2005). The physico-chemical processes will be described by a coupled set of nonlinear partial differential equations. It is seen that the nonlinear systems show extremely complex behavior such as bifurcation, deterministic chaos, self-organization and spatio-temporal chaos. Petrological complexity as observed in the field and also required to interpret the geochemical and geophysical data needs an approach as mentioned above. Several critical problems have been solved in the literature.

Acknowledgments RNS is grateful to INSA, India for the award of a Senior Scientists scheme to him. Contribution under PSC0204 (INDEX) and MLP6107-28 (AM).

References

Abramowitz M, Stegun IA (eds) (1964) Handbook of Mathematical Functions with Formulas, Graphs, and Mathematical Tables. New York: Dover Publications
Ahren JL, Turcotte DL (1979) Magma migration beneath an ocean ridge. Earth Planet Sci Lett 45:115–122
Albarede F (1995) Introduction to geochemical modeling. Cambridge University Press, Cambridge
Allegre CJ, Lewin E (1995a) Scaling laws and geochemical distributions. Earth Planet Sci Lett 132:1–13
Allegre CJ, Lewin E (1995b) Isotropic systems and stirring times in the earth's mantle. Earth Planet Sci Lett 136:629–646
Asimov PD, Hirschmann MM, Stolper EM (1997) An analysis of variations in isentropic melt productivity. Phil Trans Roy Soc Lond A355:255–281
Dobran F (2001) Volcanic eruptions: mechanisms in material transport. Kluwer Acadamic/ Plenum Publishers, New York
Ellam RM (1992) Lithospheric thickness as a control of basalt geochemistry. Geology 20:153–156
Foucher JP, PichonX Le, Sibuet JC (1982) The ocean-continent transition in the uniform lithospheric stretching model: role of partial melting in the mantle. Phil Trans Roy Soc Lond A305:27–43
Ganguly J (2005) Adiabatic decompression and melting of mantle rocks: an irreversible thermodynamic analysis. Geophys Res Lett 32:L06312. doi:10.1029/2005GL022363
Ganguly J (2008) Thermodynamics in earth and planetary sciences. Springer, New York, p 501
Ganguly J, Saxena SK (1987) Mixtures and mineral reactions. Springer, New York

Gurnis M, Davies GF (1985) Simple parametric models of crustal growth. J Geodyn 3:105–135
Kellogg JB, Jacobsen SB, O'Connell RJ (2002) Modeling the distribution of isotopic rations in geochemical reservoirs. Earth Planet Sci Lett 204:183–202
McKenzie D (1984) The generation and compaction of partially molten rock. J Petrol 25:713–765
McKenzie D (1985) 230Th–238U disequilibrium and melting processes beneath ridge area. Earth Planet Sci Lett 72:149–157
McKenzie D, Bickle MJ (1988) The volume and composition of melt generated by extension of the lithosphere. J Petrol 29:625–679
McKenzie D, O'Nions RK (1991) Partial melt distribution from inversion of rare earth element concentrations. J Petrol 32:1021–1091
McKenzie D, Jackson J, Priestley K (2005) Thermal structure of oceanic and continental lithosphere. Earth Planet Sci Lett 233:337–349
Meinbom A, Anderson DL (2003) The statistical upper mantle assemblage. Earth Planet Sci Lett 217:123–139
Ribe NM (1987) Theory of melt segregation: a review. J Volcano Geotherm Res 33:241–253
Rudge JF, McKenzie D, Haynes PH (2005) A theoretical approach to understanding the isotopic heterogeneity of mid-ocean ridge basalt. Geochim Cosmochim Acta 60:3873–3887
Shaw DM (2006) Trace elements in magmas: a theoretical treatment. Cambridge University Press, Cambridge
Sigurdsson H (1999) Melting the earth: the history of ideas on volcanic eruptions. Oxford University Press, Oxford
Slater L, McKenzie D, Gronvold K, Shimazu N (2001) Melt generation and movement beneath Theistareykir, NE Iceland. J Petrol 42:321–354
Sleep NH (1974) Segregation of magma from a mostly crystalline mesh. Bull Geol Soc Am 85:1225–1232
Steefel CI, DePaolo DJ, Lichtner PC (2005) Reactive transport modeling: an essential tool and a new research approach for the earth sciences. Earth Planet Sci Lett 240:539–558
Stracke A, Bourdon B, McKenzie D (2006) Melt extraction in the mantle: constraints from U–Th–Pa–Ra studies in oceanic basalts. Earth Planet Sci Lett 244:97–112
Turcotte DL, Schubert G (2002) Geodynamics, 2nd edn. Cambridge University Press, Cambridge
Williams RW, Gill JB (1989) Effects of partial melting on the uranium decay series. Geochi Cosmochim Acta 53:1607–1619

Geochemical Modelling of Melting and Cumulus Processes: A Theoretical Approach

K. Vijaya Kumar and K. Rathna

Abstract Mantle melting and fractional crystallization are two fundamental processes that control the compositional variations in the erupted basaltic melts. Most of the geochemical variations in the primary mantle-derived melts are generally attributed to mantle heterogeneity. With simple forward modelling techniques, it is illustrated that different melting types (batch, fractional and continuous) are capable of producing large variations in elemental abundances in the primary magmas derived from a homogeneous mantle source. There is no reason to invoke mantle heterogeneity. Similarly, variations in the cumulate and corresponding residual liquid geochemistry are demonstrated alluding to rare earth element abundances. Variable partition coefficients for cumulus phases and different amounts of intercumulus liquid depict pseudo liquid-lines-of-descent in cumulate rocks on X–Y type plots. It is documented that assimilation fractional crystallization (AFC) produces greater variations in trace element concentrations in both cumulates and residual liquids. It is found that infinitesimally small solid and liquid compositions, amended by AFC, have similar trace element abundances. Two Indian examples are cited to support the theoretical modelling of mantle melting and cumulus processes presented here.

1 Introduction

Upper mantle is the source of basalt; therefore, mantle-melting models are postulated mostly based on the composition of erupted basaltic melts. On the basis of average composition of the erupted basalt, geochemists presumed that partial melting of the mantle produces a "batch" of primary basaltic melt with uniform

---

K. Vijaya Kumar (✉) · K. Rathna
School of Earth Sciences, SRTM University, Nanded 431606, India
e-mail: vijay_kumar92@hotmail.com

S. Kumar and R. N. Singh (eds.), *Modelling of Magmatic and Allied Processes*, Society of Earth Scientists Series, DOI: 10.1007/978-3-319-06471-0_3,

composition. As a result, the compositional variations in a cogenetic suite of basalts are generally considered to be a result of fractional crystallization of this homogeneous primary magma (Bowen 1928). Erupted primary basalt (expectedly with high Mg# and compatible element concentrations), as it turned out, represents the average composition of aggregated partial melts derived from mantle. This realization, and a better understanding of the physics of mantle melting (McKenzie 1984 and 1989) and detection of crystallized products of small melt fractions (supposedly derived from a homogeneous mantle source) with a wide range of trace element compositions (Langmuir et al. 1977; Vijaya Kumar et al. 2006 and references cited therein) has dramatically changed our early thoughts on the magma generation processes within the upper mantle.

A second process that controls the compositional spectrum of igneous rocks is fractional crystallization. Cumulate successions from layered complexes record the magma chamber processes encrypted within the individual layers and minerals. Magma replenishments are well documented in the reversals of mineral compositions and intermittent disappearance of some phases (Namur et al. 2010 and references therein). Homogeneity and diversity in trace element chemistry in the erupted basalt compositions depend on whether the magma chambers are long-lived or ephemeral respectively (Meyer et al. 1989). The present contribution aims to provide a critical but generalized overview of the current models on mantle-melting and magma chamber processes, and their bearing on petrogeny.

2 Mantle Melting Mechanisms

Localization of present day volcanism indicates that mantle melts only at isolated pockets, and the composition of melts (basaltic rather than peridotitic) points out that the melting is partial. Implication of these observations is that the required conditions of mantle melting are available only locally and the accessible temperature is not sufficient to melt the mantle completely. Under normal mantle conditions, the solidus of the peridotite is much higher than the ambient mantle temperature at any given pressure; therefore, the melting does not take place. Taking a simpler view of the generation of partial melts, leaving aside complex mathematical models of mantle melting, there are at least two ways in which such melting can be induced (1) by increasing the ambient mantle temperature to intersect the solidus or alternatively decreasing the solidus temperature of the mantle peridotite to cross the geothermal gradient. The former mechanism is believed to operate when the rising mantle plume impinges upon the lithosphere. The potential temperature in the mantle plume is at least 200 °C higher than the lithospheric mantle; consequently, as the plume up-wells, melt is generated from the lithosphere due to heat released from the plume (Fig. 1a). In such isobaric melting processes, temperature increases with extents of melting; therefore only small melt fractions will be produced from the lithospheric mantle. Plume induced melting (i.e. increasing the geothermal gradient) of the lithosphere is similar to the

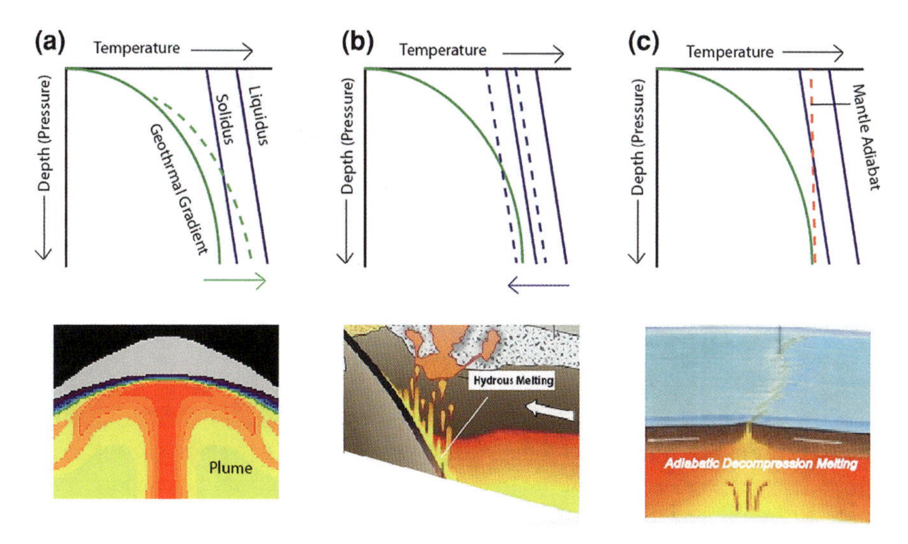

Fig. 1 Schematic sketch illustrating the initiation of melting in the *upper* mantle. Solidus temperature of peridotite mantle increases around 10–12 °C/Kbar; for an increase of 30 Kbar (~ 100 km) the solidus temperature increases by 330 °C. To initiate melting the ambient temperature of the mantle at any given pressure (expressed by geothermal gradient) should cross the solidus of the mantle peridotite. There are at least two ways by which the geothermal gradient intersects the mantle solidus: (1) either perturbation of lithosphere geotherm by a deep up-welling plume (**a**), or addition of fluids to mantle peridotite at, for example, subduction zones, thus decreasing its liquidus and solidus (**b**) and (2) adiabatic decompression melting that takes place in rising mantle diapirs (**c**)

melt production in the crust due to contact metamorphism. Of course, these minor volume melt fractions are in addition to the huge melt produced by decompression melting of plume. Decrease in the solidus temperature of the mantle peridotite may be achieved by addition of fluids (wet solidus of mantle is at least 200 °C lower than the dry solidus) which initiates mantle melting (Fig. 1b). Such a mechanism is often related to fluids derived from subducting oceanic lithosphere in subduction zones. Upwelling fluids from the deeper mantle (mantle degassing) can also be another source. Both these processes produce small amounts of melt. (2) A significant mechanism that promotes huge magmatism is adiabatic decompression melting of the upwelling mantle as (operates) in oceanic ridges, continental rifts and plumes (Fig. 1c). Due to rifting/up-warping, the overburden on the convecting asthenospheric mantle/plume is removed; as the diapir up-wells, the mantle adiabat may intersect the peridotite solidus at variable depths and initiates melting. The amount of melt produced depends upon the water content and potential temperature of the source; both have a positive impact on the amount of melt produced. The composition of melt so generated by these mechanisms depends on (1) mineralogical and chemical composition of the source, (2) residual mineralogy, (3) depth and degrees of melting and (4) type of melting process.

Before discussing the different types of mantle melting, it is worthwhile to evaluate, why rare earth or other incompatible elements are used to model the mantlemelting processes. Major element compositions of source regions for basalts are likely to be constant excepting possibly Fe abundances. Major element compositions, especially Mg and Fe, of the basaltic liquids are successfully modeled using trace element modelling approaches to decipher Fe/Mg ratios of the mantle sources (Hanson and Langmuir 1978; Langmuir and Hanson 1980; Rajamani et al. 1985, 1989; Vijaya Kumar et al. 2006). However, geochemical modelling of mantle melting processes generally uses trace elements, because their activity in the solid and liquid under equilibrium are proportional to their concentrations (i.e. obeys Henry's law); therefore, their distribution can be expressed by a partition coefficient (Kd_i). Unlike fractional crystallization, which is a liquid dominated mechanism, partial melting is a solid dominated process. Therefore, elements which preferentially partition into the solids (compatible elements with $D > 1$) are not suitable to model the types or extents of melting as their composition in the liquids will be buffered by the residual minerals. For example, Ni would preferentially partition into residual olivine during partial melting, with the result the composition of Ni in the melts formed by different extents of melting would be constant even up to very large degrees of melting ($\sim$50 %). On the contrary, the elements which preferentially partition into the liquid (incompatible elements with $D \ll 1$) would be very sensitive to different degrees of melting. Hence, we infer that mantle melting would be better modeled by incompatible elements whereas fractional crystallization processes are modeled by compatible elements. This contrasting behavior of compatible and incompatible elements is utilized to construct the process identification diagrams (Allegre and Minster 1978) and is extensively used to model melting and crystallization processes (e.g. Rajamani et al. 1985). One such diagram is illustrated in a plot between Ni versus Ce (Fig. 2). Melts formed by different extents of melting plot parallel to Ce axis (i.e. very high Ce variation), whereas melts that are related by different degrees of fractional crystallization will be parallel to Ni axis (i.e. very high Ni variation).

Additionally, the partition coefficients of elements are controlled by pressure and temperature with the effect being more pronounced on the compatible elements (Irving 1978). The effect of temperature on the Mg partition between olivine and liquid is shown by considering a set of amphibolites from the Nellore-Khammam Schist Belt in Peninsular India (Vijaya Kumar et al. 2006; Fig. 3a). The effect is much more pronounced on Ni partition between olivine and liquid (Fig. 3b). Therefore, using constant mineral/melt partition coefficients (Kds) for compatible elements during partial melting, for example in a rising mantle diapir, will lead to erroneous results.

The major mineral phases in the mantle—olivine and orthopyroxene are sterile, in the sense that they have very low partition coefficients for incompatible elements including REE. Hence, any variation in the partition coefficients due to changes in pressure and temperature will be less drastic on incompatible elements. The rare earth elements are (1) incompatible during mantle melting, (2) geochemically coherent, (3) immune to moderate amounts of fractionation, i.e. intra

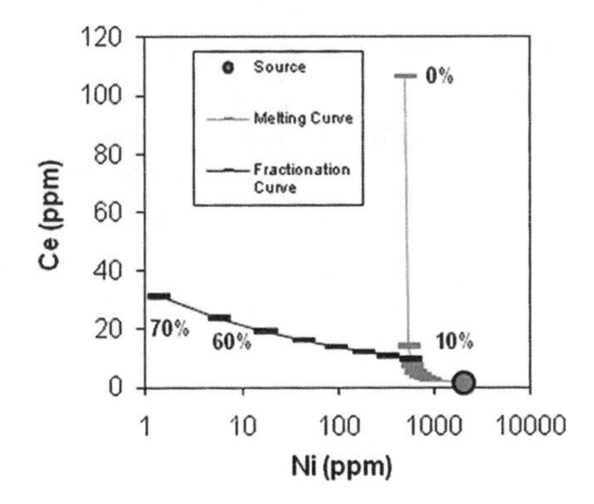

Fig. 2 Ni versus Ce plot showing the liquid compositions calculated for different extents of partial melting and fractional crystallization processes. Modal batch melting ($C_i^l/C_i^o = 1/D_i^o(1-F)+F$; Schilling and Winchester 1967) and Rayleigh fractionation ($C_i^l/C_i^o = F^{D_i-1}$; Neumann et al. 1954) equations are used for calculations. Ni and Ce concentrations of mantle peridotite source are taken as 2000 and 1.6 ppm respectively. Bulk distribution coefficients for Ni and Ce are 4 and 0.015 during melting, and 6 and 0.02 during fractional crystallization. Each *tic mark* indicates 10 % increments both on melting and fractionation curves

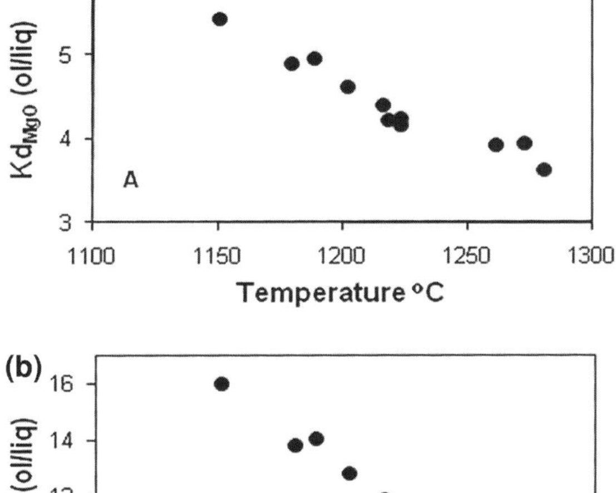

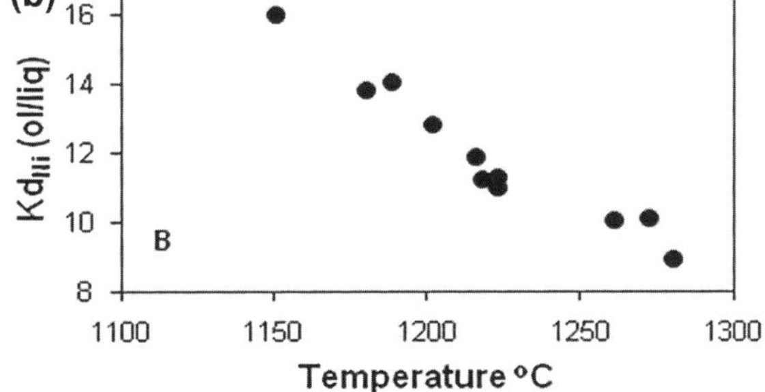

Fig. 3 Calculated $Kd_{Mg}^{(ol/liq)}$ (**a**) and $Kd_{Ni}^{(ol/liq)}$ (**b**) versus liquidus temperature for a set of amphibolites from the Nellore-Khammam Schist Belt (Vijaya Kumar et al. 2006). $Kd_{Mg}^{(ol/liq)}$ and one-atmosphere liquidus temperature are calculated using [Mg]–[Fe] program of Rajamani et al. (1985); $Kd_{Ni}^{(ol/liq)}$ is calculated using the equation $Kd_{Ni}^{(ol/liq)} = \left[3.92 \times Kd_{Mg}^{(ol/liq)} - 5.30\right]$ (Jones 1984)

REE ratios do not change with fractionation and (4) reasonably resistant to alteration; for these reasons they are the favored geochemical species to model mantle melting processes.

2.1 Types of Melting Models

Depending on the level of equilibrium between the partial melt and mantle residue, at least three types of melting models are formulated. The simplest and most favored model by the geochemists is the "batch" or "equilibrium" melting first framed by Schilling and Winchester (1967) and applied by Gast (1968). But the trace element compositional variations in basalts (Langmuir et al. 1977), clinopyroxene in basalts (Johnson et al. 1990) and ultra-depleted cumulates (Piccardo and Guarnieri 2011), and melt inclusions (Sobolev and Shimizu 1993; Gurenko and Chaussidon 1995), cannot be accounted for by equilibrium melting processes. To explain these compositional peculiarities in the basaltic melts, "fractional" melting (Bowen 1928; Presnall 1969) and "continuous" melting (Langmuir et al. 1977) processes are devised. In the following sections, each process is described and the influence of each process on the rare earth element compositions of melts is illustrated.

2.2 Batch Melting

In batch or equilibrium melting (Presnall 1969), after a certain fraction of melting, the melt is extracted from the solid phases in a single batch, and there is no further melting of the refractory source. Batch melting model corresponds to a closed system in which chemical equilibrium is achieved between solid and liquid i.e. the bulk composition of system remains unchanged until melting is completed. During melting in an adiabatically rising diapir, new fertile material is added to the melting column as different batches of partial melts are extracted to keep the source composition constant (Zou 1998).

Earlier models of mantle melting assumed that the melting is modal i.e. melting proportions of minerals are same as their original abundances in the source. But later it is realized that melting proportions of minerals are not related to the original abundances of minerals in the source, but are controlled by eutectic proportions (nonmodal melting; Shaw 1970). Since the mantle undergoes eutectic melting, nonmodal melting models provide more consistent results. All the calculations of melt compositions in the present paper are done by using nonmodal melting equations. Equations for all the melting models are given in the Appendix. If P_i is substituted by D_i, the nonmodal equations would become modal melting equations. Mantle source compositions, melting proportions and partition coefficients adopted in this study are given in Appendix.

Melts formed by different extents of batch melting of spinel-peridotite source with 2 $\times$ Chondrite REE abundances are illustrated in Fig. 4. With increasing extents of melting, fractionation of LREE over HREE decreases and the patterns would become parallel to that of the source once the degrees of melting is $\sim$20 %.

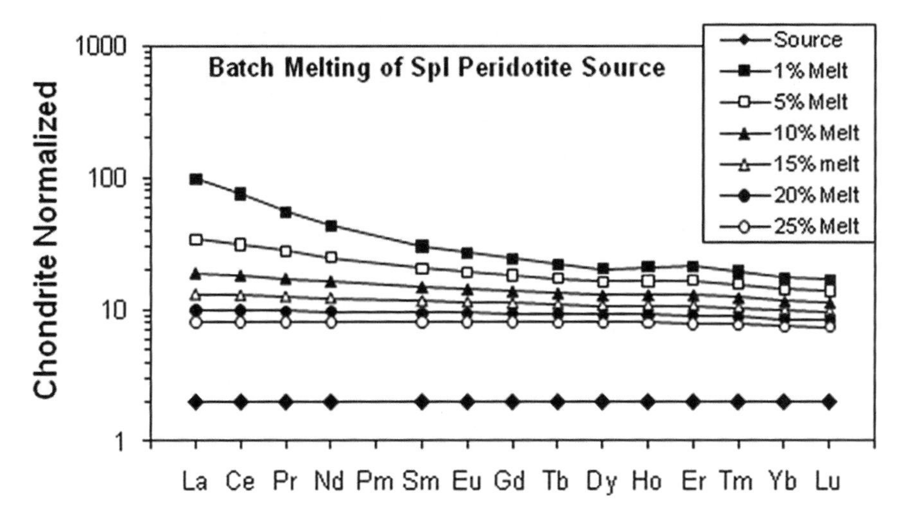

Fig. 4 Chondrite-normalized REE plot illustrating the different extents of (nonmodal) batch melting of spinel-peridotite source with 2 × Chondrite abundances. Chondrite values are from Hanson (1980). Source mineralogy, melting proportions and partition coefficients are as given in Appendix. By 20 % melting, the melt REE pattern becomes parallel to that of the source irrespective of the original mineralogy of the source

The rare earth element concentrations in the melts are basically controlled by the abundance of clinopyroxene and aluminous phases in the residue (plagioclase, spinel, garnet, phlogopite and amphibole). The alumina phases and clinopyroxene melt preferentially as the melt fraction increases (see Appendix). Therefore, any type of mantle melting would fractionate LREE over HREE when the degrees of melting are lower. Once the degree of melting is greater than 20 %, the residual mineralogy would be dominated by olivine and orthopyroxene, which have low partition coefficients for REE, as a result of which the melt composition pattern would be parallel to that of the source composition irrespective of the original source mineralogy (Fig. 4). Under any extents of batch melting, the melt composition will never be more depleted than the source composition.

2.3 Fractional Melting

In fractional melting small melt fractions are continuously and completely extracted as the melting proceeds (Presnall 1969). Only the last drop of the liquid is thought to be in equilibrium with the residue. Different extents of fractional melts from spinel-peridotite source are shown in Fig. 5. By second increment of melting (see 5 % melt pattern) $(La/Ce)_N$ ratios fall to less than one. Fractional

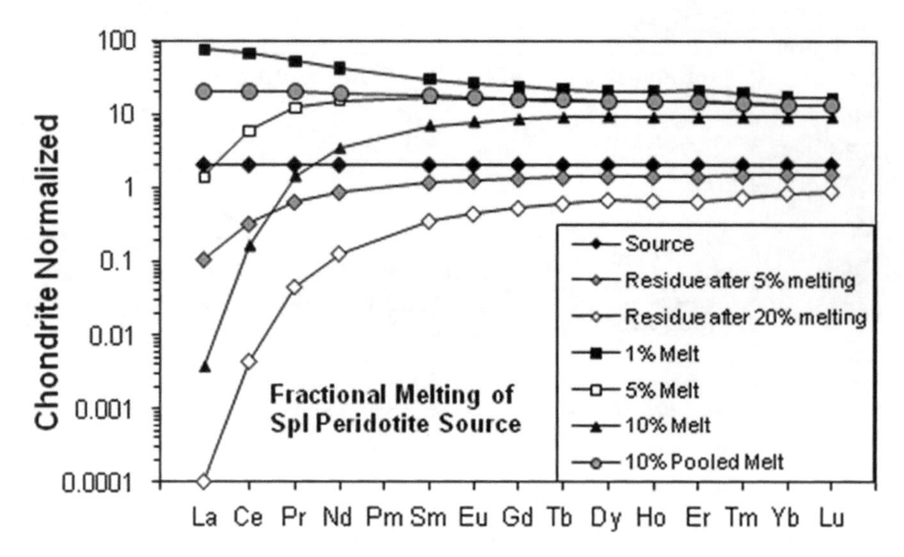

Fig. 5 Chondrite-normalized REE plot illustrating the different extents of (nonmodal) fractional melting of spinel-peridotite source with 2 × Chondrite abundances. Source mineralogy, melting proportions and partition coefficients are as given in Appendix. During melting, it is assumed that the mineral/melt partition coefficients (Kd_i) remain constant, but the source composition continuously changes with progressive melting. Residue (effective source to later melts) composition is calculated using the equation $C_s^o = C_i^l\left[\left(D_i^o - FP_i\right)/(1 - F)\right]$ and denoted as source (original source; parental to 1 % melt), residue after 5 % melting (parental to 10 % melt) and so on in the legend. The notations are the same as given in the Appendix. Mineral phases in the residue are calculated using the equation $P_o = FP_l + (1 - F)P_r$ where P_o = Mass proportion of any phase in the original source, F = % of melting, P_l = Mass proportion of phase entering the liquid and P_r = Mass proportion of the phase in the residue. Fractional melting is extremely efficient in fractionating incompatible elements and affecting the highly incompatible element ratios (e.g. La/Ce ratios). 10 % pooled melt is also shown in the Figure

melting processes are particularly effective in fractionating highly incompatible elements for small melt fractions (for example, see La/Ce ratios in Fig. 5). Although fractional melting explains the ultra depleted melt compositions (see Fig. 5) and intra-elemental fractionations among incompatible elements (e.g. fractionation of U/Th series), its general application to model partial melting is not straight forward for two reasons: (1) whenever a mineral phase exhausts during progressive fractional melting (which is much faster than the batch melting), temperature required to melt the residue will be abnormally high and may not be available at the source and (2) the source becomes alarmingly depleted in incompatible elements by fourth or fifth increment melting (assuming 2 % increments; see the source REE patterns in Fig. 5).

2.4 Pooled or Aggregated Melts

Aggregated or pooled melts can be described as the assemblage of different small melt fractions in the proportions they are obtained from the source. The pooled melt will correspond to a weighted mean of the compositions of all the instantaneous melts. Such an averaging mechanism decreases the efficacy of incompatible element fractionation (Fig. 5).

2.5 Continuous Melting

Continuous melting can be defined as an intermediate process between batch and fractional melting, where the instantaneous melt is continuously but not completely removed as the melting proceeds, so a part of the melt is always retained in the residue (Langmuir et al. 1977). In other words, continuous melting is fractional melting with residual porosity. For continuous melting, bulk composition of that part of the system, where melting is taking place, changes continuously as the instantaneous melt is extracted. This makes the effective parent to later derived melts increasingly refractory and produces melts with depleted REE patterns (Fig. 6a). Composition of continuous melts is intermediate between batch and fractional melts (see Figs. 4, 5 and 6). A review of the mathematical equations for continuous melting models is provided by Zou (1998) and Shaw (2000). Figure 6a is constructed using critical continuous melting equation proposed by Sobolev and Shimizu (1993; see Appendix).

2.6 Dynamic Melting

In reality a gamut of processes (batch, continuous and fractional melting, and zone refining) will be taking place simultaneously during the mantle melting. Liquid extracted from the mantle is an outcome of the sum effect of all these processes. Although, it is profoundly difficult to model these processes concurrently, availability of crystallized products of small melt fractions (Vijaya Kumar 2006; Vijaya Kumar et al. 2006 and references cited therein) and mantle residues resultant of sequential mantle melting (Prinzhofer and Allegre 1985) in a few localities enable us to see through the mantle melting processes. If we assume that in a rising mantle diapir melting begins in the garnet-peridotite stability field then as melting progresses, the rising mantle column passes in sequence through garnet-spinel peridotite, spinel-peridotite, spinel-plagioclase peridotite and plagioclase peridotite (Prinzhofer and Allegre 1985; McKenzie and O'Nions 1991). The diapir is supposed to maintain its chemical homogeneity while liquids are extracted (Albarède 1996). The basaltic melts are thus obtained at different depths with different extents of

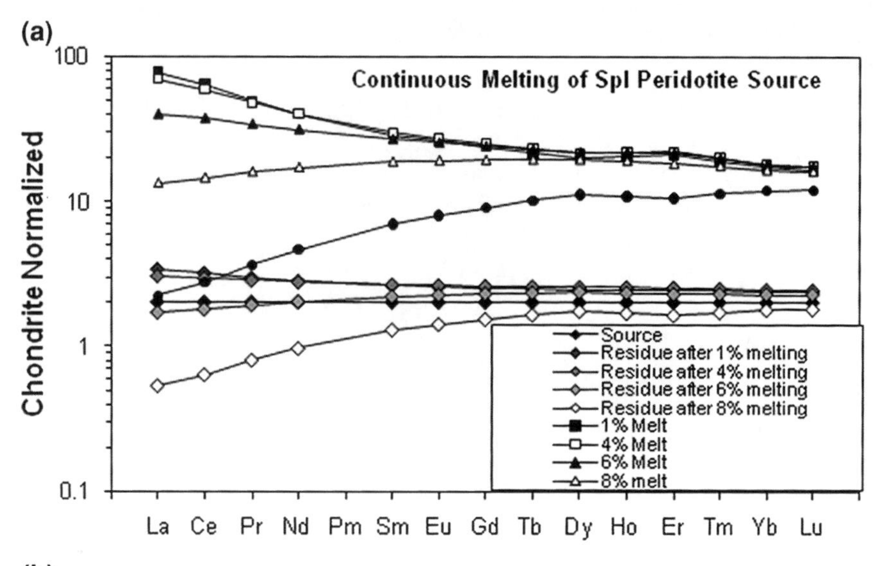
(a)
Continuous Melting of Spl Peridotite Source
Chondrite Normalized
100
10
1
0.1
Source
Residue after 1% melting
Residue after 4% melting
Residue after 6% melting
Residue after 8% melting
1% Melt
4% Melt
6% Melt
8% melt
La Ce Pr Nd Pm Sm Eu Gd Tb Dy Ho Er Tm Yb Lu

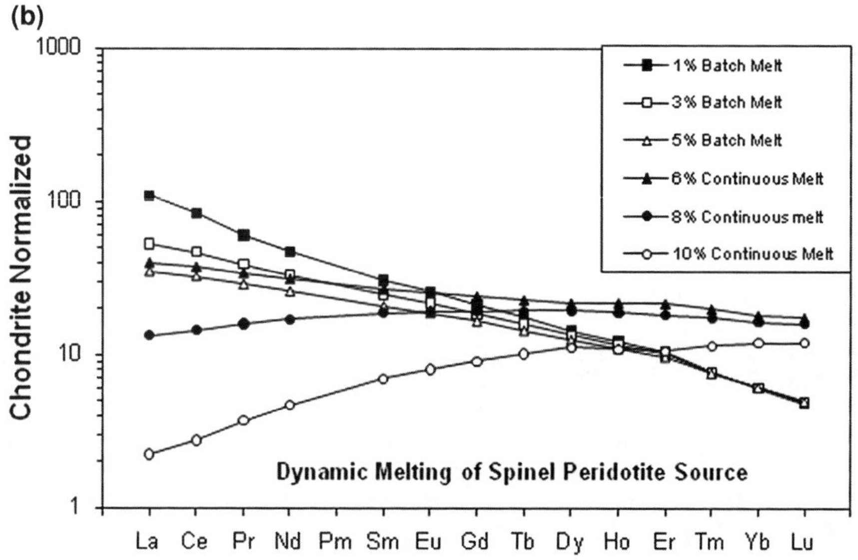
(b)
Chondrite Normalized
1000
100
10
1
1% Batch Melt
3% Batch Melt
5% Batch Melt
6% Continuous Melt
8% Continuous melt
10% Continuous Melt
Dynamic Melting of Spinel Peridotite Source
La Ce Pr Nd Pm Sm Eu Gd Tb Dy Ho Er Tm Yb Lu

◀ **Fig. 6** **a** Chondrite-normalized REE plot illustrating the different extents of (nonmodal) continuous melting of spinel-peridotite source with 2 × Chondrite abundances. Source mineralogy, melting proportions and partition coefficients are as given in Appendix. During melting it is assumed that the mineral/melt partition coefficients (Kd_i) remain constant, but the source composition continuously changes with progressive melting. Residue (effective source to later melts) composition is calculated using the equation $C_{res} = C_i^l(\alpha/\alpha + 1) + C_s(1 - \alpha/\alpha + 1)$ and $C_s = D_i^o C_i^l$ and denoted as source 1 (original source; parental to 1 % melt), residue after 1 % melting (parental to 4 % melt), residue after 4 % melting (parental to 6 % melt), and so on in the legend. The notations are the same as given in the Appendix. Continuous melting produces melt compositions that are intermediate between batch and fractional melting. For discussion, see the text. **b** Chondrite-normalized REE plot illustrating the different extents of (nonmodal) batch and continuous melting of spinel-peridotite source with 2 × Chondrite abundances. Note the crossing of REE patterns. In the rising diapir, the initial stages of melting tend to be equilibrium (batch) melting, but once the residue develops critical porosity, some partial melt will be retained in the pores of the residue, and disequilibrium (continuous) melting will set in (Prinzhofer and Allegre 1985). A composite of the melts produced by both batch (or aggregated) and continuous melting of a chemically homogeneous source is what constitutes dynamic partial melts

melting. In the rising diapir, the initial stages of melting tend to be equilibrium (batch) melting, but once the residue develops critical porosity, some partial melt will be retained in the pores of the residue, and disequilibrium (continuous) melting will set in (Prinzhofer and Allegre 1985). A composite of the melts produced by both batch (or aggregated) and continuous melting of a chemically homogeneous source is what constitutes dynamic partial melts (Fig. 6b). Dynamic partial melts characteristically show both the LREE enriched and depleted patterns with crossing REE patterns (Fig. 6b).

The dynamic melting model has been effectively used to explain the trace element variations of basaltic and related rocks from a variety of tectonic environments including tholeiites from the Mid-Atlantic ridge, Iceland, Troodos massif, Hawaii and Newfoundland, Precambrian Nellore-Khammam Schist Belt, and cogenetic komatiites and basalts from Gorgana (Langmuir et al. 1977; Wood 1979; Strong and Dostal 1980; Elliott et al. 1991; Eggins 1992; Sobolev and Shimizu 1993; Gurenko and Chaussidon 1995; Arndt et al. 1997; Vijaya Kumar et al. 2006).

An Indian example (amphibolites from the Nellore-Khammam Schist Belt, South India; Vijaya Kumar et al. 2006) is highlighted here as a specific example of the application of the concept of dynamic melting. The Nellore-Khammam Schist Belt (NKSB) amphibolites are tholeiites with highly variable incompatible trace element abundances for similar Mg#s, relatively constant compatible element concentrations, and uniform incompatible element ratios. Chondrite-normalized REE patterns of the tholeiites range from strongly LREE depleted $(La/Yb)_N = 0.19$) to LREE enriched $(La/Yb)_N = 6.95$). Constant $(La/Ce)_N$ ratios but variable $(La/Yb)_N$ values are characteristic geochemical traits of the tholeiites; the latter has resulted in crossing REE patterns especially at the HREE segment (see Fig. 7). Even for the most LREE depleted samples, the $(La/Ce)_N$ ratios are >1 and are similar to those of the LREE enriched samples. There is a systematic decrease in FeO^t, K_2O and P_2O_5, as well as Ce and other incompatible elements from the

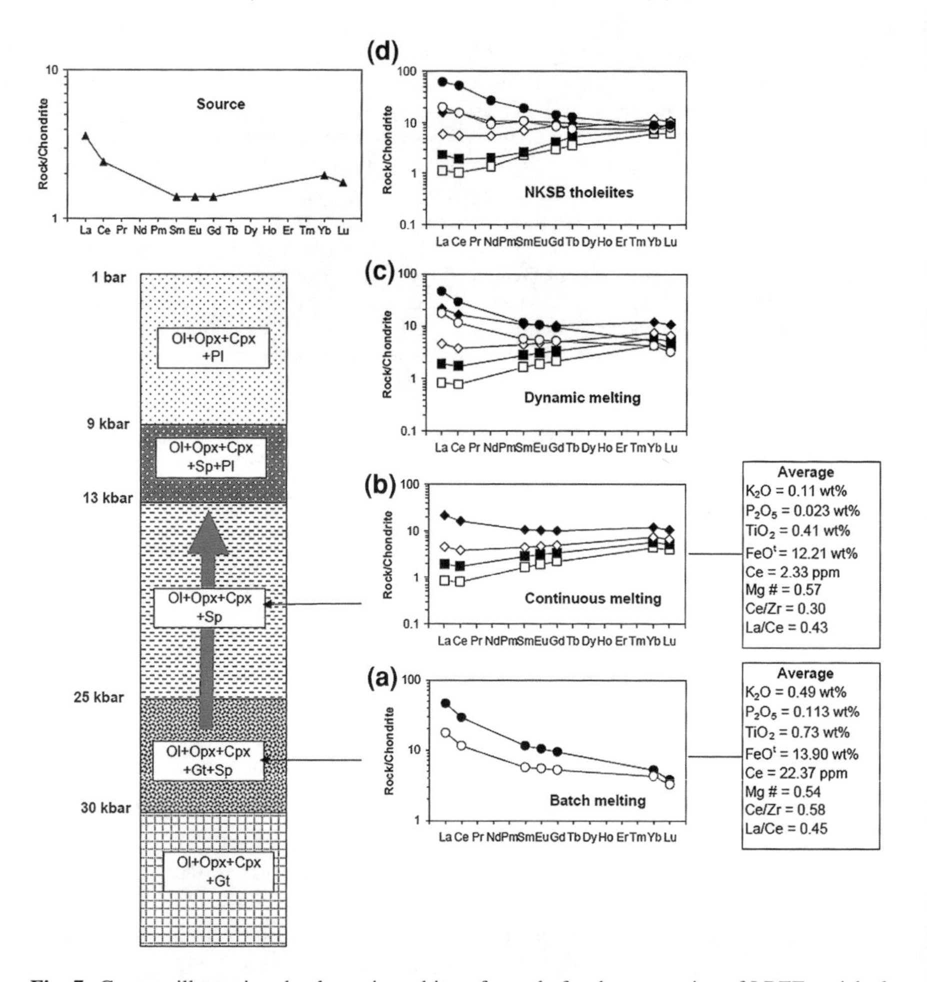

Fig. 7 Cartoon illustrating the dynamic melting of mantle for the generation of LREE enriched and depleted patterns from a homogenous source. Samples with enriched LREE and other incompatible elements (**a**) are considered to be early derived melts by batch melting within garnet-spinel stability field whereas LREE depleted melts represent the final fractions of melts produced by critical (continuous) melting of the mantle within spinel-peridotite stability field (**b**). Composite of **a** and **b** results in the dynamic partial melts (**c**). Dynamic mantle melting model impeccably explains the REE patterns (**d**) of the NKSB tholeiites. Average abundances of elements for LREE enriched and depleted NKSB amphibolites are also shown in the figure. Mantle source REE compositions are calculated by inversion modelling of the melts. Boundaries for mantle phase assemblages are taken from Vijaya Kumar et al. (2006 and references cited therein)

LREE enriched to the depleted samples without any variation in the incompatible element ratios and Mg#s (see the averages given in Fig. 7). Neither batch and fractional melting, nor magma chamber processes can explain the non-correlation between the LREE enrichment and HREE concentrations. The geochemical peculiarities of the NKSB tholeiites are modeled using dynamic partial melt models. Polybaric dynamic melting within a single mantle column with variable mineralogy is the likely mechanism for the derivation of NKSB tholeiitic melts (Fig. 7). Samples with enriched LREE and other incompatible elements (Fig. 7a) are considered to be early derived melts by batch melting within garnet-spinel stability field, whereas LREE depleted melts represent the final fractions of melts produced by critical (continuous) melting of the mantle within spinel-peridotite stability field (Fig. 7b). A composite of the melts produced by both batch and continuous melting denotes the dynamic partial melts (Fig. 7c). Dynamic mantle melting model impeccably explains the REE patterns (Fig. 7d) and also other geochemical peculiarities of the NKSB tholeiites.

3 Cumulus Processes

The term 'cumulate' is proposed as a group name for igneous rocks formed by crystal accumulation (Wager et al. 1960). After accumulation of the crystal precipitate, crystallization processes such as compaction, episodic expulsion of variably fractionated melt, reaction between cumulus grains and entrapped melt and post-cumulus growth lead to differences in the rock finally produced. Magma chamber processes are best recorded in cumulate rocks. Parental magma compositions, conditions of crystallization, an evolutionary path of the residual liquid(s) and magma additions, if any, are faithfully recorded by variations in cumulus and intercumulus mineral-chemical compositions (see Cawthorn 1996a). Since cumulate rocks are formed by accumulation of one or more minerals in a magma chamber, they do not adequately represent the melt compositions. Consequently, the whole-rock major element geochemistry of the cumulate rocks is controlled by modal mineralogy; therefore, it is only of indirect use in inferring a parental liquid composition and evolution. If the character of the intercumulus liquid is assessed adequately, then whole-rock trace element chemistry provides significant clues regarding the nature of the parental liquid (Leelanandam and Vijaya Kumar 2004 and references therein). However, it is the chemistry of individual cumulus minerals and their mineral-chemical relationships with the other cumulus and intercumulus phases that provide important constraints on the parental liquid composition, liquid-lines-of-descent and conditions of crystallization (Claeson and Meurer 2004; Charlier et al. 2010). For example, similar Mg#s but variable CaO contents in the orthopyroxene supposedly reflect more than one parental liquid to massif anorthosite cumulates (Charlier et al. 2010). Mafic and ultramafic cumulates from the layered complexes have distinct mineral-chemical characteristics depending on the magma type and tectonic environment of formation. For example, the silicate and oxide mineral compositions in Bushveld-type stratiform intrusions are distinctly different from those of Alaskan-type arc-root complexes and Alpine-type peridotite-gabbro complexes (Irvine 1967; Jan and Windley 1990; Roeder 1994; Cookenboo et al. 1997; Zhou et al. 1997; Barnes and Roeder 2001).

3.1 Parental Liquids to Cumulate Rocks from Layered Complexes

Most of our understanding of the mechanisms of basic magma differentiation comes from the studies on cumulate rocks from layered intrusions. These repositories of igneous processes have been studied extensively using a variety of methods. Despite such rigorous studies, our understanding of the "parental liquids" that produced these gigantic bodies is rather limited (Cawthorn 1996a). A broad consensus among the studious workers on the layered complexes is that at least two parental magmas are involved in producing the layered rocks. These two distinct parental magmas, designated as "U-type" and "A-type", produced ultramafic and anorthositic magma series respectively with marked crystallization sequences (see Cawthorn 1996a for detailed discussion). It is interesting to note that gabbronorites (and also some selected ultramafic rocks) could have been produced by either parental liquid (McCallum 1996). Multiple injections of these distinct melts to the magma chamber containing variable amounts of residual magma produced unique features including reversals in crystallization sequences, mineral compositions and isotopic ratios (McCallum et al. 1980; Sharpe 1985; Eales et al. 1990). A major outcome of the mixing of these two contrasting liquids is the formation of PGE deposits with cations (PGE) deriving from U-type liquid and anions (sulfur) from the A-type melt (Barnes and Naldrett 1986; Lambert and Simmons 1988).

3.2 Simple Forward Modelling of Cumulus Processes

Fractional crystallization and the formation of cumulates is a process of fundamental importance in igneous petrology. Study of continental layered intrusions allows both small and large scale cumulus and postcumulus processes to be examined. It is likely that many of the processes that operate in continental layered intrusions are also important in the formation of oceanic cumulates.

Mafic-ultramafic cumulate rocks of layered complexes define linear trends on X–Y type major element plots (Fig. 8a) and define curvilinear pseudo liquid-lines-of-descent on trace element plots (Fig. 8b). For example, olivine has higher Kds for Ni and lower Kds for Hf, therefore dunite cumulate contain high Ni and low Hf abundances, whereas pyroxene has relatively higher Kd for Hf with intermediate Kds for Ni, while plagioclase has lower Kds for both Ni and Hf. This variation in mineral/melt partition coefficients produce curvilinear trends that can be erroneously interpreted as liquid-lines-of-descent. Cumulus phases when equilibrated with trapped melt get enriched in incompatible elements with little change in their Mg#s (Barnes 1986). Barnes (1986) demonstrated that in gabbronorite cumulates, $\sim$10 % of trapped melt leads to a shift in Mg# of 0.03. Additionally during in situ differentiation, mode of the rock undergoes little change where as the concentrations of

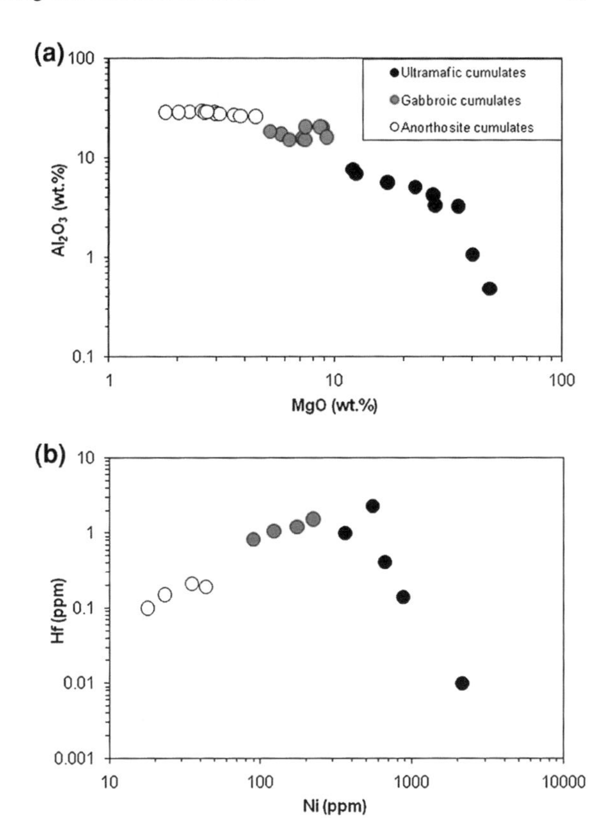

Fig. 8 **a** MgO (wt%) versus Al_2O_3 (wt%) for mafic-ultramafic cumulates from the Kondapalli Layered Complex. Linear trend is characteristic of cumulate rocks. **b** Ni (ppm) versus Hf (ppm) for mafic-ultramafic cumulates from the Kondapalli Layered Complex. Due to differential partition coefficients for trace elements for major cumulus phases and different proportions of intercumulus liquid, the cumulate rocks show pseudo liquid-lines-of-descent

incompatible elements increase with increasing formation of cumulates. Adcumulates with high incompatible elements need not be interpreted as derivatives from an evolved liquid, but can be attributed to high proportion of trapped melt (Cawthorn 1996b). For example, the elements which are essential constituents of cumulus minerals in a Plag-Opx cumulate can be explained by mixing of plagioclase and orthopyroxene with subordinate clinopyroxene/hornblende, but elements that are incompatible with cumulus phases seem to be influenced by intercumulus liquid. In the latter scenario cumulate rocks do not form a linear array in binary diagrams (Haskin and Salpas 1992). Thus, unlike the compositional variation in a liquid-lines-of-descent, the cumulate chemistry should be interpreted cautiously. The volume of trapped melt in an igneous cumulate may be controlled by the ability of the crystallizing crystal-melt system to reach textural equilibrium. One way to assess the trapped melt component is to look into the relationship between the Mg# and incompatible element composition of the cumulate rocks. Since Mg# and incompatible elements respond differently to the trapped melt, for a little variation in the Mg# there would be wide variation in the incompatible element concentrations. Additionally, the presence of cumulus accessory minerals influences the magnitude

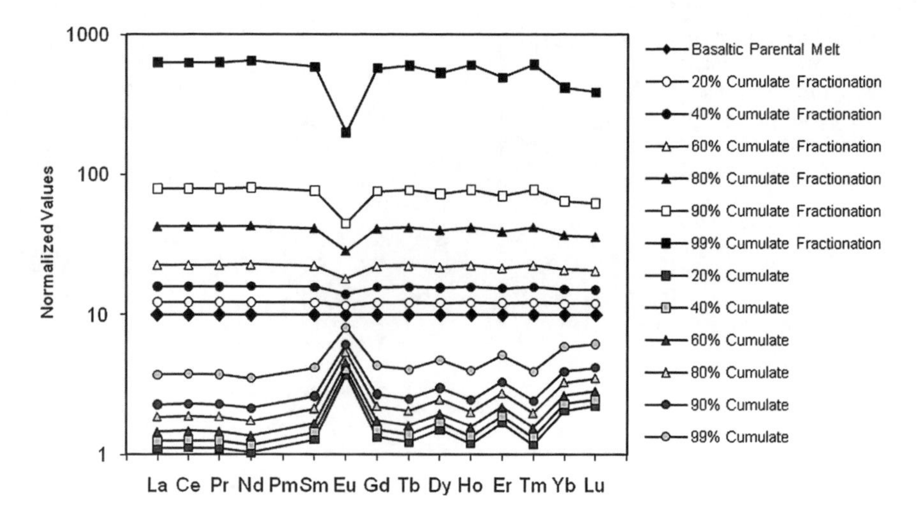

Fig. 9 Rare earth element composition of cumulates and residual liquids for different degrees of fractionation

of trace element of the cumulus rocks disproportionate to their modal abundances, and the presence of cumulus ilmenite/magnetite alters the whole rock Mg#.

Although individual cumulates are products of type and amount of cumulus minerals and trapped melt, some of the chemical parameters such as Mg#s of cumulus mafic minerals and whole rock cumulates are considered to record the evolution of the magma from which the cogenetic cumulates are formed (Tiezzi and Scott 1980). As the cumulus minerals are liquid analogues, they vary in composition as the parental magma progressively evolves. As a consequence of this, a series of genetically related cumulate rocks with little trapped melt also records changes in the parental magma composition in terms of elemental enrichment or otherwise. Depending upon the dominant cumulus phase, plagioclase or pyroxenes, the liquid becomes either more felsic or mafic. But the Mg#s of cumulus pyroxenes consequently that of whole-rocks, and the anorthite content of cumulus plagioclase decrease as the magma evolves. Mg#s of cumulates (when affected only marginally by the intercumulus liquid) may carry the signatures of magma evolution.

Both residual liquid and cumulate REE compositions are depicted through a series of plots (Fig. 9) when the parental liquid undergoes simple fractional crystallization. Liquid composition can be calculated by using the equation $C_L = C_S/D$ and published mineral/melt partition coefficients (Bedard 2001; Claeson and Meurer 2004). C_L is the concentration of element in the equilibrium liquid, C_S is the concentration of element in the solid (cumulate rock) and D is the bulk distribution coefficient ($\Sigma X_i Kd_i$); X_i is modal abundance of phase i in the solid and Kd_i is mineral/melt partition coefficient of element in phase i. Percentage of estimated intercumulus liquid is substituted as a phase in X_i with Kd_i equals to one. An olivine gabbronorite cumulate (30 % Ol + 20 % Opx + 35 % Plag + 10 % Cpx + 5 % ICL) is considered in the calculations of fractionation crystallization trends. The cumulate and residual patterns are parallel to the original parental magma with cumulates developing positive Eu anomalies and residual liquids developing negative Eu anomalies. The shape of the patterns and magnitude of the Eu anomalies depend on the relative proportions of cumulus minerals and amount of intercumulus liquid.

3.3 Effects of Assimilation Fractional Crystallization on Cumulate Rocks

Unlike the layered complexes of shallow intrusion such as Skaergaard and Kiglapait, those of deep crustal origin extensively interact with intrusion-induced partial melts of country rocks, there by camouflaging the original geochemical traits of the parental liquids. Further, very slow cooling of the magma redistributes the elements on a grain-size scale and obscures the chemistry of liquidus minerals. Distinction of the effects of crustal contamination and those of intercumulus liquid is also difficult especially when the latter equilibrates with cumulus minerals during the post-cumulus processes. Nevertheless, studying the differentiation processes in these deep magma chambers helps us in deciphering the evolutionary history of mafic magma under moderate- to high-pressure conditions in terms of fractionating mineralogy and elemental behavior under the influence of crustal melts. The effects of assimilation on the cumulate chemistry are illustrated by considering intrusion of basaltic magma into the metapelites, which induces many prograde reactional adjustments. The ones which are important include:

$$\mathrm{Bt + Sill + Qtz = Gt + Melt}$$

$$\mathrm{Spl + Bt + Sill = Gt + Melt}$$

$$\mathrm{Bt + Sill = Gt + Ti - rich\ Spl + C + Melt}$$

$$\mathrm{Bt + Sill = Gt + C + Melt}$$

$$\mathrm{Bt + Plag + Al_2SiO_5 + Qtz = Gt + K - feld + Melt}$$

The dehydration melting of biotite takes place at very early stages of pelitic melting, whereas the extensive melting may produce highly polymerized melts related to the melting of feldspars. Interaction of these crustal melts with a fractionating basaltic magma produces cumulate rocks with distinct trace element chemistry than the ones produced by simple fractional crystallization. Absence of positive anomalies in anorthosite cumulate and presence of positive anomalies in clinopyroxenite cumulate are some of the effects of assimilation fractional crystallization.

Both residual liquid and cumulate compositions are depicted through a series of plots (Fig. 10) when the parental liquid undergoes assimilation fractional crystallization with upper crust as contaminant. AFC equation $C_L = C_O F^{[(D/1 - r) - 1]} + [r / (r + D - 1)C_A(1\text{-}F^{[(D/1 - r) - 1]})]$ derived by DePaolo (1981) is used for calculation: where C_L = concentration of trace element in the contaminated magma,

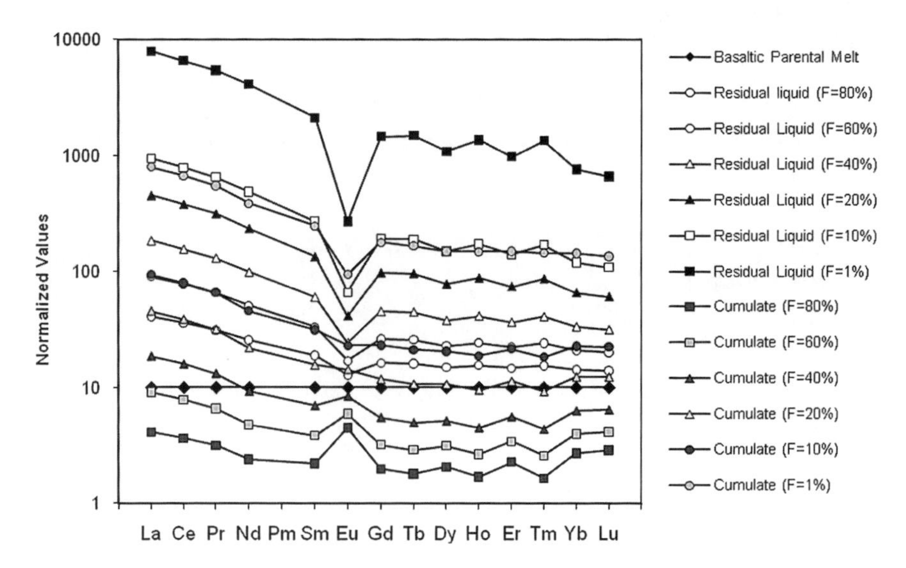

Fig. 10 Rare earth element composition of cumulates and residual liquids when the parental melt was undergoing assimilation fractional crystallization with upper crust as assimilate

C_O = concentration of trace element in the original magma, C_A = concentration of the trace element in the contaminant, F = fraction of magma remaining, D = bulk distribution for the fractionating assemblage, r = ratio of rate of assimilation to the rate of fractional crystallization (r is given a constant value of 0.2). In the modelling calculations, upper crust is considered as assimilate. Mineral/melt partition coefficients are after Bedard (2001) and Claeson and Meurer (2004). The cumulates and residual liquids affected by the assimilation of crustal material show LREE fractionated patterns with respect to flat trend of the parental magma and steepness of the trend increases with degree of fractionation. Two significant features in the cumulate rocks that are formed from a parental liquid undergoing assimilation fractional crystallization include: (1) subdued positive Eu anomalies during early stages of fractionation (see cumulate trends when F = up to 40 % in Fig. 10) and absence of Eu anomalies during the advanced stages of fractionation (see cumulate trend when F = 10 %) even in the plagioclase-rich cumulates; correspondingly residual liquids develop large negative Eu anomalies and (2) trace element chemistry of the solids (see cumulate trend when F = 1 %) and liquids (see residual liquid trend when F = 10 %) are similar. This latter observation is very significant as the cumulate chemistry can be misinterpreted as liquid composition. The importance of textural observations cannot be over emphasized.

3.4 Inversion of Cumulate Rocks

Markl and Frost (1999) have inverted the major element chemistry of cumulus minerals to calculate the equilibrium liquid composition to a massif anorthosite intrusion. But the choice of the mineral/melt Kds for major elements in such calculations depend heavily on accurate intrinsic parameters including P, T, $f(O)_2$, PCO_2 (Markl et al. 1998), and are meaningless if the cumulate rocks are altered and/or metamorphosed. Another way of obtaining the major element chemistry involves identifying a mesocumulate which is near liquid composition and correcting it for the effects of mineral fractionation or accumulation. It is akin to removing the effects of phenocrysts in a porphyritic rock. Again alteration and metamorphism make this method less dependable.

Many workers have inverted the trace and REE of the cumulus minerals and even whole rocks to get a picture of parental/equilibrium magma composition (Leelanandam and Vijaya Kumar 2004). The parental liquid calculated from the whole rock (cumulate) chemistry does not represent composition of instantaneous liquid in equilibrium with the cumulate, but only records the composition of average melt that was in equilibrium with the cumulate during the period when all the cumulus minerals crystallized.

REE compositions of a suite of ultramafic (dunite, orthopyroxenite, clinopyroxenite) and mafic (gabbronorite and anorthosite) cumulates from the Kondapalli Layered Complex are shown in Fig. 11. Continuous decrease in the Mg#s from dunite to gabbro does indicate accumulation of crystals in a continuously fractionating magma, in a way reflecting the liquid-line-of-descent of the parental melt. But the rare earth elements show wide variation as the REE concentrations depend on mineral/melt partition coefficients for the cumulus minerals and amount of intercumulus liquid. From the plot, it is evident that clinopyroxenite contains higher proportion of intercumulus liquid (Leelanandam and Vijaya Kumar 2007). To test whether ultramafic and mafic cumulate rocks derived from the same or similar parental melts, we have inverted the REE composition of clinopyroxenite and anorthosite cumulates (Fig. 11c). If a set of cumulate rocks are formed from a single parental magma, then the parental liquids calculated from different cumulate rocks are expected to show similar elemental patterns. The calculated liquids in Fig. 11c show very similar patterns irrespective of the type of cumulate from which they are calculated. Although, the absolute abundances of the trace elements calculated from the bulk compositions of the cumulate rocks are only indicative, their relative abundance patterns do suggest cogenetic nature of the cumulate rocks.

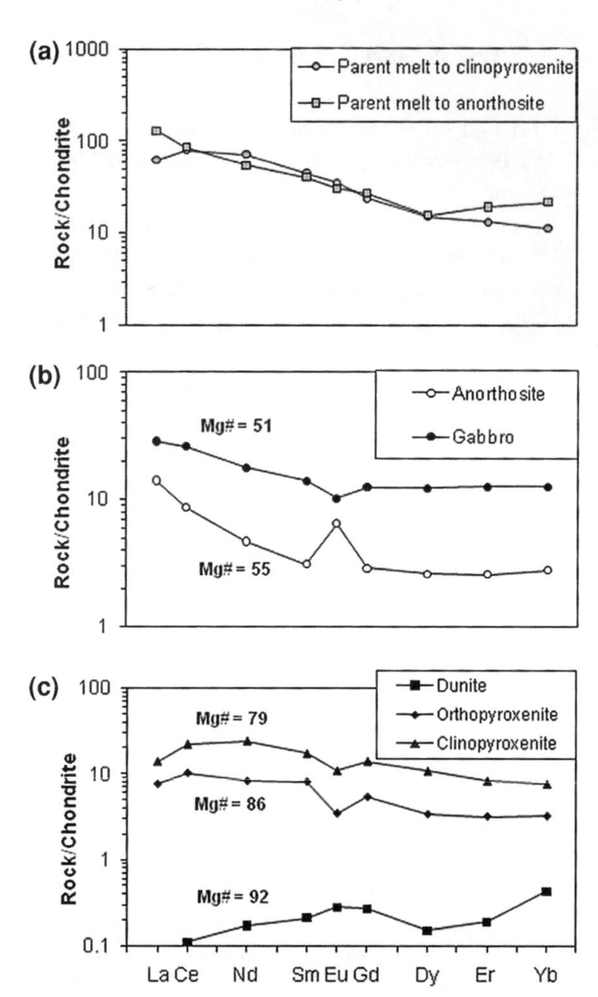

Fig. 11 **a** and **b** Rare earth element compositions of ultramafic and mafic cumulate rocks form the Kondapalli Layered Complex. **c** Calculated equilibrium melt rare earth element compositions for clinopyroxenite and anorthosite cumulate rocks. Liquid trace element compositions are calculated by using the equation $C_L = C_S/D$ and published mineral/melt partition coefficients (Bedard 2001; Claeson and Meurer 2004). C_L is the concentration of element in the equilibrium liquid, C_S is the concentration of element in the solid (cumulate rock) and D is the bulk distribution coefficient ($\Sigma X_i Kd_i$); X_i is the modal abundance of phase i in the solid and Kd_i is mineral/melt partition coefficient of element in phase i. Percentage of estimated intercumulus liquid is substituted as a phase in X_i with Kd_i equals to one

4 Concluding Remarks

There are three general models of mantle melting: batch, fractional and continuous. A combination of continuous and batch melting may be denoted as dynamic melting. Batch melting assumes that melt remains in equilibrium with the residue throughout the melting episode, whereas fractional melting assumes that small melt fraction is continuously and completely extracted from the source as it is formed and only the instantaneous melt is in equilibrium with the residue; continuous melting supposes retention of a critical fraction of melt in the mantle residue as the fractions of melt are continuously extracted. Compositional variations in the melts derived by different types and extents of melting are illustrated by using simple forward modelling techniques.

In cumulate rocks, the trace element abundances are controlled by the proportion of cumulus phases, their partition coefficients, and the amount of intercumulus component (Cawthorn et al. 1991; Bedard 1994). The parental liquid compositions for adcumulate rocks should be ideally calculated for "core" compositions of cumulus phases (Papike et al. 1995). There are major difficulties in the inversion of chemistry of a bulk cumulate rock to parental melt composition through the veil of uncertainties in cumulus mineral abundances and trapped melt component, post-cumulus alterations, and partition coefficients (Scoates and Frost 1996). The ad-cumulate rocks with little or no intercumulus liquid would be fairly good candidates to assess REE pattern shapes in associated melts at least qualitatively (Dymek and Owens 2001; Vijaya Kumar et al. 2007). Keen textural observations are mandatory and a pre-requisite for assessment of cumulus and intercumulus mineralogy and modelling of cumulus processes.

Acknowledgments Part of this article was earlier published in DST DCS News Letter by KVK on the invitation of Shri. T. M. Mahadevan. I thank Prof. C. Leelanandam for data on the Kondapalli layered complex and Prof. C. Leelanandam and Prof. Santosh Kumar for constructive reviews.

Appendix

Batch Melting

$$C_i^l/C_i^o = 1/\left[D_i^o(1-F)+F\right] \qquad \text{(Modal)}$$

$$C_i^l/C_i^o = 1/\left[D_i^o+F(1-\ P_i)\right] \qquad \text{(Nonmodal)}$$

Fractional Melting

$$C_i^l/C_i^o = \left(1/D_i^o\right)(1-F)^{[(1/D_i)-1]} \qquad \text{(Modal)}$$

$$C_i^l/C_i^o = \left(1/D_i^o\right)\left(1-\ P_iF/D_i^o\right)^{[(1/P_i)-1]} \qquad \text{(Nonmodal)}$$

Pooled or Aggregated Melts

$$C_i^l/C_i^o = \left[1-\left(1-F\ P_i/D_i^o\right)^{1/P_i}\right] \qquad \text{(Nonmodal)}$$

Continuous Melting

$$\frac{C_i^l}{C_i^o} = \frac{1}{D_i^o+(1-P_i)\,\alpha/(\alpha+1)}\left[\frac{(D_i^o+\alpha)-(P_i+\alpha)F}{(D_i^o+\alpha)-(P_i+\alpha)\alpha/(\alpha+1)}\right]^{((1-P_i)/(P_i+\alpha))} \qquad \text{(Nonmodal)}$$

C_i^l Concentration of the element i in the liquid (*current liquid in the case of continuous melting*)
C_i^o Initial concentration of the element i in the original source
D_i^o Initial solid bulk distribution coefficient ($\Sigma X_i Kd_i$)
P_i Weighted distribution coefficient of liquid
F Degree of melting
α Amount of critical melt retained with the residue.

Mantle source mineralogy and melting proportions					
Source mineralogy		*Gt-Spl peridotite*		*Spl peridotite*	
Ol		0.55		0.53	
Opx		0.22		0.24	
Cpx		0.15		0.20	
Spl		0.02		0.03	
Gt		0.06		0.00	
Melting proportions					
Ol		−0.10		−0.30	
Opx		0.25		0.40	
Cpx		0.61		0.82	
Spl		0.04		0.08	
Gt		0.20		0.00	
Mineral/melt partition coefficients					
	Ol	Opx	Cpx	Spl	Gt
La	0.000053	0.000044	0.0536	0.00002	0.01
Ce	0.000125	0.00014	0.0858	0.00003	0.021
Pr	0.000251	0.00033	0.137	0.0001	0.054
Nd	0.000398	0.00052	0.1873	0.0002	0.087
Pm					
Sm	0.00065	0.0016	0.291	0.0004	0.217
Eu	0.0008	0.0033	0.329	0.0006	0.32
Gd	0.0015	0.005	0.367	0.0009	0.498
Tb	0.0021	0.0067	0.405	0.0012	0.75
Dy	0.0027	0.0084	0.442	0.0015	1.06
Ho	0.005	0.0127	0.415	0.0023	1.53
Er	0.01	0.017	0.387	0.003	2
Tm	0.016	0.025	0.409	0.0038	3
Yb	0.027	0.033	0.43	0.0045	4.03
Lu	0.03	0.041	0.433	0.0053	5.5

Equations for Fractional Crystallization and Assimilation Fractional Crystallization

Fractional Crystallization (FC)

$$C_l = C_o \text{ x } F^{(D-1)}$$
$$C_s = C_o \left(1 - F^D\right)/(1 - F)$$

where

C_l is the concentration of an element in residual liquid
C_o is the concentration of an element in parental liquid
C_s is the concentration of an element in the solid (cumulate)
F is the percentage of liquid remaining
D is the initial solid bulk distribution coefficient ($\Sigma X_i Kd_i$).

Assimilation Fractional Crystallization (AFC)

$$C_l = C_o \times \left[F^{(D+r-1)/(1-r)}\right] + \left[r/(r+D-1) \times C_A \times \left(1 - F^{(D+r-1)/(1-r)}\right)\right]$$
$$C_s = DC_l$$

where

C_l is the concentration of an element in liquid
C_s is the concentration of an element in solid
C_A is the concentration an element in the contaminant
F is the percentage of the liquid remaining
D is the initial solid bulk distribution coefficient
r is the ratio of the rate of assimilation to the rate of fractional crystallization.

Mineral/melt partition coefficients for FC and AFC in a basaltic melt

	Plag	Ol	Opx	Cpx	Amph	Grt	Mt	Ilm	Apt	ICL[a]
Rb	0.08	0.006	0.0026	0.006	0.11	0.0007	0.001	0.09	0.2	1
Ba	0.9	0.006	0.0026	0.00068	0.43	0.0001	0.001	0.09	0.45	1
Nb	0.03	0.002	0.027	0.008	0.76	0.04	0.04	2.3	0.05	1
K	0.036	0.006	0.006	0.007	0.22	0.001	0.001	0.09	0.1	1
La	0.124	0.003	0.0019	0.0536	0.2	0.025	0.015	0.0023	12	1
Ce	0.117	0.004	0.0035	0.0858	0.43	0.04	0.016	0.0019	15	1
Pr	0.105	0.005	0.0059	0.124	0.65	0.05	0.018	0.0016	17	1
Sr	1.3	0.008	0.006	0.1283	0.49	0.03	0.022	0.7	1.4	1
P	0.075	0.008	0.02	0.13	0.25	0.075	0.024	0.002	18.7	1
Nd	0.068	0.01	0.02	0.1873	1.23	0.086	0.026	0.0012	19	1

(continued)

(continued)

Mineral/melt partition coefficients for FC and AFC in a basaltic melt										
Sm	0.058	0.02	0.055	0.291	2.03	0.4	0.024	0.0023	20	1
Zr	0.01	0.02	0.021	0.26	0.83	0.58	0.12	0.33	16	1
Eu	0.71	0.02	0.059	0.3288	1.73	0.74	0.025	0.0009	13	1
Ti	0.043	0.02	0.086	0.34	3	0.38	5	12.5	0.14	1
Gd	0.035	0.03	0.069	0.367	2.51	1.13	0.018	0.006	20	1
Tb	0.031	0.035	0.11	0.404	2.7	2.5	0.019	0.0095	19	1
Dy	0.026	0.034	0.15	0.38	3.01	4.4	0.018	0.013	18	1
Y	0.026	0.033	0.2	0.412	2.58	6	0.018	0.0045	17.5	1
Ho	0.018	0.037	0.2	0.4145	2.9	8	0.018	0.022	16.8	1
Er	0.0145	0.04	0.24	0.387	2.9	11	0.018	0.031	15.5	1
Yb	0.0097	0.04	0.39	0.43	2.96	14	0.018	0.057	13	1
Lu	0.008	0.05	0.47	0.433	2.59	16	0.018	0.07	10	1

ICL[a] = Intercumulus Liquid

References

Albarède F (1996) Introduction to geochemical modelling. Cambridge University Press, London, p 564

Allègre CJ, Minster JF (1978) Quantitative methods of trace element behaviour in magmatic processes. Earth Planet Sci Lett 38:1–25

Arndt NT, Kerr AC, Tarney J (1997) Dynamic melting in plume heads: the formation of Gorgana komatiites and basalts. Earth Planet Sci Lett 146:289–301

Barnes SJ (1986) The effect of trapped liquid crystallization on cumulus compositions in layered intrusions. Contrib Mineral Petrol 93:524–531

Barnes SJ, Naldrett AJ (1986) Geochemistry of the J-M reef of the stillwater complex, Minneapolis Adit area II. Silicate mineral chemistry and petrogenesis. J Petrol 27:791–825

Barnes SJ, Roeder PL (2001) The range of spinel compositions in terrestrial mafic and ultramafic rocks. J Petrol 42:2279–2302

Bedard JH (1994) A procedure for calculating the equilibrium distributions of trace elements among minerals of cumulate rocks, and the concentration of trace elements in coexisting liquids. Chem Geol 118:143–153

Bedard JH (2001) Parental magmas of the Nain Plutonic Suite anorthosites and mafic cumulates: a trace element modelling approach. Contrib Mineral Petrol 141:747–771

Bowen NL (1928) The evolution of the igneous rocks. Princeton University Press, Princeton, p 332

Cawthorn RG (ed) (1996a) Layered Intrusions. Elsevier, Amsterdam, p 531

Cawthorn RG (1996b) Models for incompatible trace-element abundances in cumulus minerals and their application to plagioclase and pyroxenes in the Bushveld Complex. Contrib Mineral Petrol 123:109–115

Cawthorn RG, Meyer SP, Kruger FJ (1991) Major addition of magma at the Pyroxenite Marker in the western Bushveld Complex, South Africa. J Petrol 32:739–763

Charlier B, Duchesne JC, Vander AJ, Storme JY, Maquil R, Longhi J (2010) Polybaric fractional crystallization of high-alumina basalt parental magmas in the Egersund Ogna massif-type anorthosite (Rogaland, SW Norway) constrained by plagioclase and high-alumina orthopyroxene megacrysts. J Petrol 51:2515–2546

Claeson DT, Meurer WP (2004) Fractional crystallization of hydrous basaltic "arc-type" magmas and the formation of amphibole-bearing gabbroic cumulates. Contrib Mineral Petrol 147:288–304

Cookenboo HO, Bustin RM, Wilks KR (1997) Detrital chromian spinel compositions used to reconstruct the tectonic setting or provenance: implications for orogeny in the Canadian Cordillera. J Sediment Res 67:116–173

DePaolo DJ (1981) Trace element and isotopic effects of combined wallrock assimilation and fractional crystallization. Earth Planet Sci Lett 53:189–202

Dymek RF, Owens BE (2001) Chemical assembly of Archaean anorthosites from amphibolite- and granulite-facies terranes, West Greenland. Contrib Mineral Petrol 141:513–528

Eales HV, de Klerk WJ, Teigler B (1990) Evidences for magma mixing processes within the Critical and Lower Zones of the northwestern Bushveld Complex, South Africa. Chem Geol 88:261–278

Eggins SM (1992) Petrogenesis of Hawaiian tholeiites: 2, aspects of dynamic melt segregation. Contrib Mineral Petrol 110:398–410

Elliott TR, Hakesworth CJ, Gronvold K (1991) Dynamic melting of the Iceland plume. Nature 351:201–206

Gast PW (1968) Trace element fractionation and the origin of tholeiitic and alkaline magma types. Geochim Cosmochim Acta 32:1057–1086

Gurenko AA, Chaussidon M (1995) Enriched and depleted primitive melts in olivine from Icelandic tholeiites: origin by continuous melting of a single mantle column. Geochim Cosmochim Acta 59:2905–2917

Hanson GN (1980) Rare earth elements in petrogenetic studies of igneous systems. Annu Rev Earth Planet Sci 8:371–406

Hanson GN, Langmuir CH (1978) Modelling of major elements in mantle-melt systems using trace element approaches. Geochim Cosmochim Acta 42:725–741

Haskin LA, Salpas PA (1992) Genesis of compositional characteristics of Stillwater AN-I and AN-II anorthosite units. Geochim Cosmochim Acta 56:1187–1212

Irvine TN (1967) Chromian spinel as a petrogenetic indicator: Part 2. Petrologic applications. Can J Earth Sci 7:71–103

Irving AJ (1978) A review of experimental studies of crystal/liquid trace element partitioning. Geochim Cosmochim Acta 42:743–770

Jan MQ, Windley BF (1990) Chromian spinel-silicate chemistry in ultramafic rocks of the Jijal complex, Northwest Pakistan. J Petrol 31:667–715

Johnson KTM, Dick HJB, Shimizu H (1990) Melting in the oceanic upper mantle: an ion microprobe study of diopsides in abyssal peridotites. J Geophys Res 95:2661–2678

Jones JH (1984) Temperature- and pressure-independent correlations of olivine/liquid partition coefficients and their application to trace element partitioning. Contrib Mineral Petrol 88:126–132

Lambert DD, Simmons EC (1988) Magma evolution in the Stillwater Complex, Montana: II. Rare earth element evidence for the formation of the J-M Reef. Econ Geol 83:1109–1126

Langmuir CH, Hanson GN (1980) An evaluation of major element heterogeneity in the mantle sources for basalts. Philos Trans R Soc Lond A 297:383–407

Langmuir CH, Bender JF, Bence AE, Hanson GN, Taylor SR (1977) Petrogenesis of basalts from the FAMOUS area: Mid-Atlantic ridge. Earth Planet Sci Lett 36:133–156

Leelanandam C, Vijaya Kumar K (2004) Rare earth element geochemistry of the Kondapalli layered complex, Andhra Pradesh. J Geol Soc India 64:251–264

Leelanandam C, Vijaya Kumar K (2007) Petrogenesis and Tectonic Setting of the Chromitites and Chromite-bearing Ultramafic Cumulates of the Kondapalli Layered Complex, Eastern Ghats Belt, India: evidences from the textural, mineral-chemical and whole-rock geochemical studies. IAGR Mem 10:89–107

Markl G, Frost BR (1999) The origin of anorthosites and related rocks from Lofoten Islands, northern Norway: II. Calculation of parental liquid compositions from anorthosites. J Petrol 40:61–77

Markl G, Frost BR, Bucher K (1998) The origin of anorthosites and related rocks from the Lofoten Islands, Northern Norway: I. Field relations and estimation of intrinsic variables. J Petrol 39:1425–1452

McCallum IS (1996) The stillwater complex. In: Cawthorn RG (ed) "Layered intrusions" developments in petrology, vol 15. Elsevier, Amsterdam, pp 441–483

McCallum IS, Raedeke LD, Mathez EA (1980) Investigations of the stillwater complex: Part I. Stratigraphy and structure of the Banded Zone. Am J Sci 280-A:59–87

McKenzie DP (1984) The generation and compaction of partially molten rock. J Petrol 25:713–765

McKenzie DP (1989) Some remarks on the movement of small melt fractions in the mantle. Earth Planet Sci Lett 95:53–72

McKenzie D, O'Nions RK (1991) Partial melt distributions from inversion of rare earth element concentrations. J Petrol 32:1021–1091

Meyer PS, Dick HJB, Thompson G (1989) Cumulate gabbros from the Southwest Indian Ridge, 54°S–7°16′E: implications for magmatic processes at a slow spreading ridge. Contrib Mineral Petrol 103:44–63

Namur O, Charlier B, Toplis MJ, Higgins MD, Lie Geois JP, Vander Auwera J (2010) Crystallization sequence and magma chamber processes in the ferrobasaltic Sept Iles layered intrusion. Can J Petrol 51:1203–1236

Neumann H, Mead J, Vitaliano CJ (1954) Trace element variation during crystallization as calculated from the distribution law. Geochim Cosmochim Acta 6:90–99

Papike JJ, Spilde MN, Fowler GW, McCallum IS (1995) SIMS studies of planetary cumulates: orthopyroxene from the Stillwater Complex. Montana Amer Mineral 80:1208–1221

Piccardo GB, Guarnieri L (2011) Gabbro-norite cumulates from strongly depleted MORB melts in the Alpine-Apennine ophiolites. Lithos 124:200–214

Presnall DC (1969) The geometrical analysis of partial fusion. Am J Sci 267:1178–1194

Prinzhofer A, Allegre CJ (1985) Residual peridotites and mechanisms of partial melting. Earth Planet Sci Lett 74:251–265

Rajamani V, Shivkumar K, Hanson GN, Shirey SB (1985) Geochemistry and petrogenesis of amphibolites, Kolar schist belt, South India: evidence for komatiitic magma derived by low percentage of melting of mantle. J Petrol 26:92–123

Rajamani V, Shirey SB, Hanson GN (1989) Fe-rich Archaean tholeiites derived from melt-enriched mantle sources: evidence from the Kolar schist belt, South India. J Geol 97:487–501

Roeder PL (1994) Chromite: from the fiery rain of chondrules to the Kilauea Iki Lava Lake. Can Mineral 76:827–847

Schilling J-G, Winchester JW (1967) Rare-Earth fractionation and magmatic processes. In: Runcorn SK (ed) Mantles for the Earth and Terrestrial Planets. Wiley Interscience, New York, pp 267–283

Scoates JS, Frost CD (1996) A strontium and neodymium isotopic investigation of the Laramie anorthosites, Wyoming, USA: implications for magma chamber processes and the evolution of magma conduits in Proterozoic anorthosites. Geochim Cosmochim Acta 60:95–107

Sharpe MR (1985) Strontium isotope evidence for preserved density stratification in the Main Zone of the Bushveld Complex. Nature 316:119–126

Shaw DM (1970) Trace element fractionation during anatexis. Geochim Cosmochim Acta 34:237–243

Shaw DM (2000) Continuous (dynamic) melting theory revisited. Can Mineral 38:1041–1063

Sobolev AV, Shimizu N (1993) Ultra-depleted primary melt included in an olivine from the Mid-Atlantic Ridge. Nature 363:151–154

Strong DF, Dostal J (1980) Dynamic melting of Proterozoic upper mantle: evidence from rare earth elements in oceanic crust of Eastern Newfoundland. Contrib Mineral Petrol 72:165–173

Tiezzi LJ, Scott RB (1980) Crystal fractionation in a cumulate gabbro, Mid-Atlantic Ridge, 26°N. J Geophys Res 85:5438–5454

Vijaya Kumar K (2006) Mantle melting models: an overview. Deep Cont Stud India Newsl 16:2–10

Vijaya Kumar K, Narsimha Reddy M, Leelanandam C (2006) Dynamic melting of the Precambrian mantle: evidence from rare earth elements of the amphibolites from the Nellore-Khammam schist belt, South India. Contrib Mineral Petrol 152:243–256

Vijaya Kumar K, Frost CD, Frost BR, Chamberlain KR (2007) The Chimakurti, Errakonda, and Uppalapadu plutons, Eastern Ghats Belt, India: an unusual association of tholeiitic and alkaline magmatism. Lithos 97:30–57

Wager LR, Brown GM, Wadsworth WJ (1960) Types of igneous cumulates. J Petrol 1:73–85

Wood DA (1979) Dynamic partial melting: its application to the petrogenesis of basalts erupted in Iceland, the Faeroe Islands, the Isle of Skye (Scotland) and the Troodos Massif (Cyprus). Geochim Cosmochim Acta 43:1031–1046

Zhou M-F, Lightfoot PC, Keays RR, Moore ML, Morrison GG (1997) Petrogenetic significance of chromian spinels from the Sudbury Igneous Complex, Ontario, Canada. Can J Earth Sci 34:1405–1419

Zou H (1998) Trace element fractionation during modal and nonmodal dynamic melting and open-system melting: a mathematical treatment. Geochim Cosmochim Acta 62:1937–1945

Parameterized Mantle Convection Analysis for Crustal Processes

R. N. Singh and A. Manglik

Abstract Thermal convection is considered as the main heat transport mechanism in the mantle that brings heat from earth's interior to the base of the lithosphere. Many large-scale geological and tectonic processes such as initiation of plate tectonics and its persistence throughout the geological history, formation and stability of cratons, generation of komatiites in Archaean, etc. are controlled by heat. These processes and the presence of radioactive elements support the view that the early earth would have been at a much higher temperature compared to the present state indicating that the mantle convection would have been more vigorous during the early Archaean. Therefore, reconstruction of evolution of average mantle temperature and the geotherm of the convecting mantle through the geological history is important for the understanding of the geological processes. This chapter deals with simplified treatment of mantle convection, so called parameterized model of thermal convection. From the energy balance for the mantle, an equation for the average temperature is derived using Nusselt and Rayleigh number relationship and temperature dependent viscosity. Solution of this non-linear equation, given evolution of core heat flux and decay of radioactive elements, yields the cooling history of average mantle temperature. Implications of the mantle cooling history on the generation komatiites, initiation of plate tectonics and craton stability are discussed.

1 Introduction

Mantle provides heat and material to the crust. Heat generated by freezing of inner core driving core convection provides heat at the base of mantle. Some feeble concentration of radioactive elements is present in the mantle. Further, mantle has

R. N. Singh (✉) · A. Manglik
CSIR-National Geophysical Research Institute, Uppal Road, Hyderabad, India
e-mail: rnsingh@ngri.res.inrishiprema@gmail.com

S. Kumar and R. N. Singh (eds.), *Modelling of Magmatic and Allied Processes*,
Society of Earth Scientists Series, DOI: 10.1007/978-3-319-06471-0_4,

been hotter in the past compared to the present state. This heat is transported to the surface of the earth and eventually lost to earth's environment. Heat is transported within mantle by thermal convection process. In this chapter, we are interested in the heat flow to crust from mantle during the geological history. To understand how this has taken place over the vast expanse of geological history of the regions, it is necessary to know how the mantle temperature has changed over time. For instance, igneous activities since the Archaean times have shown that concentration of MgO has progressively decreased over time. This can be explained in terms of decreasing mantle temperature with time. It would be desirable to know both spatial and temporal behavior of mantle temperatures, but it is difficult to reconstruct the past from the present state with requisite space-time resolution as both discretization and model errors would grow. It is, however, possible to reconstruct the past average temperature of the mantle, by ignoring spatial variations and focusing on time variation only. We focus on simplified treatment of mantle convection, so called parameterized model of thermal convection. The vast and active field of theory of three dimensional mantle convection is not addressed in this chapter.

2 Energy Balance of the Mantle

Mantle convection is modeled as the Rayleigh-Benard convection in which the growth of top and bottom boundary layers destabilizes the mantle fluids. Growth of the top boundary layer underlies the plate tectonics and the bottom boundary layer to genesis of mantle plumes. In between these boundary layers, the mantle temperature follows adiabatic gradient of mantle materials. Mantle adiabatic gradient depends upon the mineral compositions and prevailing thermal and pressure states. Using thermodynamic data of mantle forming minerals or modal compositions, one can calculate the adiabatic gradient. This gradient can be tied to either lower boundary or upper boundary to estimate the mantle geotherm compatible with ongoing thermal convection. As earth had initial high heat due to accretionary process during its formation and added with heating of outer core due to freezing of inner core, mantle temperatures would have been higher in the past. Thus, it would be interesting to obtain the evolution of average temperature of mantle with heat balance equation and then use the knowledge of the adiabatic gradient to construct the geotherm of the convecting mantle. Energy balance equation for mantle can be written as:

Change in stored heat = heat flux from core – heat flux at the Moho + rate of radiogenic heat generation within mantle.

In mathematical form, it is written as:

$$Mc\frac{dT}{dt} = Dq_c - Aq_s + MH. \tag{1}$$

Here, T is average temperature of mantle with mass M and heat capacity c, and t is time. $q_c(q_s)$ is heat flux from core into mantle (mantle into crust), D(A) is surface area of core (mantle), and H is radiogenic heat production per unit mass of mantle.

3 Heat Flux at Core-Mantle Interface

Heat flow from core depends upon working of core convection and the efficiency of mantle to remove heat from the core-mantle boundary. This is given by (Labrosse et al. 1997):

$$q_c = q_0 - rt, \tag{2}$$

where $q_0 = 0.06 - 0.09\ \mathrm{Wm^{-2}}$ and $r = 2.10^{-19}\ \mathrm{Wm^{-2}\,s^{-1}}$. These values have been determined by consideration of energy and entropy balance in the core based on generation of main geomagnetic field by dynamo action.

4 Heat Flux from Mantle

As mantle is under convection, the heat flux depends upon the efficiency of heat transfer within the mantle. This relationship is frequently written in terms of two non-dimensional numbers, Nusselt number (Nu) and Rayleigh number (Ra):

$$Nu = \frac{q}{KdT/dz} = \frac{qd}{K(T - T_s)}, \tag{3a}$$

$$Ra = \frac{\rho\alpha g(T - T_s)d^3}{\kappa\mu}. \tag{3b}$$

Here, T_s is the surface temperature. $\rho, \alpha, \kappa, \mu, d, g,$ and K are density, coefficient of thermal expansion, thermal diffusivity, viscosity, thickness of the fluid layer, gravity acceleration and thermal conductivity, respectively. Many experimental and theoretical studies have shown that:

$$Nu = aRa^{\beta}. \tag{4}$$

The value of β has been determined using experiments and analytical and numerical models of Rayleigh-Benard convection. In the field of mantle convection, Turcotte and Oxburgh (1967) used boundary layer model of thermal convection to get $\beta = \frac{1}{3}$ for an isoviscous fluid. A very small value of 0.01 was derived numerically by Christensen (1984) using rigid lid as boundary condition at the top of mantle. Taking the top boundary condition as free slip, Gurnis (1989) obtained its value as 0.3. Similar approach was followed by Solomatov (1995) and Reese et al. (1999) for stagnant lid and sluggish lid boundary conditions and its values determined. Next development took place by considering temperature dependent viscosity and value of β was found to be 0.3 by Korenaga and Jordan (2002) and Korenaga (2003). A value of $2/7 \approx 0.28$ was determined using experimental approach by Castaing et al. (1989). Similar value was obtained numerically by Bercovici et al. (1989, 1992). Wolstencroft et al. (2009) conducted numerical experiments taking three dimensional spherical geometry for both basal heating and internal heating with free slip boundary condition on the top of mantle. Their values are 0.294 ± 0.004 for basal heating and 0.337 ± 0.009 for internal heating.

From the above relationship, it is shown that surface heat flux is related to average mantle temperature as:

$$q_s = c(T - T_s)^{1+\beta} v^{-\beta}, \tag{5}$$

where v is kinematic viscosity and c is a constant. Viscosity is a function of composition, pressure and temperature distributions in the mantle. Assuming a thermally activated rheology, the kinematic viscosity is given by:

$$v = bT\exp\left(\frac{E + PV}{RT}\right). \tag{6}$$

Here, $b, P, E,$ and V are constant depending upon stress, pressure, activation energy and activation volume, respectively. From this, Davies (1980) derived the following form of temperature dependence of viscosity:

$$v = v_0(T/T_0)^{-n}, \tag{7}$$

where v_0, T_0 are reference viscosity and temperature, respectively. The value of n is given by:

$$n = -\frac{\partial \log v}{\partial \log T} = \frac{E + PV}{RT} - 1. \tag{8}$$

We get from Eqs. (5) and (7) (Davies 1980):

$$q_s = q_{s0}(T/T_0)^{1+\beta(1+n)}, \tag{9}$$

where q_{s0} is the surface heat flux at mantle temperature T_0.

In the above scaling relation no considerations of plate tectonic processes have been included. Korenaga (2006) has included also the effects of variable lithospheric thickness as a result of changes in the mantle temperature and also composition leading to dehydration stiffening of plates at higher mantle temperatures. The relationship between mantle heat flux and temperature in this case is given as:

$$Aq_s = a\left[\frac{C_{pe}\alpha\rho T^3 Dh}{C_{vd}^m v_m + C_{vd}^l v_l\left(\frac{h}{R}\right)^3}\right]^{1/2}. \tag{10}$$

Here $D, h, v_m, v_l,$ and R are thickness of mantle, plate thickness, kinematic viscosity of mantle, kinematic viscosity of plate and radius of curvature of plate (200 km). The constant a is determined by using the present value of mantle heat flow and temperature. Other constants are:

$$C_{pe} \sim 1/\sqrt{\pi}; \quad C_{vd}^m \sim 3\left(\frac{L}{D} + 2.5\right); \quad C_{vd}^l \sim 2.5. \tag{11}$$

Here, L/D is the aspect ratio of mantle convection at the large scale. In this, viscosity and plate thickness are a function of temperature and also mantle composition. Korenaga (2006) has shown that the plate thickness, which normally decreases with the rise of mantle temperature, will increase instead if the effects of compositional buoyancy are also included. This formulation includes the effects of decompression melting below mid oceanic ridges, dehydration effects in mantle rheology and viscous dissipation during plate bending on subduction.

Effects of cell size and also presence of continent on Nu–Ra relationship has been determined (Grigne et al. 2005; Jaupart and Mareschal 2011) as:

$$Nu = \frac{1}{(2\pi)^{2/3}} Ra^{1/3} \sqrt{\left(1 - \frac{a}{L}\right)} \left\{(L/D)^2 + (L/D)\right\}^{-1/3}. \tag{12}$$

Here, a is the width of the continental lid. This relationship has been derived using the model of plate tectonics operative during the last about 200 Ma. For this period, the generation of oceanic lithosphere by which mantle is cooling now is well constrained. This model then can be extrapolated to the past processes involved in mantle cooling. Silver and Behn (2008) have used the following parameterization to include the vigor of the plate tectonic processes. Here, the expression for mantle heat flow is multiplied by a term, $p(t)$, representing vigor of plate tectonics as:

$$q_{s1} = q_{s0} p(t) q_s. \tag{13}$$

This formulation shows that for $q_s = 0$, we have $q_{s1} = 0$. To avoid this situation Korenaga (2008a) has extended this formulation as:

$$q_{s1} = q_{s0} p(t) q_s + (1 - (p(t)) q_{min}, \tag{14}$$

where q_{min} refers to heat flow for stagnant lid approximation with no plate tectonics.

In case mantle behaves as Newtonian rheology, the temperature dependence of viscosity is given by:

$$\nu = a\exp\left(\frac{E}{RT}\right). \tag{15}$$

Here, temperature dependence in pre-exponential factor has been ignored as it is small compared to that in exponential term. The following version is frequently used:

$$\nu = \nu_{ref}\exp\left(-b\left(T - T_{ref}\right)\right), \tag{16}$$

where ν_{ref} and T_{ref} are reference viscosity and temperature, respectively.

Another form of Eq. (15) is used in the literature:

$$\nu = \nu_0\exp\left(g\frac{T_m}{T}\right), \tag{17}$$

where T_m is melting temperature and g, a constant.

5 Transient Radiogenic Heat

In the mantle, heat is generated due to decay of U, Th and K. Based on geochemical considerations the concentrations of these elements have been determined. The heat generated in the past was higher and it has been decreasing exponentially. There are four isotopes $U^{238}, U^{235}, Th^{232}$ and K^{40}. Representing their concentrations as C^U, C^{Th} and C^K, we can write the heat generation rate as (Turcotte and Schubert 2002):

$$H(t) = aC^U H^{U^{238}} + (1-a)C^U H^{U^{235}} + C^{Th}H^{Th} + bC^K H^{K^{40}}. \tag{18}$$

In the above equation, a is fraction of natural uranium isotope U^{238} and b is fraction of natural potassium isotope K^{40}. $H's$ are the rate of decay of corresponding isotopes. If the initial concentration is denoted by C_0's, and decay constants by $\tau_{1/2}$, then we get the radiogenic heat in the past as:

$$\begin{aligned} H(t) =& aC_0^U H^{U^{238}} \exp\left(\frac{(t_0 - t)ln2}{\tau_{\frac{1}{2}}^{U^{238}}}\right) + (1-a)C_0^U H^{U^{235}} \exp\left(\frac{(t_0 - t)ln2}{\tau_{\frac{1}{2}}^{U^{235}}}\right) \\ &+ C_0^{Th} H^{Th} \exp\left(\frac{(t_0 - t)ln2}{\tau_{1/2}^{Th}}\right) + bC_0^K H^{K^{40}} \exp\left(\frac{(t_0 - t)ln2}{\tau_{1/2}^{K^{40}}}\right). \end{aligned} \tag{19}$$

Here t_0 is the age of the earth. This has been approximated for the rate of radiogenic heat changes with time as (Jackson and Pollack 1984):

$$H = H_0 \exp(-\lambda t). \tag{20}$$

6 Equation for Mantle Temperature

Substituting Eqs. (2), (9) and (20) into Eq. (1), we get the following equation for the average temperature of mantle:

$$Mc\frac{dT}{dt} = MH_0 e^{-\lambda t} + D(q_0 - rt) - Aq_{s0}(T/T_0)^{1+\beta(1+n)}. \tag{21}$$

This is a nonlinear ordinary differential equation whose numerical solution can be easily obtained. This equation was formulated by Davies (1980) without core heat flux terms. There have been developments assuming mantle as layered: upper and lower mantle. Such models have been developed to accommodate geochemical arguments for having one reservoir for mid-oceanic basalts (MORB) and other for ocean island basalts (OIB). Phase transformation of mantle minerals and limiting of subducting plate to the base of upper mantle also supported this layering of mantle. However, with time all arguments for layered mantle convection have been shown as not convincing due to uncertainties of the data and whole mantle convection has been shown as a reasonable model to explore thermal and mechanical evolution of mantle (Korenaga 2008b). We shall thus not review the literature on the parameterized layered mantle convection.

7 Initial Temperature

It is now believed that earth started hot. This hot state was arrived at due to heating during accretion processes, core formation and radiogenic heating. The effects of accretion, core formation and short decaying radioactive heat would have been lost during a few million years and the temperature distribution would have been determined by long decaying radiogenic heat and processes of heat transport and material phase transformation in the earth. The initial temperature of upper mantle is taken as 3,500 °C, lower mantle as 4,500 °C and at earth's centre as 6,070 °C (Davies 2011). Equation (21) can be solved from initial condition to the present condition as a forward problem and can also be solved backward from the present condition.

7.1 Special Cases

Davies (1980) presented solutions of Eq. (21) for some special cases.

Case I

In this case there are no heat sources and core heat flux, we have:

$$Mc\frac{dT}{dt} = -Aq_{s0}(T/T_0)^m, \tag{22}$$

where $m = 1 + \beta(1 + n)$. For $m = 1$, the solution of this equation is:

$$T = T_0\exp(-t/\tau), \tag{23}$$

where $\tau = T_0Mc/(Aq_{s0})$. The earth's heat in this case decays exponentially from its hot initial condition with a time constant of about 15 Ga, more than three times the age of earth.

Case II
For $m \neq 1$, the solution of Eq. (21) is:

$$\frac{1}{T^{m-1}} = 1/T_0^{m-1} + (m-1)tAq_{s0}/(McT_0^m). \tag{24}$$

In this case, the earth would have a thermal catastrophe at the following time in the past:

$$t = -McT_0/((m-1)Aq_{s0}). \tag{25}$$

Davies (1980) found this catastrophe at 1.5 Ga ago for m = 10. These results change when the effects of radiogenic heat sources is included.

Case III
For constant heat sources and for $m = 1$, Eq. (21) is reduced to:

$$\frac{dT}{dt} = \frac{H}{c} - T/\tau. \tag{26}$$

The solution of the above equation is given by:

$$T = \frac{H}{c} + (T_m - H/c)\exp(-t/\tau). \tag{27}$$

Here also, the decay rate is same as in the case without radiogenic heat. Davies (1980) performed numerical computations for $m \neq 1$ and showed that the decay rate gets reduced by a factor $1/(m-1)$.

Christensen (1985) linearized Davies (1980) equation for mantle temperature as:

$$Mc\frac{dT}{dt} = MH - Aq_{s0}(1 + m(T - T_0)/T_0). \tag{28}$$

For $\theta = T - T_0$, we have (Jaupart et al. 2007):

$$Mc\frac{d\theta}{dt} = MH - Aq_{s0}(1 + m\theta/T_0). \tag{29}$$

We then have decay rate of temperature perturbation for constant H as:

$$\tau_1 = \frac{McT_0}{mAq_{s0}}. \tag{30}$$

This rate comes out as ~0.8 Ga (Christensen 1985). Thus, more efficient thermal convection (larger values of β) gives faster decay of mantle temperature than less efficient convection. For time varying radiogenic heat as given by Eq. (20), perturbation in mantle temperature follows (Jaupart et al. 2007; Jaupart and Mareschal 2011):

$$\begin{aligned}\theta =&\theta_0 \exp\left(-\frac{t}{\tau_1}\right) + \frac{Aq_{s0}\tau_1}{Mc}(\exp(-t/\tau_1) - 1)\\ &+ \frac{H_0\tau_1}{T_0Mc(1-\lambda\tau_1)}\left(\exp(-\lambda t) - \exp(-\frac{t}{\tau_1})\right).\end{aligned} \tag{31}$$

The above analytical solutions provide good estimates of thermal decay time under various assumptions as mentioned above. However, for the estimation of mantle temperature history for geological purposes we would require solution for the full nonlinear Eq. (21) (Schubert et al. 2001). This has been achieved using numerical approach. We will discuss some of the results which satisfy all available constraints provided by theory and geochemical and geophysical observations.

8 Model of Mantle Thermal History

One of the diagnostics for mantle temperature evolution is that it should satisfy present value of convective Urey number defined as:

$$U_r(0) = \frac{MH(0)}{Aq_{s0}}. \tag{32}$$

It has value of 0.22 based on best geochemical considerations and surface heat flow measurements (Korenaga 2008b). Davies (1980) model gave this value for present as 0.5 and subsequently corrected to 0.9 (Richter 1984). It is interesting to recall that originally Urey number was thought to be 1 (Urey 1955) by fully balancing the heat flux and radiogenic heat in the earth. We present recent results of mantle thermal history using small Urey number.

Korenaga (2006) presented a mantle history model which is in agreement with the knowledge of thermal convection scaling, geochemical and geophysical data. Figure 1 shows the results from Korenaga (2008b) using heat flow scaling given by

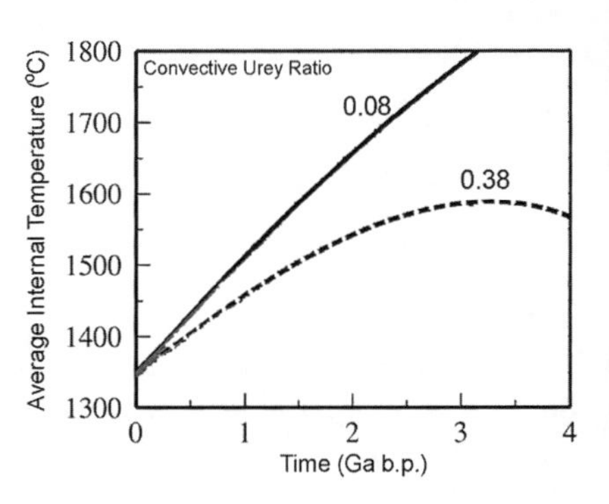

Fig. 1 Evolution of average internal temperature through the geological history starting from the present (time = 0) value of 1,350 °C for two values of convective Urey ratio (after Korenaga 2008b)

Eq. (12) for two values of present convective Urey ratio: 0.08 and 0.38. This will cover all available estimates of convective Urey ratio for the present age. In the first case the convective heat flux and continental heat production are 36 and 10 TW, respectively. In the second case, these values are 37 and 7 TW, respectively. The mantle temperature equation was integrated backward with present mantle temperature taken as 1,350 °C. For viscosity, Arrhenius form was used. The mantle temperatures increase with age, but the rate of increase declines with age.

Recently Tirone and Ganguly (2011) have calculated mantle thermal history by combining parameterized energy equation with equilibrium thermodynamic calculations. They assumed mantle as MAS ($MgO - Al_2O_3 - SiO_2$) system. Gibbs energy of the mineral assemblages was minimized at pressure and temperatures at depths, providing isentrope for the mantle. The viscosity function was estimated using experimental diffusion data for mineral assemblages. Figure 2 shows their results for the mantle thermal history for two values of core heat flux $(0.04, 0.08)\,\mathrm{Wm}^{-2}$. They provide estimate of average temperature of the mantle, temperature at the base of mantle adiabat and at the core-mantle boundary. The average mantle temperature can be used to estimate normal behavior of mantle as seen in mid-ocean ridge basalts whereas core-mantle boundary temperature can be useful in finding out the enhanced temperature of plumes originating from core-mantle boundary. In contrast to Korenaga's model above, in this model rate of cooling is faster in the past than in the present.

Loyd et al. (2007) have included the effects of both cell size and the continental lid size using Nu–Ra relationship given by Eq. (12). Their model of mantle thermal history is given in Fig. 3. The variation of wavelength with time $(l(t) \equiv L/D)$ has been assumed as (Grigne et al. 2005):

$$l(t) = l_{max} - l_{min} + (l_{max} - l_{min})cos\left(2\pi\frac{t}{\tau} + b\right). \tag{33}$$

For periodicity τ and $l_{max}(l_{min})$ as $500\,\mathrm{Ma}$ and 7(3), the value of $b = -53.4$. Loyd et al. (2007) have also given the mantle cooling model for $l_{max}(l_{min})$ as 6(4). These results are shown in Fig. 3. The cooling model of Korenaga (2006) lies within these cooling histories with Urey ratios as 0.28 and 0.54. There have been

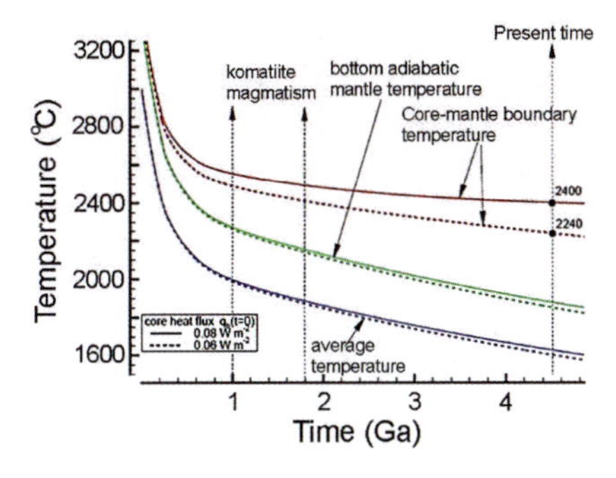

Fig. 2 Mantle thermal history by combining parameterized energy equation with equilibrium thermodynamic calculations (after Tirone and Ganguly 2011, personal communication)

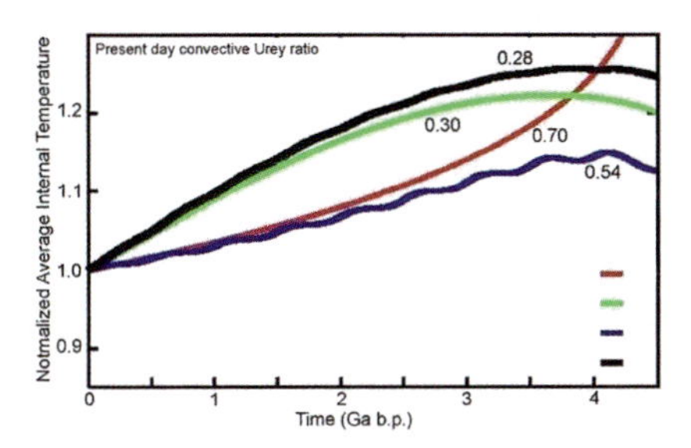

Fig. 3 Mantle thermal model due to Loyd et al. (2007) (Copyright (2007) National Academy of Sciences, U.S.A.). As wavelength has periodicity of 500 Ma, mantle temperature shows periodic behavior. *Red curve* shows unrealistic mantle temperature evolution without inclusion of plate tectonic processes. *Green curve* shows Korenaga (2006) model incorporating the variation in lithospheric thickness due to changing rheology of the evolving mantle. *Black* and *blue curves* show the result for $l_{max}(l_{min})$ as 6(4) and 7(3) respectively

several studies taking more complex layered model of mantle to build mantle cooling history. However, we have limited our discussion to whole mantle convection models.

9 Initiation of Plate Tectonic Processes

It is certain that plate tectonic processes have operated during the last ~200 Ma. It is seen in marine magnetic anomalies. Types of rocks, igneous and metamorphic, which are formed in subduction and emplaced on the continents, are studied to find the operation of plate tectonics prior to ~200 Ma. Age of subduction, thickness of plates and melting of mantle wedge material by subducting plates all depend upon mantle temperatures. One view has been that as mantle temperatures have been high in the Archaean, plate velocities would have been large, and plate size and thickness would have been small with temperature dependent viscosity. Thus, the style of plate tectonics would not have been the same. But with higher mantle temperatures, partial melting would have started deeper with increased thickness of depleted mantle and thicker basaltic layer. With dehydration the thicker depleted mantle would be stiffer, so that the thickness of lithosphere would not have been lesser. However, the plate would be spreading slower. Thus, with both temperature and composition dependent viscosity, the style of plate tectonics may have changed from the present day plate tectonics, but would have been operating in the Archaean (Korenaga 2008b).

10 Mantle Stress History and Craton Stability

Convective stress in the mantle is scaled to Rayleigh number of the mantle. It depends upon mantle flow velocity and viscosity. As viscosity depends upon temperature which declines with time, vigor of convection will decline with time. Thus mantle velocity will decline with time. Decline in velocity is offset by increase in the viscosity with time, so the mantle convective stress will increase with time. Sandu et al. (2011) have estimated increase of the mantle stress with time and have considered the stability of cratonic roots of continental lithosphere. In spite of the increase in the mantle stress, the roots are not deformed and these roots are well viscously coupled with subjacent mantle.

11 Generation of Komatiites

Komatiites were formed in the Archaean and are not formed now. The pressure and temperature estimated for generation of melts leading to emplacement of komatiites have been studied. Plume materials with deep melting have been ascribed as source of komatiite in absence of water. In case water is present, it can be generated at shallower depths. Melting mantle materials rising along adiabat will intersect the solidus at a particular pressure and temperature and terminate at shallower levels. This melting zone will determine the chemical composition of komatiites. We need mantle temperatures enhanced by 200–300 °C in the Archaean. This range of temperature is given by the models discussed above.

12 Concluding Remarks

The evolution of mantle temperature over the geological history of earth has manifested itself in geological records. For example, the abundance of MgO-rich komatiites in the Archaean and their absence in subsequent times is a good petrological indicator for the mantle temperature in the Archaean. Coupled with thermodynamic models of melt generation, these observations help in quantifying the mantle temperature of the early earth and, thus, the cooling rate of the mantle which in turn helps in understanding the evolution of the plate tectonic processes. Therefore, petrological constraints are important to investigate thermal evolution of earth and the convective regime of the mantle throughout the geological history.

Acknowledgments RNS is grateful to INSA, India for the award of a Senior Scientists scheme to him. Contribution under PSC0204 (INDEX) and MLP6107-28(AM).

References

Bercovici D, Schubert G, Glatzmaier GA, Zebib A (1989) Three-dimensional thermal convection in a spherical shell. J Fluid Mech 206:75–104. doi:10.1017/S0022112089002235

Bercovici D, Schubert G, Glatzmaier GA (1992) Three-dimensional convection of an infinite-Prandtl-number compressible fluid in a basally heated spherical shell. J Fluid Mech 239:683–719. doi:10.1017/S0022112092004580

Castaing B, Gunaratne G, Heslot F, Kadanoff L, Libchaber A, Thomae S, Wu X-Z, Zaleski S, Zanetti G (1989) Scaling of hard thermal turbulence in Rayleigh–Benard convection. J Fluid Mech 204:1–30

Christensen UR (1984) Heat transport by variable viscosity convection and implications for the Earth's thermal evolution. Phys Earth Planet Inter 35:264–282

Christensen UR (1985) Thermal evolution models for the Earth. J Geophys Res 90:2995–3007

Davies GF (1980) Thermal histories of convective earth models and constraints on radiogenic heat production in the Earth. J Geophys Res 85:2517–2530

Davies GF (2011) Mantle convection for geologists. Cambridge University Press, Cambridge

Grigne G, Labrosse S, Tachley PJ (2005) Convective heat transfer as a function of wavelength: implications for the cooling of the Earth. J Geophys Res 110:B03409. doi:10.1029/2004JB003376

Gurnis M (1989) A reassessment of the heat transport by variable viscosity convection with plates and lids. Geophys Res Lett 16:179–182

Jackson MJ, Pollack HN (1984) On the sensitivity of parameterized convection to the rate of decay of internal heat sources. J Geophys Res 89:10103–10108

Jaupart C, Mareschal JC (2011) Heat generation and transport in the earth. Cambridge University Press, Cambridge

Jaupart C, Labrosse S, Mareschal J-C (2007) Temperatures, heat and energy in the mantle of the Earth. In: Schubert G (ed) Treatise on geophysics, vol 1. Elsevier, NewYork

Korenaga J (2003) Energetics of mantle convection and the fate of fossil heat. Geophys Res Lett 30:1437. doi:10.1029/2003GL016982

Korenaga J (2006) Archean geodynamics and the thermal evolution of Earth. In: Benn K, Mareschal J-C, Condie K (eds) Archean geodynamics and environments. AGU, Washington, pp 7–32

Korenaga J (2008a) Comment on "Intermittent plate tectonics?". Science 320:1291

Korenaga J (2008b) Urey ratio and the structure and evolution of Earth's mantle. Rev Geophys 46:1–32. doi:10.1029/2007RG000241

Korenaga J, Jordan TH (2002) Onset of convection with temperature- and depth-dependent viscosity. Geophys Res Lett 29:1923. doi:10.1029/2002GL015672

Labrosse S, Poirier J-P, Le Mouël J-L (1997) On cooling of the Earth's core. Phys Earth Planet Inter 99:1–17

Loyd SJ, Becker TW, Conrad CP, Lithgow-Bertelloni C, Corsetti FA (2007) Time variability in Cenozoic reconstructions of mantle heat flow: plate tectonic cycles and implications for Earth's thermal evolution. Proc Nat Acad Sci 104:14266–14271

Richter FM (1984) Regional models for the thermal evolution of the Earth. Earth Planet Sci Lett 68:471–484

Reese CC, Solomatov VS, Moresi L-N (1999) Non-Newtonian stagnant lid convection and magmatic resurfacing on Venus. Icarus 139:67–80

Sandu C, Lenardic A, O'Neill CJ, Cooper CM (2011) Earth's evolving stress state and the past, present, and future stability of cratonic lithosphere. Int Geol Rev. doi:10.1080/00206814.2010.527672

Silver PG, Behn MD (2008) Intermittent plate tectonics? Science 319:85–88

Schubert G, Turcotte DL, Olson P (2001) Mantle convection in the Earth and Planets. Cambridge University Press, New York

Solomatov VS (1995) Scaling of temperature- and stress-dependent viscosity convection. Phys Fluids 7:266–274

Tirone M, Ganguly J (2011) A viscosity model for the mantle based on diffusion in minerals and constrained by the thermal history and melting of the mantle. DI11A-2132 presented at 2011 Fall Meeting, AGU, San Francisco, Calif. 5–9 Dec

Turcotte DL, Oxburgh ER (1967) Finite amplitude convection cells and continental drift. J Fluid Mech 28:29–42

Turcotte DL, Schubert G (2002) Geodynamics, 2nd edn. Cambridge University Press, Cambridge

Urey HC (1955) The cosmic abundances of potassium, uranium, and thorium and the heat balances of the Earth, the Moon, and Mars. Proc Natl Acad Sci USA 41:127–144

Wolstencroft M, Davies JH, Davies DR (2009) Nusselt-Rayleigh number scaling for spherical shell Earth mantle simulation up to a Rayleigh number of 10. Phys Earth Planet Int 176:132–141

Modelling Paleogeotherms in the Continental Lithosphere: A Brief Review and Applications to Problems in the Indian Subcontinent

R. N. Singh and Jibamitra Ganguly

Abstract In this work, we have reviewed the fundamental aspects of heat transfer theory that underlie the modelling of steady state and transient paleogeotherms in the continental lithosphere. We reviewed different types of models of paleogeotherms involving one or two layers, a combination of rheological and thermal boundary layers, and the stagnant-lid convection model in which convection is allowed in the lower part of a thermal boundary layer. We also reviewed the studies on the calculation of steady state paleogeotherm in the *Dharwar* cratons during the Proterozoic period that have been constrained by the heat flow data and the inferred P-T conditions of the mantle xenoliths in the Proterozoic kimberlite pipes. Several transient thermal models have also been discussed. These include the transient effects of CO_2 fluxing from the mantle, crustal melting and melt focusing, and overthrusting. For the latter we considered both single and double thrust sheets derived from the same and different places. An interesting conclusion that emerges from this study is that a metamorphic field profile (MFP), which defines the locus of temperature maxima experienced by exhuming parcels of rocks in response to erosion, is not significantly affected by the different thrusting scenarios.

R. N. Singh (✉)
CSIR-National Geophysical Research Institute, Uppal Road,
Hyderabad 500007, India
e-mail: rnsingh@ngri.res.in

J. Ganguly
Department of Geosciences, University of Arizona, Tucson, AZ 95721, USA
e-mail: ganguly@email.arizona.edu

S. Kumar and R. N. Singh (eds.), *Modelling of Magmatic and Allied Processes*,
Society of Earth Scientists Series, DOI: 10.1007/978-3-319-06471-0_5,

1 Introduction

Knowledge of the steady state and transient thermal structures of different regions of the Earth is of fundamental importance in a wide spectrum of problems in Earth sciences, especially petrology, geophysics and geodynamics. The paleo-thermal structures help us understand the formation of metamorphic and igneous rocks, rheological properties of, and processes within the Earth's interior, thickness of the lithosphere, subsidence of ocean floor, formation of sedimentary basins and passive margins, orogenesis etc. In this work we present a brief review of the methodologies used in the calculation of steady state and some transient paleo-geotherms in the continental lithosphere and their applications to selected problems in the Indian subcontinent.

The lithosphere of the Earth is bounded by atmosphere/ocean on the top and convecting mantle at the bottom and consists of the crust and part of the upper mantle. Lithosphere is broken up into a number of plates that translate on the convecting mantle as rigid units. Within the lithosphere the heat transport takes place primarily via heat conduction, and in some cases by a combination of heat conduction and advection. Heat within the lithosphere comes from the heat given off by the secular cooling of the convecting mantle and the decay of radioactive elements that are primarily concentrated within the crust in the continental regions. Heat flows from the lithosphere to the environment, which is at lower temperatures. In oceanic region the crust is very thin and the thickness of lithosphere increases with its age as it moves away from the spreading centers, thus causing the ocean floor to sink progressively with distance from the latter. Lithosphere below ocean is less than hundred kilometers whereas in continental regions it is $\sim$200 km or even thicker (Ganguly et al. 1995; Rudnick et al. 1998). Lithosphere below ocean undergoes magmatic accretion at the ridges and subduction at the trenches whereas in continental region, the lithosphere is subjected to uplift/erosion, tension/compression, orogenesis, metamorphism and magmatic accretions.

2 Modelling Lithospheric Paleogeotherm: Mathematical Formulation

Since the heat transport within the lithosphere is primarily via conduction and advection, the geotherm, or the variation of temperature with depth, is calculated by solving the following partial differential equation:

$$\frac{\partial(\rho C_p T)}{\partial t} = \frac{\partial}{\partial z}\left(K\frac{\partial T}{\partial z}\right) + \frac{\partial(\rho C_p T v)}{\partial z} + A(Z) \tag{1}$$

It is simply an equation of continuity in a Cartesian coordinate system. Here ρ is the density, C_P is the isobaric specific heat capacity, K is the thermal conductivity, v is the advection velocity, taken as positive upwards, and $A(Z)$ is the radioactive heat production rate per unit volume that varies as a function of depth. A may also vary as a function of time, but the time dependence of A is usually neglected because of the very long half-lives of the heat producing nuclides in the Earth. It is easy to see from dimensional considerations that the quantity ($\rho C_P T$) represents the heat content per unit volume. In a static medium, the advection term appears when a reference layer within the medium moves up or down with respect to the surface ($Z = 0$) due to erosion or accumulation of material at the top.

Commonly, the above one-dimensional (1-D) equation is simplified by assuming that the material properties and velocity are independent of depth, in which case we have

$$\frac{\partial T}{\partial t} = k\frac{\partial^2 \mathrm{T}}{\partial Z^2} + v\frac{\partial T}{\partial z} + A(Z), \tag{2}$$

where k is the thermal diffusivity and is defined as $k = K/\rho C_P$. The boundary conditions that could be used for the solution of the 1-D heat transport equation are summarized below.

$$\mathrm{T} = \mathrm{T_s} \quad \text{at } Z = 0 \quad \text{and} \quad K\frac{dT}{dZ} = Q_s \quad \text{at } Z = 0, \tag{3a}$$

$$\mathrm{T} = \mathrm{T}_s \quad \text{at } Z = 0 \quad \text{and} \quad K\frac{dT}{dZ} = Q_L \quad \text{at } Z = L, \tag{3b}$$

$$\mathrm{T} = \mathrm{T}_s \quad \text{at } Z = 0 \quad \text{and} \quad T = T_L \quad \text{at } Z = L, \tag{3c}$$

$$\mathrm{T} = \mathrm{T}_L \quad \text{at } Z = L \quad \text{and} \quad K\frac{dT}{dZ} = Q_L \quad \text{at } Z = L, \tag{3d}$$

where Q stands for the heat flux. In (3a), the temperature and heat flux at the earth's surface are held constant; in (3b) the surface temperature and the heat flux at the base of the lithosphere are held constant; in (3c), the surface temperature and the basal temperature of the lithosphere are held constant, and finally in (3d), the temperature and the heat flux are specified at the base of the lithosphere. Any one of these set of boundary conditions yield a unique geotherm within the lithosphere, depending on the explicit nature the boundary conditions.

For a 3-D problem in a Cartesian coordinate system, the heat transfer equation involving conduction, advection and heat production assumes the following form.

$$\frac{\partial(\rho C_p T)}{\partial t} = \nabla \cdot (K\nabla T) + \nabla(\rho C_p T v) + A(Z), \tag{4}$$

where ∇ is the gradient operator ($\nabla = i(\partial/\partial x) + j(\partial/\partial y) + k(\partial/\partial x)$, where i, j and k are three orthogonal unit vectors).

The radioactive heat production in the Earth varies as a function of depth, with most of the radioactive elements being concentrated within the crust. A widely used form of this depth dependent variation of A is

$$A(Z) = A_s e^{-\frac{Z}{D}}, \tag{5}$$

where A_s is the volumetric heat production rate at the surface and D is a scaling depth within which the radioactive elements are primarily concentrated (when $Z = D$, $A(Z) = A_s/e \sim A_s/3$). The basis for the Eq. (5) lies in the fact that it leads to the observed linear relationship between the surface heat flux, Q_s, and A_s in a given region. This linear relation was first observed by Birch et al. (1968) and Lachenbruch (1968) and has been known as the Birch-Lachenbruch relation (it is, however, also possible to derive the observed Q_s versus A_s relation by a stepwise decrease of A as a function of depth). The B-M relation is a consequence of differential uplift and erosion that all continental regions have been subjected to. Considering its derivation using Eq. (5), one finds the slope of the B-M relation to be equal to D. Thus, the B-M relation is commonly expressed as

$$\mathrm{Q_s = Q_m + A_s D}, \tag{6}$$

where Q_m is the heat flux from the mantle. The value of D in a given region did not change with time after formation of crust from the melting and differentiation of the Earth's mantle. Using the above relation to the observed heat flux versus heat production of surface rocks, the scaling depth D has been found to vary between ~7 and 11 km (Pollack and Chapman 1977).

In general, the surface and basal heat fluxes of a slab of thickness L are related according to

$$Q_s = Q_L + \int_0^L A(Z)dZ. \tag{7}$$

Using the exponential decay model of A(Z) with depth, Eq. (5), we then have (Singh and Negi 1979):

$$Q_s = Q_L + D\left(1 - e^{-\frac{L}{D}}\right). \tag{8}$$

In a given region, the lithospheric paleo-temperature could change significantly not only as a function of depth, but also as function of the horizontal directions due to change of thermal conductivity resulting from the presence of magmatic bodies, convection effects from fluid circulation, and variation of the lithospheric thickness. Thus, one needs to exercise caution in deriving geological conclusions from 1-D modelling, but useful geological information could be extracted from such modelling if care is taken to evaluate the above problems on the basis of observational data.

3 Steady State Paleogeotherm

3.1 Single Layer Models

After the transient effects have subsided, a geotherm gradually returns to the steady state condition. Because of its very old age, the steady state condition is commonly achieved in the continental lithosphere. In the absence of any internal flow, the steady state geotherm may be calculated from Eq. (1) or (4) by imposing v = 0, and the steady state condition, $\partial T/\partial t = 0$. For 1-D problem, the steady state equation is then given by

$$\frac{\partial T}{\partial Z}\left(K\frac{\partial T}{\partial Z}\right) + A(Z) = 0. \tag{9}$$

Commonly used boundary conditions for the solution of Eq. (9) to construct continental paleogeotherms are given by Eq. (3a) (i.e. constant surface temperature and constant surface heat flux). Assuming K to be constant, solution of this equation is then given by:

$$T(Z) = T_s + \frac{(Q_s - A_s D)Z}{K} + \frac{A_s D^2}{K}\left(1 - e^{-\frac{Z}{D}}\right). \tag{10}$$

If L_M is taken as the depth to Moho, and the heat flux at Moho, $Q_{L(M)}$, is used as a boundary condition along with fixed surface temperature, then the solution of Eq. (9), using the exponential model of radiogenic heat production rate versus depth (Eq. (9)), is given by:

$$T(Z) = T_s + \frac{Q_{L(M)}z}{K} + \frac{A_s D^2}{K}\left(1 - e^{-\frac{Z}{D}} - \frac{Z}{D}e^{-\frac{L_M}{D}}\right). \tag{11}$$

For $L_M >> D$, the last term within the parentheses becomes negligible, and Q_L becomes the mantle heat flux, Q_m. Russell et al. (2001) presented an analytical solution of Eq. (9) with the boundary conditions given by Eq. (3d), but treating K as a function of temperature. They used the transformation:

$$U = \int_{T_0}^{T} \frac{K(T')}{K_0}dT', \tag{12}$$

so that Eq. (9) reduces to:

$$\frac{\partial^2 U}{\partial Z^2} = \frac{A}{K_0}, \tag{13}$$

where K_0 is the thermal conductivity at a reference temperature T_0. The thermal conductivity is expressed as a function of temperature according to:

$$K(T) = K_0[1 + B(T - T_0)], \tag{14}$$

where B is a constant. We thus get for U:

$$U = (T - T_0)\left[1 + \frac{B}{2}(T - T_0)\right], \tag{15}$$

In this case the boundary conditions Eq. (3d) are transformed as:

$$U = T_L - T_0 + \frac{B}{2}(T_L - T_0^2) \quad \text{at } Z = L, \tag{16}$$

$$K_0 \frac{dU}{dz} = Q_L \quad \text{at } Z = L. \tag{17}$$

The right hand side of Eq. (16) becomes zero if the reference temperature is taken as T_L. We then have for constant value of radiogenic heat

$$U(Z) = \frac{Q_L}{K_0}(Z - L) - \frac{A}{2K_0}(Z - L)^2. \tag{18}$$

Here K_0 is thermal conductivity at Moho $(Z = L, T = T_L)$. From values of $U(Z)$, the values of $T(Z)$ can be obtained from the following relationship which follows from Eq. (15) upon substitution of T_L for T_0 (this exercise yields a quadratic equation in $(T - T_L)$):

$$T(Z) = T_L + \frac{-1 \pm (\sqrt{1 + 2UB})}{B}. \tag{19}$$

Only positive values of T are physically meaningful. Russell et al. (2001) modeled the observed P(Z)-T data according to the above relation to get optimal values of surface heat flow, surface heat generation and the value of B, using parameter fitting methodology.

3.2 Two Layer Models

Russell and Kopylova (1999) calculated lithospheric paleogeotherm using a two-layer model, each characterized by its own radiogenic heat and thermal conductivity. They fitted the analytical solution of the problem to the metamorphic P-T arrays keeping surface heat flow and thickness of first layer as unknowns. The second layer was taken as an infinite half-space. The radiogenic model used by Russell and Kopylova (1999) is:

$$A(Z) = \begin{cases} A_s \exp\left(-\frac{Z}{d}\right) & \text{for } 0 \leq Z \leq d \\ A_m & \text{for } Z \geq d \end{cases}. \tag{20}$$

where d is the thickness of the top layer.

Russell and Kopylova (1999) obtained analytical solutions for T versus Z for both layers. For the first layer the solution is the same as given by Eq. (11), whereas for the bottom layer the solution is:

$$T(Z \geq d) = T_s + Q_s\left[\frac{d}{k} - \frac{d - Z}{K_2}\right] - \frac{A_m(d - Z)^2}{2K_2} + A_s d\left[\frac{0.632(d - Z)}{K_2} - \frac{0.3679d}{K}\right], \tag{21}$$

where K_1 and K_2 are, respectively, the thermal conductivities for the top and bottom layers. Russell and Kopylova (1999) retrieved Q_s and d by fitting metamorphic P-T array. They also developed thermal model using step function model of radiogenic heat in the lithosphere. The analytical solution derived in this case is:

$$T(Z \leq d) = T_s + \frac{Q_s Z}{K} - \frac{A_s d^2}{2K}, \tag{22}$$

$$T(z \geq d) = T_s + Q_s\left[\frac{d}{k} - \frac{d - Z}{K_2}\right] - \frac{A_s d^2}{2K} - \frac{A_m(d - Z)^2}{2K_2} + A_s\frac{(d - Z)}{K_2}. \tag{23}$$

Here too, Q_s and d have been varied to fit the data for metamorphic P-T arrays.

3.3 Thermal/Rheological Boundary Layer Models

The rheology of minerals and rocks forming the lithosphere suggests that the lower part of the lithosphere could undergo slow deformation. To address this problem, Parsons and McKenzie (1978) divided the lithosphere into two parts. The base of the top part is defined by a mechanical boundary layer (MBL), while that of the lower part by a thermal boundary layer (TBL). Richter and McKenzie (1981) developed a parameterization of convection below the mechanical lithosphere using numerical methods and obtained models of both heat flow from thermal boundary layer and its thermal structure. This model was used in McKenzie and Bickle (1988) and McKenzie et al. (2005) to get a consistent thermal model of both mechanical and thermal lithosphere in which there is continuity of temperature and heat flow across the base of thermal boundary layer. Based on these theoretical

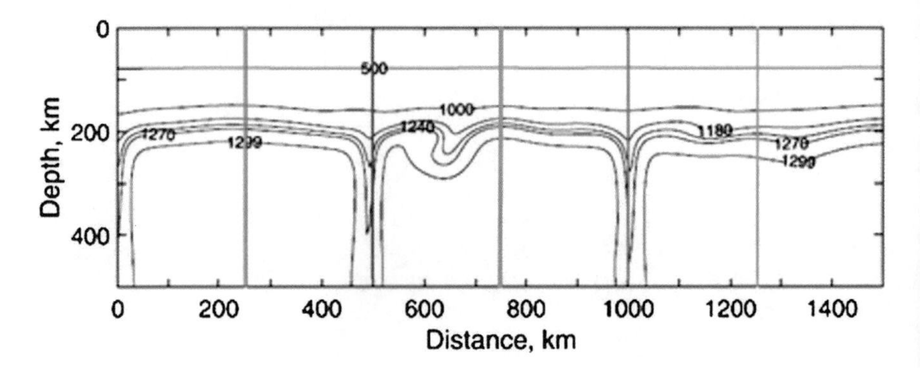

Fig. 1 Stagnant-lid convection pattern beneath a normal craton (before imposition of plume material) showing vigorous internal downwelling at ~500 and ~1000 km, and initiation of downwelling at ~650 km. The downwellings are balanced by broad upwellings. (Sleep 2003) (copyright permission from Wiley)

developments, McKenzie has developed software FITPLOT, which is presented and discussed in detail in Mather et al. (2011).

In several shield areas, the thermobaromatric data of mantle xenoliths in kimberlites are sometimes found to be scattered and also to show two distinct P-T regimes leading to kinks in the overall thermal profiles. In an attempt to explain these data, Sleep (2003) studied spatial and temporal changes of thermal profiles underlying cratons from where the mantle xenoliths were derived by kimberlite magmas. By solving thermal convection equation for nonlinear rheology, Sleep (2003) constructed geotherms permitting convective flow within lower part of a thermal boundary layer under cratons. This formalism, which is called stagnant lid convection model (Solomatov 1995, Solomatov and Moresi 2000), yields geotherms for upwelling and downwelling mantles, and also for areas that have mantle plumes. The upwelling and downwelling regions are found to be separated by several hundred kilometers (Fig. 1). The spatial variation of geotherms (Fig. 2) in such regions could explain some of the scatter in the P-T data obtained from xenoliths across strikes of kimberlite exposures.

Sleep (2003) also computed the effects of hot plume material ponding below the cratons. The ponding of plume material thins the lithosphere to some extent and may yield a linear array extrapolating to the shallower geotherm without any significant kink. The results show that the basal part of thermal lithosphere can be further heated within tens of millions of years since ponding of hot material. Sleep (2003) found that although the xenolith P-T array of the Lesotho kimberlite pipe, South Africa, define a linear trend, the deepest segment of the array lies at higher temperature than the MORB adiabat. Thus, he suggested that these deeper xenoliths probably sampled plume material.

The sheared xenoliths from the Jericho kimberlite pipe, Slave province, Canada, which represents a single eruptive event at 172 Ma, are found to define P-T array that intersects the regional conductive geotherm at high angle from the high

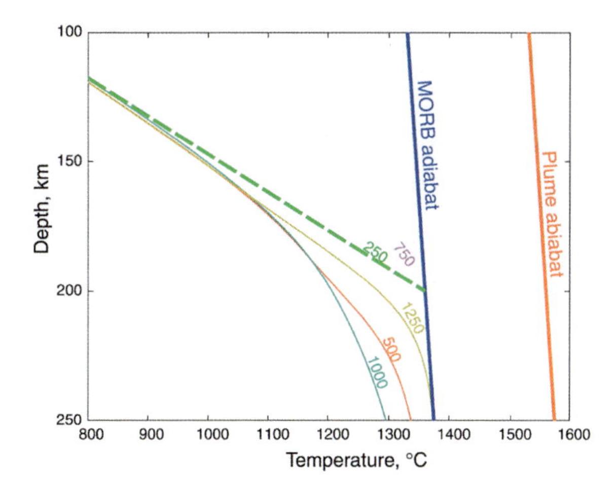

Fig. 2 Computed geotherms at 250 km interval from Fig. 1 along with MORB and plume adiabats (0.3 K/km) that intersect the surface at 1300 and 1500 °C, respectively. The *dashed* (*green*) *line* is the expected laterally averaged conductive geotherm. After Sleep (2003) (copyright permission from Wiley)

temperature side. In this case, Sleep (2003) argued that local heating within a deep stockwork rather than plume related heating to be the likely cause of the high temperature of the sheared Jericho xenoliths.

3.4 Dharwar Craton, Southern India

Substituting Q_m for ($Q_s - A_sD$) according to the B-M relation (Eq. 6) in Eq. (10), Ganguly et al. (1995) calculated the steady state Proterozoic mantle geotherm in the Dharwar craton, southern India. The calculated geotherm was constrained by the thermobarometric data of Proterozoic mantle xenoliths in the kimberlite pipes in Wajrakarur region, southern India. The thermal conductivity was varied as a function of depth according to the data of Schatz and Simmons (1972) by dividing the lithosphere into thin shells and implementing the above equation in a finite difference scheme such that K varied stepwise, being constant within each shell, and the bottom temperature of one shell became the surface temperature for the next shell below.

The scanty heat flow versus heat production data of the surface rocks of the Indian shield yields $D = 11.5$ km and $Q_m = 23$ mW/m^2 (Gupta et al. 1991). A value of A_s (1 Ga) $= 2.22 \times 10^{-6}$ W/m^3 was obtained from the present day heat production data after accounting for the decay of radioactive elements since the Proterozoic time. Substitution of the values of D, Q_m and A_s (1 Ga) in Eq. (9) and use of the numerical scheme discussed above yields a steady state paleogeotherm that is $\sim$100 °C higher than the P-T array defined by the thermo-barometric data of the kimberlite xenoliths (Ganguly and Bhattacharya 1987). Thus, instead of retrieving Q_m from the B-M relation, as has been the common practice, Ganguly et al. (1995) treated it as an adjustable parameter such that the calculated geotherm satisfies the P-T data of the xenoliths. This procedure yields $Q_m = 20.92$ mW/m^2

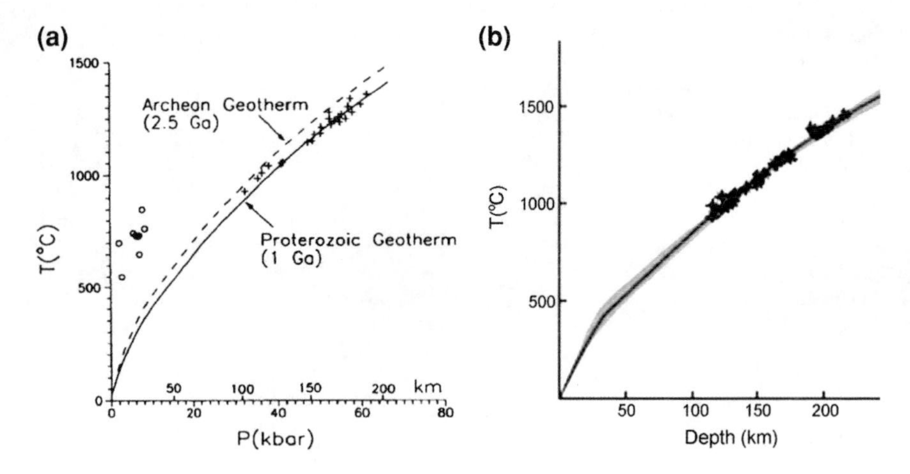

Fig. 3 Steady state lithospheric geotherms in the southern Indian shield. **a** Ganguly et al. (1995) (copyright permission from Wiley): *solid line* 1 Ga, *dashed line* 2.5 Ga (*dashed line*), *open circles* P-T data for the Archean granulites/charnockites. The 1 Ga (Proterozoic) geotherm has been constrained by the thermobarometric data (crosses) of Proterozoic mantle xenoliths in kimberlite pipes in the Wajrakarur area, southern India. **b** Roy and Mareschal (2011) (copyright permission from Wiley): the geotherm is constrained by xenolith P-T, surface heat flux and crustal heat production data. The successful models are confined within the *shaded area*

and a Proterozoic geotherm of the southern Indian shield area, as illustrated in Fig. 3a. The work of Ganguly et al. (1995) represents the first study that combined present day surface heat production rate and xenolith P-T data to calculate a paleogeotherm and mantle heat flux, Q_m. Subsequently the xenolith P-T data were combined with heat flow data in many other regions to construct steady state paleogeotherm. A recent summary of results obtained in various terrains are given in Artemieva (2011).

In order to compare with the P-T data for the Archean granulites in the southern Indian shield, and thus quantify the magnitude of thermal perturbation during granulite facies metamorphism, the calculated geotherm was adjusted to 2.5 Ga by using a cooling model of the upper mantle, as discussed in Ganguly et al. (1995). The adjusted geotherm is shown in Fig. 3a by a dashed line, which lies ~200 °C below the P-T data of the granulites that are illustrated by open circles.

Patel et al. (2009) interpreted the P-T arrays for xenoliths from the Narayanpet kimberlite pipe, Easter Dharwar craton (India) on the basis geotherms calculated by Pollack and Chapman (1977) and found good match with the generic geotherm calculated for $Q_s = 45$ mW m^{-2} and D = 8 km. However, unlike the work of Ganguly et al. (1995), the results of Patel et al. (2009) were not based on models of radiogenic heat and likely surface heat flow or mantle heat flux for the Dharwar craton region.

Recently Roy and Mareschal (2011) recalculated the Proterozoic geotherm of the southern Indian shield by constraining it with the xenolith P-T data, as was done by Ganguly et al. (1995), and also imposing additional constraints. In addition to the data of Ganguly and Bhattacharya (1987), they also used subsequent xenolith P-T data from Nehru and Reddy (1989). Roy and Mareschal (2011) argued that small uncertainties in the temperature gradient in the depth range of 150–200 km resulting from uncertainties in the xenolith P-T data could lead to large uncertainties in the Moho temperatures. Thus, they used the data on surface heat flux, Q_s and crustal heat production rate, A_s, as additional constraints to limit the range of Moho temperatures. Their successful paleogeothermal models yield Moho temperatures of 375–475 °C and $Q_m = 14–20$ mW/m^2. The upper limit of the acceptable Q_m value is similar to the Q_m value inferred by Ganguly et al. (1995). Additionally, the calculated paleogeotherms, which are confined within a very narrow range of temperature defined by the shaded area in Fig. 3b, are not significantly different from the Proterozoic geotherm deduced by Ganguly et al. 1995 (Fig. 3a).

4 Transient Thermal Conditions in the Continental Lithosphere

4.1 Thermal Perturbation of Steady State Geotherm by Fluid Advection and Crustal Melting

In order to explain the thermal perturbation required to produce the granulites in southern India, the thermometric data for which are ~200 °C higher than the temperatures defined by the steady state geotherm (Fig. 3a), Ganguly et al. (1995) calculated the transient thermal effects due to advection of hot CO_2 vapor from the mantle. In this case, Eq. (1) was modified, as follows, to account for the fact that the density and heat capacity of CO_2 is different from that of the rock through which it had moved.

$$\frac{\overline{\rho C_p} \partial T}{\partial t} = K_e \frac{\partial^2 T}{\partial Z^2} + (\rho C_p)_f (V\varphi)_f \frac{\partial T}{\partial Z} + (1 - \phi) A(Z) \tag{24}$$

where $\overline{\rho C_p}$ is the weighted average of the volumetric heat capacity of the rock and pore fluid, K_e is the effective thermal conductivity of the rock-fluid system, ϕ is the porosity of the rock and the subscript f stands for fluid. The term $(V\varphi)_f$ is the volumetric fluid flux (i.e. fluid volume passing through unit area per unit time) or the so called Darcy velocity.

Maximum effects due to fluid flow may be obtained from the solution of the steady state form of the last equation:

$$0 = K_e \frac{\partial^2 T}{\partial Z^2} + (\rho C_p)_f (V\varphi)_f \frac{\partial T}{\partial Z} + (1 - \phi) A(Z) \tag{25}$$

This approach has been used extensively in hydrological literature (Bredehoeft and Papadopulos 1965; Anderson and Mary 2005). For the boundary conditions given by Eq. (3b), the solution of Eq. (25) without the radiogenic heat term is given by

$$T = T_s + \left(\frac{Q_L}{(\rho C_p)_f (V\varphi)_f}\right) \exp(bL)(1 - \exp(-bZ)) \quad (26)$$

In the absence of radiogenic heat, the solution of Eq. (25) with boundary conditions given by Eq. (3c) is:

$$T = T_0 + \frac{(T_L - T_0)\left[1 - \exp\left(-\frac{bz}{L}\right)\right]}{\left[1 - \exp\left(-\frac{b}{L}\right)\right]} \quad (27)$$

where

$$b = (\rho C_p)_f \frac{(V\varphi)_f}{K_e} \quad (28)$$

For problems in metamorphic petrology, Brady (1988) and Ganguly et al. (1995) used the solution of Eq. (25) with the boundary conditions given by Eq. (3c). In the presence of radiogenic heat that varies as a function of depth according to the exponential model, Eq. (5), the steady state solution is:

$$\frac{(T - T_0)}{(T_L - T_0)} = \frac{[1 - \exp(-bZ/L)]}{[1 - \exp(-b/L)]} + \frac{c}{(1 - bD)}\left[(1 - \exp(-Z/D) - \frac{(1 - \exp(-bZ/L))(1 - \exp(-L/D))}{1 - \exp(-b)}\right] \quad (29)$$

where

$$c = \frac{A_s(1 - \phi)}{T_L - T_0} K_e \quad (30)$$

However, for determining time dependent perturbations of temperature, full transient solutions are required, which could be obtained analytically for some special cases, as in and Ganguly et al. (1995).

Using the boundary conditions in Eq. (3d), with L = 90 km, $T_L = 1123$ K (850 °C) and $T_s = 298$ K, Eq. (24) was solved by Ganguly et al. (1995) to obtain a relation between T versus t at different depths for specified values of the Darcy velocity. They also calculated quasi-steady state perturbation of the ambient geotherm due to CO_2 flux from different depths and different values of $V\varphi$. As illustrated in Fig. 4, quite satisfactory matches between the perturbed quasi-steady state geotherms and the thermobarometric data of the granulites could be obtained

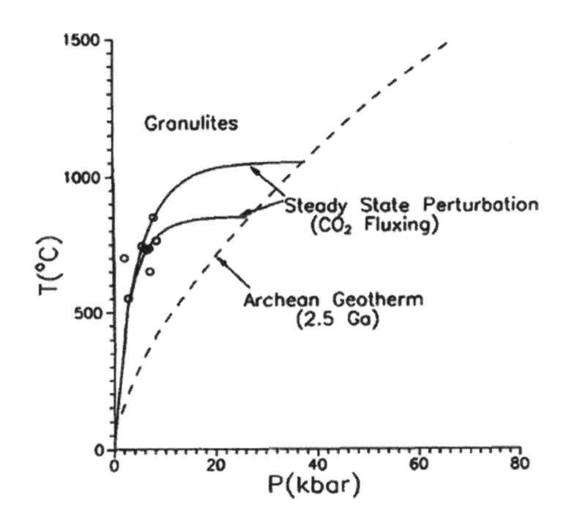

Fig. 4 Steady state perturbation (*solid lines*) of the ambient Archean geotherm due to degassing of CO_2 from depths of 90–125 km with a Darcy velocity of 0.3 cm/year. After Ganguly et al. (1995) (copyright permission from Wiley)

by setting the depth of CO_2 degassing at 90–105 km and $V\varphi = 0.3$ cm/year. On the basis of these results, Ganguly et al. (1995) have discussed if CO_2 fluxing from the mantle could have been a viable mechanism for the thermal perturbation required for the granulite facies metamorphism in the Dharwar craton. They concluded that a combination of the thermal effects of CO_2 fluxing and over-thrusting to be the most likely cause for the formation of these granulites. An interesting feature of the perturbed geotherms in Fig. 4 is the near isothermal condition around the depth of CO_2 degassing and very rapid change of temperature at shallower depths.

Depine et al. (2008) considered crustal melting and melt focusing as a plausible reason for thermal perturbation of steady state geotherms to generate P-T conditions for granulite facies metamorphism. Figure 5 illustrates their results that show the best match between a global collection of granulite P-T conditions and quasi-steady state perturbed geotherm that is reached after 25.7 Myr. The shape of the quasi-steady state geotherm resulting from melt focusing is similar to that calculated by Ganguly et al. (1995) for CO_2 fluxing (Fig. 4). Depine et al. (2008) assumed an initial temperature of 1250 °C at the crust-mantle interface, which is equivalent to positioning the asthenosphere directly below the crust. In this model, heat flux from the mantle triggers melting at the base of the crust, and thermal buffering causes the Moho to evolve from 1250 °C to the crustal solidus temperature. After exceeding a threshold of 10 vol %, the melt migrates to 13 km depth, thereby advecting heat upward, and causing a downward displacement of the crust below to satisfy mass balance. The heat transfer equation solved numerically by Depine et al. (2008) is essentially the same as Eq. (24), but an additional term was added to the right to account for the latent heat effect due to melting.

Long (2010) used the formalism of Depine et al. (2008) to interpret xenolith P-T data. Here the variables have been the mean crustal radioactive heat production, basal temperature of lithosphere and thermal properties near the bottom of the

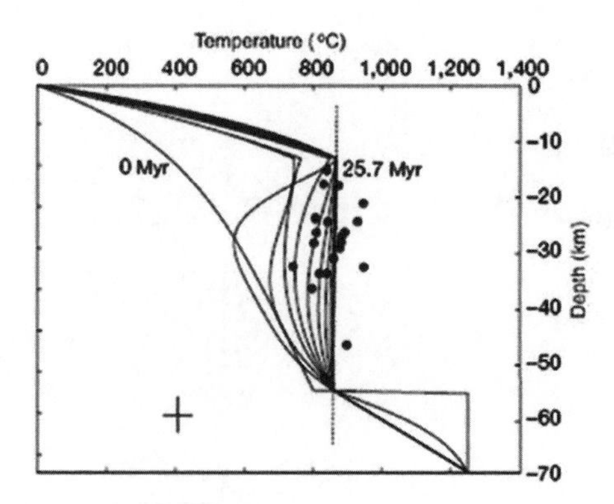

Fig. 5 Numerical model showing evolution of crustal geotherms as function of time at 3 M year steps as a result of thermal perturbation of an initial geotherm (0 M year) at 55 km with a temperature of 1250 °C. The quasi-steady state of the perturbed geotherm is reached at 25.7 year. The solidus of the crust (thin dashed vertical line) is assumed to have been defined by the dehydration melting of hornblende + quartz ± plagioclase. The filled symbols indicate a global collection of granulite facies P-T condition. The *cross* shows the P-T uncertainty of the thermobarometric data. Reprinted by permission from Macmillan Publishers Ltd: Nature, Depine et al. (2008), copyright (2008)

lithosphere. Enhanced temperature at the base of the lithosphere is ascribed to the presence of plume.

Bickle and McKenzie (1987) calculated the effects of advection and conduction in layer of constant thickness for three cases, as follows, all characterized by constant temperature boundary conditions and absence of radiogenic heat.

(a) A layer having an initial temperature profile increasing linearly form the upper surface is perturbed with uniform upward fluid flow with no changes in the thermal boundary conditions. The geotherms become progressively nonlinear with increase in the fluid advection velocity.
(b) The next case dealt with a layer for which bottom temperature is suddenly raised leading to the onset of advection.
(c) The third case involves the thermal evolution of a layer having uniform initial temperature distribution and cooling at both boundaries by sudden application of zero temperature boundary conditions and also onset of advection by fluids.

The nature of transient geotherms reflects the nature of perturbations. The fluid flow also disturbs the distribution of geochemical anomalies along with thermal perturbations. Bickle and McKenzie (1987) also solved for the distribution of geochemical anomalies as equations and boundary conditions are formally the same for the calculation of temperature and concentration changes. They worked out conditions in which geochemical perturbations would take place due to advection by fluids, without significantly affecting the spatial variation of temperature.

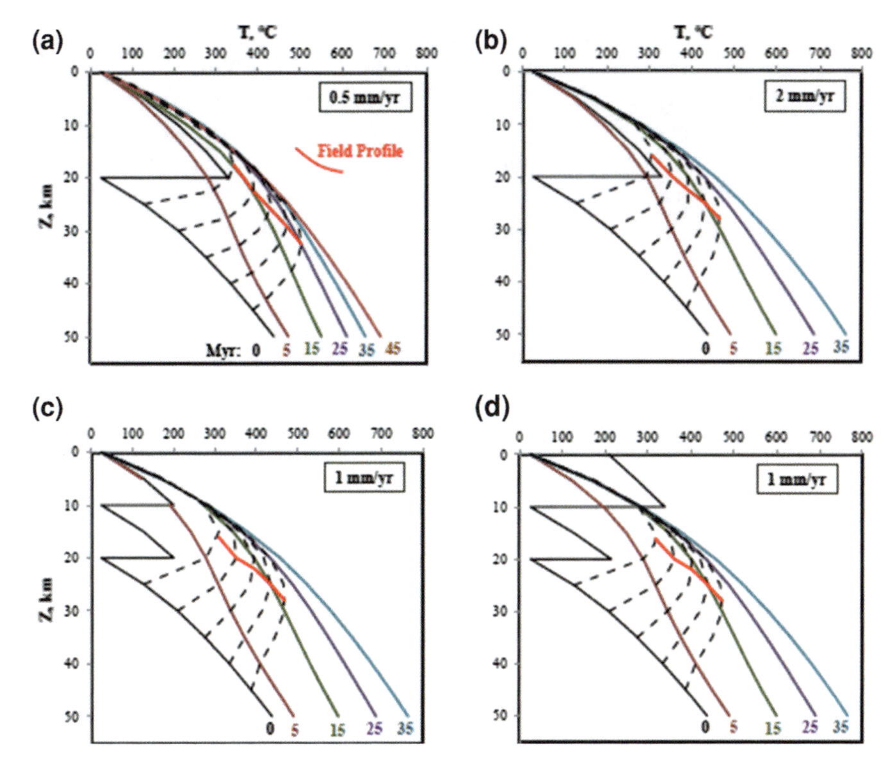

Fig. 6 Transient thermal profiles after thrusting, followed immediately by erosion, the corresponding Z-T paths of rocks exhuming from different depths and the Metamorphic Field Profiles. Each *panel* shows the erosion rate. The *curved dashed lines* show the T-Z trajectory of the material parcels exhuming from their initial locations on the "saw-tooth" geotherm. *Panels* **a** and **b**: single 20 km thrust sheet; **c** and **d**: two thrust sheets, 10 km each. In (**c**), both thrust sheets have the same radiogenic heat production rate at the surface and variation with depth, whereas in (**d**), both sheets have been derived from the same place by successive thrusting events, without any significant time lag, so that the bottom temperature and radiogenic heat production rate of the first thrust sheet define the surface conditions of the second one

4.2 Transient Geotherms and Z(P)-T Paths of Rocks During Exhumation in an Overthrusted Region

In an overthrusted region, the initial geotherm assumes a "saw tooth" shape. In this case Eq. (2) is solved numerically using the boundary condition (3b) and a finite difference/element scheme to evaluate the transient effects of overthrusting on the thermal profile. Figure 6a–d illustrate the results of such calculation, assuming that erosion had taken place immediately after thrusting. The calculations were carried in this study out using the COMSOL finite element package. The erosion rate was assumed to be 0.5 mm/year for the Fig. 6a, 2 mm/year for Fig. 6b, and 1 mm/year for Fig. 6c and d. In all cases, the boundary conditions are defined by constant surface temperature of 298 K and basal heat flux of 23 mW/m^2, as inferred by Ganguly et al. (1995) for the Dharwar craton. The initial geotherm prior to thrusting and other parameters (i.e. A(Z), D, Cp, ρ), except the thermal conductivity, K, are also taken to be the same as used for the calculation of Proterozoic geotherm in the Dharwar craton (Fig. 3a). For simplicity, we have used a constant value of K of 2.25 W/m-K in these calculations whereas Ganguly et al. (1995) used a depth dependent K, as discussed above, in the calculation of Fig. 3a.

Figure 6a and b illustrate the effect of erosion rate on the transient thermal profiles following thrusting of a single block of 20 km thickness. In Fig. 6b and c, there are two thrust sheets, each of 10 km thickness. Both sheets have the same characteristics in Fig. 6c so that the surface temperature and radiogenic heat production rate versus Z are the same for both, as could happen for thrust sheets derived from different domains in the same area. In Fig. 6d, on the other hand, the thrust sheets were derived from the same place so that the surface temperature and surface heat production rate of the second sheet, the top block in Fig. 6d, were defined by the bottom conditions of the first (earlier) one prior to thrusting.

Figure 6a–d also show the Z-T path of parcels of rocks exhuming from different depths in response to erosion that follows the thrusting event without any significant time lag. In the P-T space, the exhumation paths would appear as clockwise trajectories, except for rocks in the upper block where the initial temperature is greater than the transient temperatures for some period of time, requiring the P-T paths to become anti-clockwise in the beginning. The locus of the thermal maxima (LTM) of the rocks exhumed from different depths is known as the "metamorphic field gradient", but it is more appropriate to call it a "metamoprhic field profile" since it describes the position, and not just the gradient, of the thermal maxima in the P-T space. Thermobarometric studies of regionally metamorphosed rocks typically reflect this field profile, and not the steady state or transient geotherms.

The MFF-s for the different scenarios of thrusting and erosion used in the calculations of Fig. 6a–d are illustrated in Fig. 7. It is interesting to note that there is no significant difference among the MFF-s for the different thrusting and erosion scenarios explored in this study, especially considering the numerical errors in the simulation.

Analytical solution of Eq. (2) for the simple case of a semi-infinite slab with constant values for the source term (A), advection velocity (v), surface temperature, T_s, and a linear initial geotherm is given by Carslaw and Jaeger (1959, Eq. 7, p. 388). (The term semi-infinite implies a slab in which the temperature at a sufficiently large distance from the boundary remains unaffected by heat transport.) Although geological problems are, in general, more complicated, this solution has been found to be quite useful for getting an approximate idea of the transient thermal states in the lithosphere. In this solution, T is expressed as a function of Z, v and t, but in geological problems, these three variables are also linked by the relation $Z = Z_0 - vt$, where Z_0 is the initial depth of the exhuming rock parcel. Thus, the transient Z-T path of exhumation parcel of rock may be calculated simply by fixing Z_0 and v.

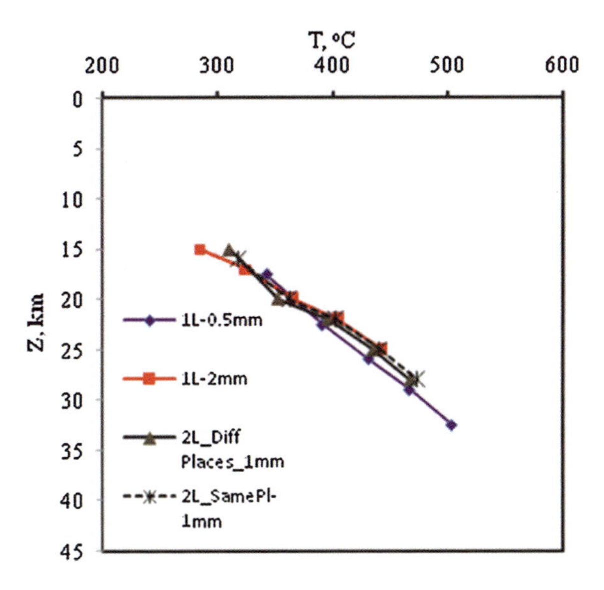

Fig. 7 Comparison of the Metamorphic Field Profiles for the different thrusting and erosion scenarios shown in Fig. 6. *Symbols* 1L-0.5 mm means 1 thrust layer and the erosion rate is 0.5 mm/year; 2L_DiffPlaces_1 mm means 2 successive thrust layers from different places and 1 mm/year erosion rate, and similarly for others

Ganguly et al. (2000) used the solution of Eq. (2) given by Carslaw and Jaeger (1959) to calculate the exhumation history of a section of Sikkim Himalayas near the South Tibetian detachment system or STDS, satisfying the constraints imposed by the mineralogical reactions that had taken place during exhumation, as determined by petrographic observations, and the retrograde compositional zoning in garnet. They considered the variation of A with depth according to the exponential form (Eq. 5) by incorporating it in a finite difference scheme in which the total depth was divided into a large number of layers and A was assumed to be constant within each layer, equaling its average value within the layer. The results of Ganguly et al. (1995) are reproduced in Fig. 8. It shows a two stage exhumation, a very rapid initial stage ($\sim$15 mm/year) followed by a slow stage at a rate of $\sim$2 mm/year. The rapid initial stage was needed to satisfy the peak P-T condition of a parcel of rock from near the STDS and the fact that the petrographic examination shows retrograde reactions Grt + Sill → Spnl + Qtz and Grt + Sill + Qtz → Cord. The slower second stage describes a P-T path that leads to the observed retrograde zoning of divalent cations in garnet. The pattern of P-T path shown in Fig. 8 has been borne out in a recent study by Sorcar et al. (2014) using several samples from near STDS and judicious thermobarometric studies.

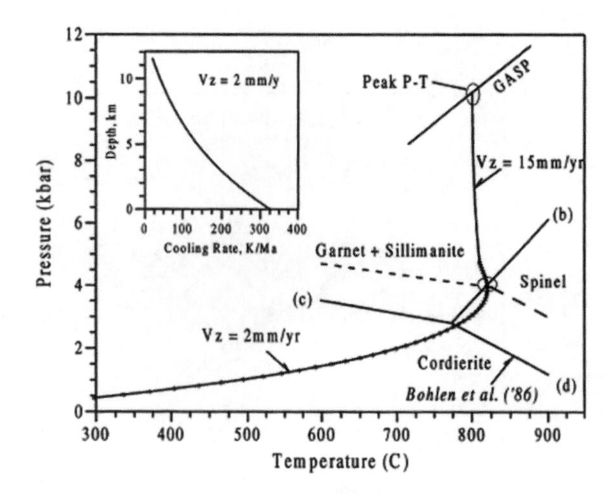

Fig. 8 Two step exhumation history of a section of the Sikkim Himalaya (*solid* and *patterned lines*) and the corresponding exhumation velocities, as reconstructed from phase equilibrium constraints ($V_z = 15$ mm/year) and retrograde compositional zoning of garnet ($V_z = 2$ mm/year). The *solid lines* labeled (*b*) (*c*) and (*d*) denote the equilibrium reaction boundaries in the $FeO–Al_2O_3–SiO_2$ system, and the corresponding *dashed lines* represent the calculated displacement of the equilibria for the observed mineral compositions. Reprinted from Ganguly et al. (2000), Copyright (2000) with permission from Elsevier

Acknowledgments RNS is grateful to Indian National Science Academy, New Delhi, for award of a Senior Scientist position.

References

Anderson MP (2005) Heat as a ground water tracer. Ground Water 43:951–968

Artemieva I (2011) The lithosphere: an interdisciplinary approach. Cambridge University Press, Cambridge, p 794

Bickle MJ, McKenzie D (1987) The transport of heat and matter by fluids during metamorphism. Contrib Mineral Petrol 95:384–392

Birch F, Roy RF, Decker ER (1968) Heat flow and thermal history on New York and New England. In: Zen E, White WS, Hadley JB, Thompson JB Jr (eds) Appalachian geology: northern and maritime. Interscience, New York, pp 437–451

Brady JB (1988) The role of volatiles in the thermal history of metamorphic terranes. J Petrol 29:1187–1213

Bredehoefh JD, Papadopoulos IS (1965) Rates of vertical groundwater movement estimated from the Earth's thermal profile. Water Resour Res 1:325–328

Carslaw HS, Jaeger JC (1959) Conduction of heat in solids, 2nd edn. Oxford University Press, London, p 510

Dapine VD, Andronicos CA, Phipps-Morgan J (2008) Near-isothermal conditions in the middle and lower crust by melt migration. Nature 456:80–83

Ganguly J, Bhattacharya PK (1987) Xenoliths in proterozoic kimberlites from southern India: petrology and geophysical implications. In: Nixon PH (ed) Mantle xenoliths, Springer, New York, pp 248–265
Ganguly J, Singh RN, Ramana DV (1995) Thermal perturbation during charnockitization and granulite facies metamorphism in southern India. J Metamorphic Geol 13:419–430
Ganguly J, Dasgupta S, Cheng W, Neogi S (2000) Exhumation history of a section of the Sikkim Himalayas, India: records in the metamorphic mineral equilibria and compositional zoning of garnet. Earth Planet Sci Lett 183:471–486
Gupta ML, Sunder A, Sharma SR (1991) Heat flow and heat generation in the Archean Dhanvar cratons and implications for the Southern Indian Shield geotherm and lithosphere thickness. Tectonophysics 194:107–122
Lachenbruch AH (1968) Preliminary geothermal model of the Sierra Nevada. J Geophys Res 73:6877–6989
Long AM (2010) Xenoliths: insights from upper mantle petrology. Unpublished MS Thesis, Cornell University
Mather KA, Pearson DG, McKenzie D, Kjarsgaard BA, Priestley K (2011) Constraints on the depth and thermal history of cratonic lithosphere from peridotite xenoliths, xenocrysts and seismology. Lithos 125:729–742
McKenzie D, Bickle MJ (1988) The volume and composition of melt generated by extension of the lithosphere. J Petrology 29:625–679
McKenzie D, Jackson J, Priestley K (2005) Thermal structure of oceanic and continental lithosphere. Earth Planet Sci Lett 233:337–349
Nehru CE, Reddy AK (1989) Ultramafic xenoliths from Wajrakarur kimberlites, India. In: Ross J et al (eds) Kimberlites and related rocks. Geol Soc Aust, Spec Publ no 14:745–759
Parsons B, McKenzie D (1978) Mantle convection and the thermal structure of the plates. J Geophys Res 83(B9):4485–4496
Patel SC, Ravi S, Anilkumar Y, Naik A, Thakur SS, Pati JK, Nayak SS (2009) Mafic xenoliths in proterozoic kimberlites from eastern Dharwar craton, India: mineralogy and P-T regime. J Asian Earth Sci 34:336–346
Pollack HN, Chapman DS (1977) Regional variation of heat flow, geotherms, and lithospheric thickness. Tectonophysics 38:279–296
Richter FM, McKenzie DP (1981) On some consequences and possible causes of layered mantle convection. J Geophys Res 86:1738–1744
Roy S, Mareschal J-C (2011) Constraints on the deep thermal structure of the Dharwar craton, India, from heat flow, shear wave velocities, and mantle xenoliths. J Geophys Res 116:B0 2409. doi:10.1029/2010JB007796
Rudnick RL, McDonough WF, O'Connell RJ (1998) Thermal structure, thickness and composition of continental lithosphere. Chem Geol 145:395–411
Russell JK, Kopylova M (1999) A steady state conductive geotherm for the north central Slave, Canada: inversion of petrological data from the Jericho Kimberlite pipe. J Geophys Res 104(B4):7089–7101
Russell JK, Dipple GM, Kopylova MG (2001) Heat production and heat flow in the mantle lithosphere, Slave Craton, Canada. Phys Earth Planet Inter 123:27–44
Singh RN, Negi JG (1979) A reinterpretation of the linear heat flow and heat production relationship for the exponential model of the heat production in the crust. Geophys J Roy Astron Soc 57:741–744
Schatz J, Simmons G (1972) Thermal conductivity of earth materials at high temperatures. J Geophys Res 77:6966–6983
Sleep NH (2003) Geodynamic implications of xenolith geotherms. Geochem Geophys Geosyst 4(9):1079. doi:10.1029/2003GC000511
Solomatov VS (1995) Scaling of temperature- and stress-dependent viscosity convection. Phys Fluids 7:266–274

Solomatov VS, Moresi LN (2000) Scaling of time-dependent stagnant lid convection: application to small-scale convection on earth and other terrestrial planets. J Geophys Res 105:21795–21817

Sorcar N, Hoppe U, Dasgupta S, Chakraborty S (2014) High temperature cooling histories of migmatites from the High Himalayan Crystallines in Sikkim, India: rapid cooling unrelated to exhumation? Contrib Mineral Petrol 167:957. doi:10.1007/s00410-013-0957-3

Accessory Phases in the Genesis of Igneous Rocks

Igor Broska and Igor Petrík

Abstract An overview of the significance and application of most common accessory minerals in igneous systems, mainly in granitic rocks is presented in two parts: (1) General description and definition of the most important accessory phases are given, and (2) a case study from Western Carpathians is dealt which unravels the granite typology. A short account of structure and composition of principal accessory phases is also discussed along with their occurrences, usage of isotopes and thermodynamic constraints which reveal the P-T evolution of parental igneous bodies.

1 Introduction

The importance of accessory mineral assemblage, their crystal-chemical compositions, time of precipitation has long attracted the attention of petrologists. Generally, the geochemistry of trace elements in igneous rocks is controlled by accessory mineral saturation and fractionation. Paragenesis of accessory mineral carries plenty of petrogenetic information about the behaviour of trace elements as well as their isotopes, which can be potentially used for the discrimination and characterization of geotectonic setting of magmatic rocks. Most accessory minerals act as principal geochronometers, both isotopic and chemical dating which underline the importance of their scientific knowledge. This includes the diffusion characteristics of major elemental compounds in the melt, their solubility and resistance.

The major goal of this paper is to characterize the accessory assemblages observed in granitoids and their mineralogy, and then we focus on typomorphic

I. Broska (✉) · I. Petrík
Geological Institute of Slovak Academy of Science, Dúbravská cesta 9,
840 05 Bratislava, Slovak Republic
e-mail: igor.broska@savba.sk

S. Kumar and R. N. Singh (eds.), *Modelling of Magmatic and Allied Processes*,
Society of Earth Scientists Series, DOI: 10.1007/978-3-319-06471-0_6,

features of accessory minerals enabling to understand the parental host magma evolution and their co-magmatic affiliation, if any, within a magmatic suite. The typomorphic features of accessories carry significant information of temperature, redox conditions, REE fractionation during magma evolution, its crystallization age, as well as subsolidus processes such as breakdown and replacement phenomena. The second part provides examples of the most important granitoid accessory minerals, as have been characterised in the Wester Carpathians, see Broska et al. (2012).

According to genetic preferences, we prefer to group the accessory phases suitably into (1) primary, (2) secondary, (3) relic, (4) hydrothermal and (5) of metamorphic origin. Accessory minerals from igneous rocks however do not contribute to rock classification as their quantity is usually below 1 wt% in the bulk rock. Exceptionally high concentrations (titanite, hornblende, apatite etc.) of phases may occur because of extreme fractionation and accumulation processes in the magma chamber.

2 Characteristics and Typomorphic Features of Accessory Minerals

2.1 Zircon

Zircon is a tetragonal mineral with general formula ATO_4 in which high field strength elements (Zr, Hf, U, Y etc.) occupy larger eight-coordinated structural A-sites, whereas T-site are represented by silicon oxide tetrahedra. Structural chains of alternating Si tetrahedra and Zr polyhedra along C-axis form the zircon prismatic habit. The ZrO_8 dodecahedron can be described as two interpenetrating ZrO_4 tetrahedra with different Zr–O distances, which are at high pressure conditions shortened in a different way (Finch and Hanchar 2003). The structure of zircon is relatively open with some voids between SiO_4 tetrahedra and ZrO_8 polyhedra allowing incorporation of various impure elements along channels parallel to (001). Due to presence of actinides and their α-decay in zircon, its structure is commonly metamictized leading to amorphous and isotropic stage. The metamict zircon significantly changes optical properties, colour and also lowers the density. For the structural damage α-particles are responsible, which cause displacements of atoms. The metamictization of crystalline zircon is characterized by its transformation from an initial stage where isolated amorphous domains are surrounded by slightly disordered crystalline material to a damage level, where crystalline remnants occur in an amorphous matrix. The radiation damage in zircon is not uniform it rather forms small haloes inside (Nasdala et al. 2005). Metamict zircon shows often porous texture providing low sum in microprobe analysis because the extent of radiation damage in crystal causes loss of variable amounts of Zr, Hf, REE, U, Th and Pb (Nasdala et al. 2009; Geisler et al. 2003).

Recrystallization of the amorphous phase in metamict zircon is strongly enhanced under hydrothermal conditions (Geisler et al. 2003). Breakdown of U leading to formation of fission tracks is used for dating events related to temperatures lower than ca 310 °C (FT method).

Zircon is transparent to translucent with variety of colours such as pink, rose, purple, hyacint, brown, redbrown, but is also colourless. Due to high refractive index some crystals can bear gemstone quality. The colour of zircon is related to the concentration of trace elements, especially rare earths and radiation damage. Zircon gradually accumulates colour through electron displacement driven by α-decay of U and Th. The source of coloration in zircon is poorly known but europium and REEs are considered as the most important contributors (Fielding 1970). Garver and Kamp (2002) have however described decoloration of zircon during orogenesis in the Southern Alps and New Zealand. Colour removal zone has formed at 400 °C, below the fission track partial annealing zone.

Zircon is a ubiquitous accessory mineral phase crystallising in many magmatic, particularly granitic rocks where supersaturation of the melt by Zr was achieved and occasionally precipitating on older inherited cores. Zircon begins to crystallize as an early magmatic phase and its abundance decreases with fractionation. The I-type granitoids are richer in zircon as compared to S-type granites, the latter variety commonly contains more inherited cores. The importance of zircon in crustal evolution studies is underscored by its use in U–Th–Pb geochronology and in many cases single zircon crystal contains a record of multiple geologic events (Hoskin and Schaltegger 2003). Zircon is also used for chronology and petrological purposes using Sm–Nd and Lu–Hf isotopic systems. In alkaline and peralkaline magmas zircon is soluble and hence Zr forms more complex silicates.

Changes in magma composition result in different habits of zircon. Pupin (1980) proposed a method of evaluation of zircon morphology using parameter dependant on *T* (IT; index of temperature) and alkalinity (I.A; index of agpaicity), which were included in a typological diagram (Fig. 1) as a useful scheme for understanding genesis and evolution of granites (sensu lato). It was shown that the zircon pyramid {211} usually forms in a hyperaluminous and hypoalkaline environment; pyramid {101} indicates hyperalkaline and hypoaluminous system and pyramid {301} is a characteristic of potassium-rich alkaline environment. Mean points of zircon populations plotted in the Pupin typogram may be used for the affiliation of granite suites (S-, I-, and A-type) and zircon typological data gathered from granitic batholiths enable wider regional correlations (e.g. Broska and Uher 1991; Belousova et al. 2006). However evaluation of morphological parameters has showed that the estimated temperature is relative rather than absolute scale, and zircon exhibits evolution within subtypes of the typology. Vavra (1990) concluded that growth rates are controlled by $ZrSiO_4$ supersaturation and concentration of distinct trace elements. Vavra (1990, 1994) further suggested that morphology of pyramids {211} versus {101} is determined by foreign trace elements, whilst prisms indicate degree of Zr supersaturation. Benisek and Finger (1993) revealed a U growth-blocking effect for {110}-type faces and therefore is a cause the domination of {110} faces in some samples. Vavra (1996) proposed

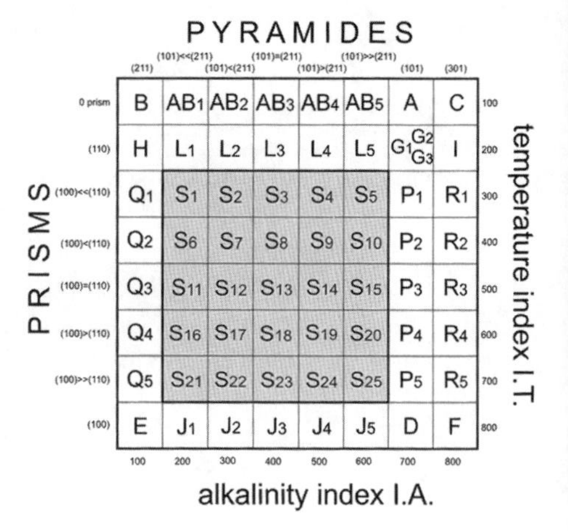

Fig. 1 Typological scheme of zircon morphological types: prisms versus pyramids (Pupin 1980). I.A. and I.T. indexes indicate relative degree of alkalinity and temperature of host granitoids

an assessment of evolution of zircon growth not only by indexing the faces of crystals but also of their inner zones.

Typical oscillatory zoning of zircon varies from sub-micron to millimetre scale where chemically and optically distinct bands have grown parallel along crystal faces. Oscillatory zoning may be characterised by its fractal dimension (Halden and Hawthorne 1993) and is observed in back scatter electron (BSE) images and more sensitively under cathodoluminiscence (CL). Brighter parts of images mean higher content of U and Th but may often contain also the REEs. Oscillatory zoning is most likely an indicative of igneous origin but sector or patchy zoning is also described especially from zircons of metamorphic origin (Hoskin and Schaltegger 2003; Corfu et al. 2003). No two zircons with the same oscillatory zoning have been described within one individual sample. Therefore, the oscillatory zoning cannot be understood as a purely magmatic phenomenon. Eventually, Hoskin (2000) proposed a model of oscillatory zoning, which is generated in the process of self-organisation. Fowler et al. (2002) pointed out different zoning in crystals, the growth of which cannot be a simple reflection of changes within bulk system. Irregular zoning patterns according to Hoskin and Black (2000) can arise by recrystallization of older zircons, by dissolution/reprecipitation or by annealing.

The main role of zircon in granitic melt is control of the distribution of Zr and Hf as a main carrier of these elements. The Zr/Hf ratio of zircon is close to that of whole-rock. Hafnium concentration typically ranges from 0.5–3 wt% with mean 1.8 wt% (Belousova et al. 2002). The HfO_2 versus Y_2O_3 diagram (Pupin 2000) can be used for distinguishing orogenic and anorogenic granitoids. The mean Zr/Hf ratio of calc-alkaline rocks reaches 45, and decreases in the aluminous leucogranites to less than 35 (Pupin 2000). In alkaline rocks, the increase of Zr/Hf ratio (often above 60) is result of alteration enhanced by fluid regime rather than magmatic signature (Kempe et al. 2004). The structure of zircon incorporates preferentially medium and heavy rare earth elements or Y with P according to the xenotime type substitution: $REE^{3+} + P^{5+} = Zr^{4+} + Si^{4+}$ and in more evolved granites also petulite type substitution: REE for Sc ($ScPO_4$ molecule). Sum of REE commonly exceeds P which is indicating of various other substitution mechanisms. Hanchar et al. (2001), experimentally proved that P exceeds REE in a range from La to Nd, from Sm to Gd the REE are generally equal to P, and from Tb to Lu the REE exceed P suggesting significance of charge neutrality and ionic diameter. Hafnium is present in zircon as isotope ^{177}Hf. Beta decay of ^{176} Lu results in formation of other isotope ^{176}Hf and the ratio $^{176}Hf/^{177}Hf$ is very suitable for determining the timing of zircon precipitation. Experimentally it was shown that zircon Hf closure temperature is by ca 200 °C higher than that of Pb and its diffusion is much lower than Pb (Cherniak and Watson 2000), so the Hf system remains closed during superimposed thermal effects. Hafnium isotopes also indicate crustal or mantle origin of zircon. High values of $^{176}Hf/^{177}Hf$ indicate involvement of crustal sources because of the fact that during mantle melting Hf enters the melt preferentially to Lu. The zircon Hf isotope data worldwide imply that at least 60–70 % of existing continental crust separated from mantle before 2.5 Ga and the continental crust was generated continuously even though with some gaps (Belousova et al. 2010). Pupin (2000) and Belousova et al. (2002) proposed the use of zircon Hf and Y for discrimination of granite types.

The zircon structure easily adopts actinides, mainly U and Th and their concentration serves as an indicator of magmatic or metamorphic origin. According to Heaman et al. (1990) the Th/U ratio is generally higher for metamorphic than for magmatic zircons. The low diffusion rates for most of the elements mean a relatively high stability of zircon under different geological conditions.

Oxygen isotope of zircon can be used for determining the crustal characteristics. Oxygen $\delta^{18}O$ values higher than ~6.3 ‰ are typical of crustal rocks (Mojzsis et al. 2001). Zircon from oceanic crust (gabbro and serpentinized peridotite) preserves lower primitive $\delta^{18}O$ values ~5.3 ± 0.8 ‰ (Cavosie et al. 2009). Elevated values of oxygen isotope ratios generally result from partial melting and contamination of metasediments in the deep crustal environment (Valley et al. 1994).

Hydrothermal experiments carried out by Watson and Harrison (1983) have defined the saturation temperature of zircon in crustal derived anatectic melts. The saturation depends on temperature and peraluminosity of the melt expressed by parameter M [cationic ratio (Na + K + 2Ca)/(Al*Si)]. The thermometer can be applied to intermediate to felsic magmas except for peralkaline and dry (<1.5 wt% H_2O) melts. The higher is T, the higher must be Zr concentration to achieve the saturation level, e.g. in a subaluminous melt (molar A/CNK = 1; M = 1.6) at 800 °C ca 220 ppm of Zr is needed to reach the saturation. With increasing M the saturation T also increases. Zirconium used for calculation should correspond to the real concentration in the melt; therefore, inherited zircon cores as well accumulated crystals increase the obtained saturation temperatures. Bulk rock Zr concentrations give minimum T if magma was undersaturated and maximum T if magma was saturated (Miller et al. 2003). These authors have shown that inheritance-poor granites yield higher saturation T than inheritance-rich granites due to the higher Zr concentrations of the former and the lower T of generation for the latter.

2.2 Apatite

The composition of apatite can be described by the general formula of the apatite supergroup $^{IX}M1_2^{VII}M2_3(^{IV}TO_4)_3X$ (Pasero et al. 2010). The Ca cation is located at different M1 and M2 crystallographical positions expressed by relation $Ca(M1)_4Ca(M2)_6(PO_4)_6X_2$. In granitoids, there are three end-member compositions: fluorapatite $Ca_5(PO_4)_3F$, chlorapatite $Ca_5(PO_4)_3Cl$ and hydroxylapatite $Ca_5(PO_4)_3(OH)$, but fluorapatite occurs abundantly. Apatite is microporous mineral due to its "honeycomb" structure (White et al. 2005) remarkably tolerant to structural distortion and chemical substitution, and consequently exhibits diverse in composition. The apatite structure allows for numerous substitutions, including many cations (e.g. K, Na, Ba, Sr, Mn, Fe, Y, REEs, U etc.) that substitute for Ca in the structure, and anionic complexes (i.e. $SO_4{}^{2-}$, $SiO_4{}^{4-}$, $CO_3{}^{2-}$ etc.) replacing $PO_4{}^{3-}$ (Hughes and Rakovan 2002; Pan and Fleet 2002).

Understanding the behaviour of apatite in granite system is possible through Ca and REE distribution but mainly by P geochemistry showing contrasting behaviour within S-, I- and A- type granite suites (Bea et al. 1992; Chappell and White 1998), which are delineated by different apatite compositions (Sha and Chappell 2000; Belousova et al. 2001). The variability in the abundances of P in granite is related to the higher P solubility in peraluminous melts, P becoming progressively more abundant in the felsic S-type melts during fractionation. London et al. (1990) attributed the higher solubility of P in peraluminous felsic melts, and its relatively high concentrations in the alkali feldspar, at elevated ASI [(ASI = aluminium saturation index, molar $Al_2O_3/(Na_2O + K_2O + CaO)$]. This is because of the fact that the solubility of apatite in metaluminous and peralkaline magmas is a function of melt composition and temperature as experimentally shown elsewhere (Watson and Capobianco 1981; Green and Watson 1982; Harrison and Watson 1984). Considerably higher apatite solubility in peraluminous melts (ASI $>$ 1.1) was observed by Pichavant et al. (1992), and Wolf and London (1994). The low solubility of monazite in peraluminous magma leads to the removal of REE and hence apatite crystallization in the granites with S-type character is delayed. On the other hand, the higher CaO content and greater solubility in metaluminous I-type magmas leads to early crystallization of apatite (Fig. 2).

Simpson (1977) synthesized Al–P feldspars with the general formulae $NaAl_2PSiO_8$ and KAl_2PSiO_8 and postulated that P could be accommodated into alkali feldspars under suitable geologically conditions. Later, London (1992) demonstrated that the incorporation of P into feldspars involves coupled substitution of $Al^{3+} + P^{5+} = 2\,Si^{4+}$. This exchange is known as berlinite substitution, where berlinite, $AlPO_4$, is isostructural with quartz or the Si_2O_4 framework component of feldspars. Relatively high P contents have been recorded in alkali feldspars (Af) from many different peraluminous granitic suites, and London (1992) reported about 60 localities where the P contents in Af, including potassium feldspar (Kfs) and sodic plagioclase (Pl), ranging from 0.1 to 0.8 wt%. In general, Kfs are richer in P, and so an increase of P in Kfs with differentiation appears to be

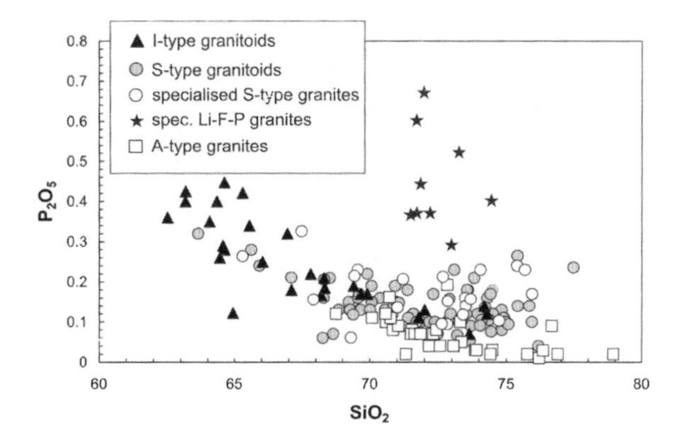

Fig. 2 Variation of P_2O_5 versus SiO_2 (wt. %) in granitoid rocks shows the differences among main genetic types. Phosphorus distribution is irregular in S- and specialised S-type granites where berlinite substitution is more extensive (example from the Western Carpathians; Broska and Uher 2001)

a general phenomenon. The highest values of P in Af from a felsic magmatic system were reported from the Alburquerque batholith in Spain, where P_2O_5 concentrations reached 2.6 wt% (London et al. 1999), and from the Podlesí granite in the Krušné Hory Mts. (Erzgebirge) with up to 2.5 wt% P_2O_5 (Frýda and Breiter 1995).

Saturation temperature of apatite derived by Harrison and Watson (1984) for metaluminous granitic rocks is not suitable for peraluminous compositions where it provides too high temperatures. Therefore, corrections were suggested by Pichavant et al. (1992), which modifed the equations for peraluminous melts and provided more realistic temperatures in these systems. Apatite, as an important carrier of volatiles in igneous rocks, may be used for estimation of primary concentration of F and Cl in the melt (Piccoli and Candela 1994).

2.3 Fe–Ti oxides

Fe–Ti oxides (cubic ulvöspinel Fe_2TiO_4, cubic magnetite Fe_3O_4, rhombic ilmenite $FeTiO_3$, tetragonal rutile TiO_2, trigonal hematite Fe_2O_3, rhombic ferropseudobrookite $FeTi_2O_5$, rhombic pseudobrookite Fe_2TiO_5) are important accessory constituents of granitic rocks. Their abundance and proportions depend mostly on temperature *T* and oxygen fugacity (fO_2) of the magma or postmagmatic fluids. The Fe–Ti oxides are characteristic of I-type granite suites giving possibilities to define the primary oxygen or water activities (fO_2, fH_2O). Moreover, pre-eruptive magmatic conditions (*T* and fO_2) in silicic volcanic rocks may be estimated from Fe–Ti oxide chemistry (Ewart et al. 1975; Ghiorso and Sack 1991) using the oxygen barometer. Such data give important information about the nature of melt in a storage chamber prior to the eruption. In granitic rocks Fe–Ti oxides are a window to the processes that ocurred during late magmatic or early subsolidus stage depending on the behaviour of O and H_2O.

Because of high Fe contents in Fe–Ti oxides their stability is interrelated with that of biotite and hornblende. Magnetite is related to biotite through a well-known reaction (e.g. Wones and Eugster 1965):

$$\underset{\text{Ann}}{KFe_3AlSi_3O_{10}(OH)_2} + 1/2\ O_2 = \underset{\text{Mt}}{Fe_3O_4} + \underset{\text{Kfs}}{KAlSi_3O_8} + H_2O \quad (1)$$

The reaction (1) indicates that with increasing f_{O2} magnetite becomes stable at the expense of annite component in biotite. The presence or absence of magnetite is a basis for distinguishing magnetite versus ilmenite series granitoids (Ishihara 1977; Kumar 2010). Magnetite and ilmenite are typical of I-type granitoids, whearas sulfide and ilmenite are for S-type granitoids (Whalen and Chappell 1988).

At higher magmatic temperatures Ti enters magnetite as ulvöspinel component to form Ti-rich magnetite. During cooling of magma the ulvöspinel component is not stable due to common late oxidation and the early Ti-rich magnetite undergoes oxidation-induced exsolution (oxyexsolution) producing magnetite and ilmenite (Fig. 3).

The oxidation promotes the Ti diffusion to (111) octahedral planes:

$$6\ \underset{\text{Usp}}{Fe_2TiO_4} + O_2 = 2\ \underset{\text{Mt}}{Fe_3O_4} + 6\ \underset{\text{Ilm}}{FeTiO_3} \quad (2)$$

The products of reaction (2) are actually solid solutions of ulvöspinel in magnetite and hematite in ilmenite. The exchange reaction between them is temperature dependent:

$$\underset{\text{Usp}}{Fe_2TiO_4} + \underset{\text{Hem}}{Fe_2O_3} = \underset{\text{Ilm}}{FeTiO_3} + \underset{\text{Mt}}{Fe_3O_4} \quad (3)$$

Thus, the coexisting oxides form a very important thermobarometer (Buddington and Lindsley 1964; Anderson and Lindsley 1988; Ghiorso and Sack 1991; Sauerzapf et al. 2008; Ghiorso and Evans 2009). The exsolution process may be prevented only in rapid-ascent eruption products, e.g. in pyroclastic fall deposits. The nonexsolved Ti-rich magnetite may be used for identification of the rapid ascent of the host rock (Turner et al. 2008).

On cooling the ulvöspinel and hematite components decrease being partly replaced by magnetite and ilmenite. The two-oxide thermobarometer, if the proper assemblage is preserved and accurate analyses of the cubic (magnetite) and rhombic (ilmenite) phases are available, provides both T and f_{O2} of the last re-equilibration of the phases. The exsolved (BSE-dark) hem-ilm lamellae form characteristic trellis intergrowths in (111) planes common in slow-ascent volcanic rocks. In deep-seated magmatic rocks, such as granites, the oxidation is slower producing thicker exsolutions—lenses of ilmenite$_{ss}$ within (BSE-bright) magnetite$_{ss}$ host. The prolonged exsolution is indicated by the formation of progressively smaller ilmenite lenses in magnetite, and occasionally tiny magnetite lamellae occurs in the earlier formed larger ilmenite lenses (Fig. 3). If proportion of Ti is higher the exsolution relations

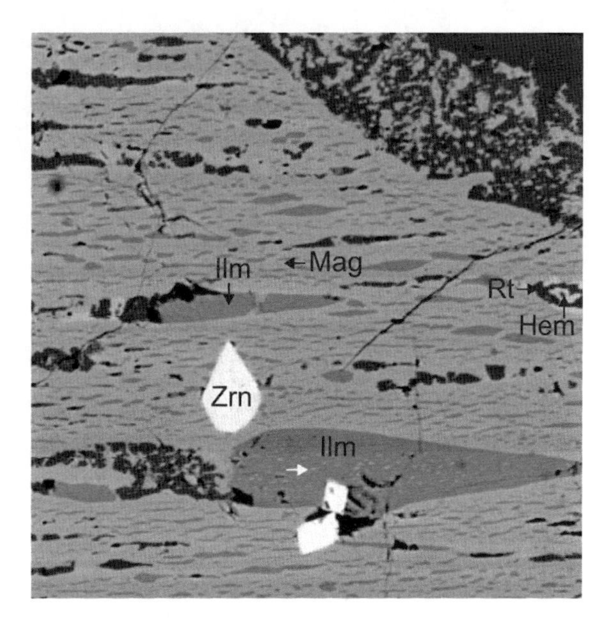

Fig. 3 BSE image of a former Ti-magnetite grain showing exsolution of darker lenses of ilmenite (*Ilm*) in magnetite (*Mag*). Rims are oxidized to rutile (*Rt*) and hematite (*Hem*). Note the tiny magnetite rods, which exslove from the large ilmenite spindle (*white arrow*). *Zrn* enclosed zircon

are reversed: magnetite forms the lenses in the ilmenite host. Ilmenite commonly unmixes to the borders or forms independent grains (e.g. Buddington and Lindsley 1964). The oxidation-induced exsolution in granitoid having Ti-rich magnetite is mostly a product of magmatic process as evident from temperatures of 800–700 °C obtained for coexisting phases (e.g. Ghiorso and Evans 2009). The finest measurable spindles from Western Carpathian granites provide temperature of ca. 670–650 °C using the above calibration, which means that exsolution ceased at or near the solidus (for granites cooled at mid-crustal depths). However, oxidation continues below sub-solidus region with operation of different reactions. The most common reaction is oxidation of the ulvöspinel component:

$$\underset{\text{Usp}}{Fe_2TiO_4} + {}^1\!/_2\, O_2 = \underset{\text{Hem}}{Fe_2O_3} + \underset{\text{Ru}}{TiO_2} \tag{4}$$

Many titanomagnetite grains have their ilmenite rims replaced by hematite and rutile (Fig. 3). The exsolved titanomagnetites from Ca-rich tonalitic rocks react with plagioclase and biotite to form titanite and pure magnetite (e.g. Broska et al. 2007):

$$\begin{aligned} &\underset{\text{An}}{3CaAl_2Si_2O_6} + \underset{\text{Ann}}{3KFe_3AlSi_3O_{10}(OH)_2} + \underset{\text{Usp}}{3Fe_2TiO_4} + 5/2O_2 \\ &= \underset{\text{Ttn}}{3CaTiSiO_5} + \underset{\text{Mu}}{3KAl_2AlSi_3O_{10}(OH)_2} + \underset{\text{Mt}}{5Fe_3O_4} + \underset{\text{Qtz}}{3SiO_2} \end{aligned} \tag{5}$$

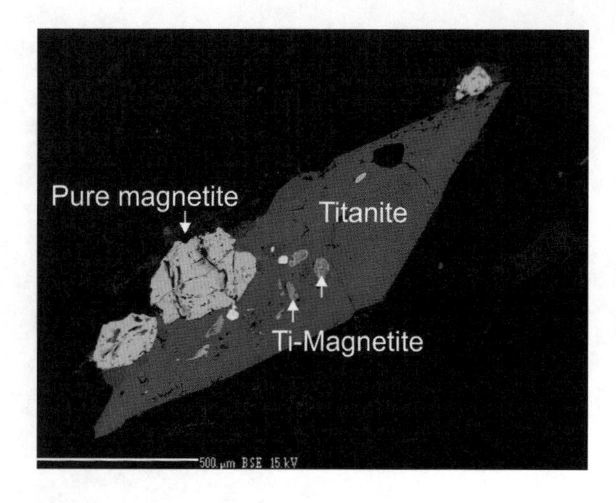

Fig. 4 BSE image of euhedral zoned titanite enclosing small grains of primary Ti-magnetite and on the rim intergowing with late pure magnetite. Host rock is I-type biotite tonalite from Tribeč Mts. (Western Carpathians)

In more mafic granitoids the Ca source may include amphibole, e.g.:

$$\underset{\text{Usp}}{4Fe_2TiO_4} + \underset{\text{Fe-act}}{2Ca_2Fe_5Si_8O_{22}(OH)_2} + 3O_2 = \underset{\text{Mt}}{6Fe_3O_4} + \underset{\text{Ttn}}{4CaTiSiO_5} + \underset{\text{Qtz}}{12SiO_2} + 2H_2O \quad (6)$$

In hornblende-free I-type tonalites from the Western Carpathians the products of reaction (5) strongly prevail and Ti-rich magnetite grains are only preserved as being enclosed in a large titanite coexisting with dominant pure magnetites (Fig. 4).

The reactions (5, 6) occur in postmagmatic conditions. For example, in Tribeč tonalite the stability curves of reactions (1, 5) (Thermocalc 3.21; Holland and Powell 1998) give an intersection at ca 590 °C and a high f_{O2} ($\Delta NN = 2.3$). The oxidation continues towards lower temperatures while magnetite oxidizes to hematite: $2Fe_3O_4 + ½O_2 = 3Fe_2O_3$; this process alters mostly magnetite rims and internal cracks (e.g. Bónová et al. 2010). The main reason for late-stage oxidation can be due to preferential escape of H_2 from dissociation of water (Buddington and Lindsley 1964; Sato and Wright 1966; Czamanske and Wones 1973). Borodina et al. (1999) suggested that magnetite usually formed by oxidation of biotite and hornblende only in epizonal granitoids. In deep-seated, catazonal granitoids magnetite does not form and oxidation results in increased Fe^{3+}/Fe^{2+} silicate ratios. However, in shallower intrusions the oxidation continues towards lower temperatures during oxidation of magnetite to hematite: $2Fe_3O_4 + ½O_2 = 3Fe_2O_3$; this process alters mostly the rims and internal cracks of magnetite (e.g. Bónová et al. 2010). Ilmenite from syenite, related to A-type granite, shows a different oxidation mechanism: it first changes to ferropseudobrookite and ultimately to hematite and rutile (Sakoma and Martin 2002). Ilmenite in hypergene conditions oxidizes to pseudorutile $\left(Fe_2^{3+}Ti_3O_9\right)$, which is unstable and again breaks down to hematite and rutile (Grey and Reid 1975).

Granitic magnetites are usually pure with small amounts of Cr_2O_3, V_2O_5 and MgO in contrast to magnetites from ultramafic rocks containing up to (wt%) 2

V_2O_5, 0.2 Cr_2O_3, 1 Al_2O_3, 1.5 MgO and 0.4 MnO (Klemm et al. 1985). Ilmenite exhibits a more variable chemistry. In calc-alkaline granite magmas the most common component is pyrophanite ($MnTiO_3$), MnO in ilmenite reaching 5–7 wt%. However, in oxidizing environments MnO in ilmenite may increase from 3 wt% in monzonite to 15–20 wt% in granodiorite, and up to 30 wt% in granite corresponding to ca 7, 32 and 63 mol.% pyrophanite, respectively (Czamanske and Mihálik 1972). Rutile, which is almost pure in calc-alkaline granites, may contain appreciable amounts of Nb, Ta in specialized F–P granites (10–15 wt%Nb_2O_3, 15–25 wt%Ta_2O_5).

2.4 Titanite

Titanite [monoclinic, $CaTiSiO_4(O,OH,F)$] is an accessory mineral occurring especially in mesocratic to melanocratic granitoids with I-type characteristics. In volcanic rocks, titanite is less common suggesting its instability in low pressure melts (Nakada 1991). Its chemical variation is controlled by the coupled substitution $Al + OH \leftrightarrow Ti + O$, which leads to hypothetical end-member $CaAlSiO_4OH$. The naturally occurring mineral of this composition (vuagantite) has a different structure so they do not form a solid solution (e.g. McNear et al. 1976; Harlov et al. 2006). Increased content of the $Ca(Al,Fe^{3+})SiO_4(OH)$ component in titanite is favoured by lower temperatures (Enami et al. 1993). On the other hand, the alternative isostructural $Al + F \leftrightarrow Ti + O$ substitution may reach up to 55 mol.% in high-grade metamorphic terrains. Aluminium substituting for Ti is accompanied by Fe^{2+} and Fe^{3+} and F is accompanied by OH: $[(Al, Fe^{3+})(F,OH)Ti_{-1}O_{-1}]$. Other important substitutions are Ca and Ti for Y, REE, U, Th, Mn coupled with Fe^{2+}: $Fe^{2+}REE_2Ca_{-2}Ti_{-1}$ (Ribbe 1980). Decreasing the ratio Fe/Al from volcanic rocks ($Fe/Al \approx 1$) to granitoids with $Fe/Al = 1–0.5$ and metamorphic rocks with $Fe/Al < 0.5$ (Nakada 1991) implies that Al is dependent on pressure. Actually, Franz and Spear (1985), Carswell et al.(1996) and Tropper et al. (2002) suggested that the above substitution $AlFTi_{-1}O_{-1}$ extends the stability of titanite (+ ru + carbonate) to ultrahigh pressures which is however also constrained by F and CO_2 fugacities.

Titanite accompanied by magnetite (Fig. 4) indicates crystallization from magma with increased f_{O2} where it is involved in buffering assemblages TMQH and TMQA (Wones 1989):

$$\underset{\text{Hed}}{3CaFeSi_2O_6} + \underset{\text{Ilm}}{3FeTiO_3} + O_2 = \underset{\text{Ttn}}{3CaTiSiO_5} + \underset{\text{Qtz}}{3SiO_2} + \underset{\text{Mt}}{Fe_3O_4} \quad (7)$$

and

$$\underset{\text{Fe-act}}{6Ca_2Fe_5Si_8O_{22}(OH)_2} + \underset{\text{Ilm}}{12FeTiO_3} + 7O_2 = \underset{\text{Ttn}}{12CaTiSiO_5} + \underset{\text{Mt}}{14Fe_3O_4} + \underset{\text{Qtz}}{36SiO_2} + 6H_2O \quad (8)$$

The latter is relevant for amphibole-bearing granitoids. Contrary to earlier estimates (Noyes et al. 1983), the T- f_{O2} calculated using the database of Holland and Powell (1998) indicates that it is more reducing following quite closely to the buffer FMQ (Thermocalc 3.21). Titanite may commonly form due to oxidation of earlier ulvöspinel (reaction 6, Fig. 4 "Ti-magnetite") or ilmenite (reaction 8) either in magmatic or post-magmatic stage. Ilmenite enclosed by titanite may come from a more reduced mafic magma in composite mafic-felsic systems (Robinson and Miller 1999). On the other hand during retrogression (re-hydration) titanite may be replaced by ilmenite $\pm$ biotite $\pm$ allanite indicating more reducing conditions (Broska et al. 2007). The secondary titanite (sagenite) may also exsolves from an earlier Ti-rich biotite. Reactions such as (8) show that stability of titanite depends on both f_{O2} and f_{H2O} (Harlov et al. 2006).

Titanite having much higher abundance in rock (compared to monazite or allanite), along with allanite, is a major repository for the REE in more Ca-rich granitoid types (Gromet and Silver 1983; Bea 1996). The early magmatic titanite shows core to rim fractionation of REE similar to that in allanite. This is considered as either depleting REEs during fractionating residual melt or due to decreasing REE partitioning coefficients. The latter enables more REE to remain in the melt which may become less polymerised under influence of rising activity of F (Sawka et al. 1984).

In contrast to early magmatic titanite, a late titanite exsolved under subsolidus conditions from earlier mafic silicates is poor in REE and does not influence the whole rock evolution.

2.5 Epidote

Epidote composition ranges between two end-members: monoclinic clinozoisite [$Ca_2Al_3Si_3O_{12}(OH)$] and "pistacite" [$Ca_2Al_2Fe^{3+}Si_3O_{12}(OH)$]. The proportion of the latter (ps) is used for characterization of the solid solution. Epidotes with Fe $>$ 1 per formula unit are generally not considered magmatic (Schmidt and Poli 2004). Epidote$_{ss}$ is a characteristic accessory mineral of Ca-rich granitoid rocks where it may occur both as primary magmatic but more often as subsolidus or metamorphic mineral.

Epidote structure enables a number of substitutions leading to many mineral species (Franz and Liebscher 2004). In granitoid rocks, the most important are in M positions: $Al_{-1}Fe^{3+}$ (epidote—clinozoisite), and in AM positions: $Ca_{-1}Al_{-1}$ $REEFe^{2+}$ to allanite, $Ca_{-1}Al_{-1}REEMg$ to dissakisite (see below). Although epidote$_{ss}$ is more characteristic of metamorphic rocks, magmatic epidote is also common in granitoids, usually associated with Ca-rich plagioclase, hornblende and allanite (tonalites, granodiorites, trondhjemites) (Zen and Hammarstrom 1984; Davidson et al. 1996; Schmidt and Poli 2004). Its occurrence is used as an indication of elevated pressure of crystallization: generally >500 MPa, although the increased ps content lowers the epidote stability to pressure ca. 300 MPa at granite solidus (Schmidt and Poli 2004). In H_2O saturated granitoid melts epidote is generally stable in T range from ca. 770 °C to solidus.

Subsolidus granitoid $epidote_{ss}$ commonly originates by hydration and oxidation reactions among Ca- and Fe-bearing phases. For example amphibole reacts with epidote and biotite by hydration (Schmidt and Thompson 1996, Thermocalc):

$$\begin{aligned} &\underset{\text{Fe-act}}{2Ca_2Fe_5Si_8O_{22}(OH)_2} + \underset{\text{An}}{4CaAl_2Si_2O_8} + \underset{\text{Kfs}}{2KAlSi_3O_8} + O_2 + 2H_2O \\ &= \underset{\text{Ep}}{4Ca_2FeAl_2Si_3O_{12}(OH)} + \underset{\text{Ann}}{2KFe_3AlSi_3O_{10}(OH)_2} + \underset{\text{Qtz}}{12SiO_2} \end{aligned} \tag{9}$$

The oxidation reaction without biotite produces epidote and magnetite:

$$\begin{aligned} &\underset{\text{Fe-act}}{Ca_2Fe_5Si_8O_{22}(OH)_2} + \underset{\text{An}}{2CaAl_2Si_2O_8} + O_2 \\ &= \underset{\text{Ep}}{2Ca_2FeAl_2Si_3O_{12}(OH)} + \underset{\text{Mt}}{Fe_3O_4} + \underset{\text{Qtz}}{6SiO_2} \end{aligned} \tag{10}$$

Similar reaction was suggested by Zen and Hammarstrom (1984) for Kfs, Mt and Qtz assemblage as melt phases to produce magmatic epidote. Experimental reactions like (9, 10) suggest that magmatic epidote is stable on the higher P and lower T side while magnetite at higher T and lower P side (Schmidt and Thompson 1996). Due to its high Fe^{3+} content epidote is sensitive to oxidation-reduction regime. The elevated f_{O2} increases epidote stability to higher T and lower P: at granite solidus from 500 to 300 MPa for NNO and HM buffers, respectively (Liou 1973).

Alumina-rich end-member clinozoisite (ps < 0.2) is also common in granitoids where it replaces plagioclase by hydration (saussuritization):

$$\underset{\text{An}}{4CaAl_2Si_2O_8} + \underset{\text{Ksp}}{KAlSi_3O_8} + 2H_2O = \underset{\text{Mu}}{KAl_2AlSi_3O_{10}(OH)_2} + \underset{\text{Czo}}{2Ca_2Al_3^{+}Si_3O_{12}(OH)} + \underset{\text{Qtz}}{2SiO_2} \tag{11}$$

Clinozoisite in association with phengitic muscovite and grossular rich garnet may indicate breakdown of plagioclase due to increased pressure—e.g. tectonic burial of the granite pluton.

2.6 Allanite

Allanite [monoclicnic $CaREEAl_2FeSi_3O_{11}(O,OH)$] is an important REE-bearing accessory mineral of the epidote group. In granitoid rocks it occurs usually as allanite-(Ce) with Ce dominating among the REEs. Its high CaO content suggests that allanite is stable mainly in granodiorites and tonalites where it is found as eu-hedral to subhedral, strongly pleochroic crystals, commonly zoned and with

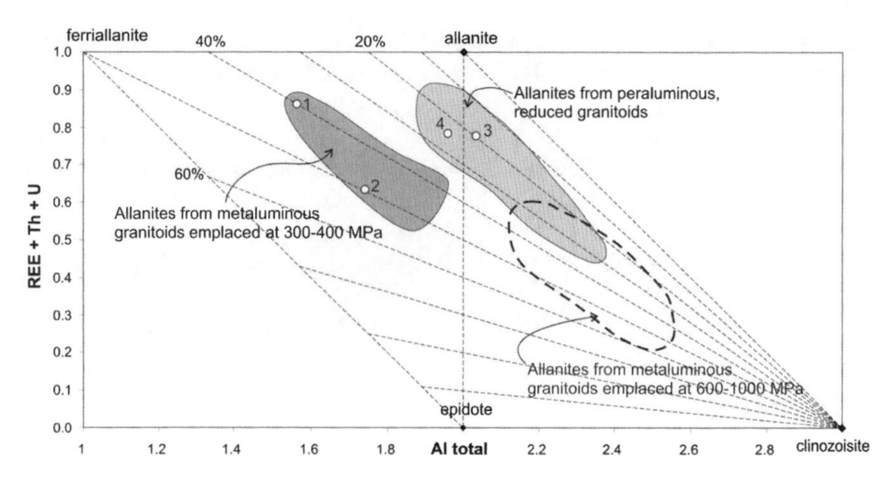

Fig. 5 Allanite compositions in terms of Al_{total} versus REE + Th + U (Petrík et al. 1995) showing fields corresponding to metaluminous (partly oxidised), peraluminous (reduced) West Carpathian granitoids compared with high pressure allanites compiled by Beard et al. (2006). The numbers refer to the analysed points in Figs. 6 and 8

complicated internal structure. The MgO- bearing analogue– dissakisite (Grew et al. 1991) is found mostly in Mg rich metamorphic rocks, e.g. amphibolites, metacarbonates (e.g. Bačík and Uher 2010). Allanite composition may be derived from epidote [$Ca_2Fe^{3+}Al_2Si_3O_{11}O(OH)$] by main substitution:

$$Ca^{2+} + Fe^{3+} = RE^{3+} + Fe^{2+} \text{ and } Ca^{2+} + Al^{3+} = RE^{3+} + Fe^{2+}$$

which join epidote—allanite and epidote—ferriallanite [$CaREFe^{3+}Fe^{2+}AlSi_3O_{11}O(OH)$]. The latter exchange is identical also with allanite—clinozoisite join. Most of allanite compositional variations can be explained by these exchanges (Petrík et al. 1995; Gieré and Sorensen 2004).

The end-member allanite with dominant Fe^{2+} is rare and majority of granitoid allanite compositions have maxima either between allanite, epidote and ferriallanite (Fig. 5) or allanite, epidote and clinozoisite (Gieré and Sorensen 2004) reflecting their variable Al –REE contents. The fact that the group of Al-poor allanites comes from low pressure plutons (e.g. Western Carpathians, emplaced at 300–400 MPa, Petrík and Broska 1994), whereas the group of Al-rich magmatic allanites and REE-rich epidotes comes from medium to high pressure plutons (e.g. Bell Island Pluton, 600–1,000 MPa, Beard et al. 2006) suggests that allanite, similar to as observed for hornblende, has a pressure-sensitive composition and may presumably be used for relative pressure estimates in having similar bulk rock compositions. Ferriallanite is found in e.g. anorogenic granites (Poitrasson 2002), carbonatites (Ripp et al. 2002) or oxidized skarns (Holtstam et al. 2003).

BSE dark domains of late origin commonly occur in primary allanite (Fig. 6) creating complicated patterns and replacing magmatic zoning (Petrík et al. 1995; Poitrasson 2002). These authors suggested that it is due to the penetrating

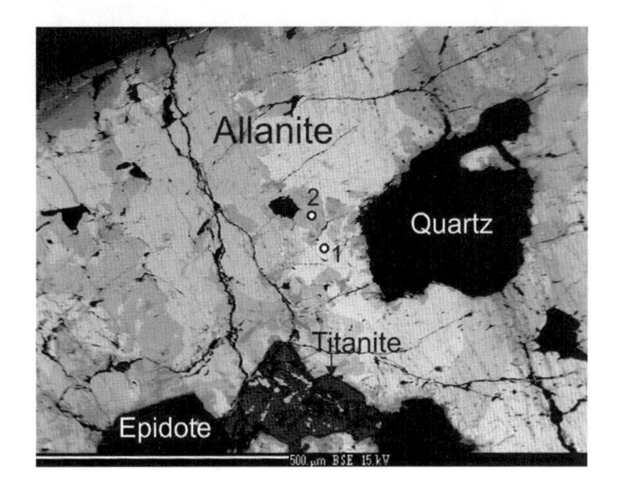

Fig. 6 Portion of a large allanite from oxidised biotite tonalite showing late oxidation (irregular BSE darker domains), numbers refer to Fig. 5

oxidizing hydrothermal fluids, which may have either postmagmatic or granite unrelated metamorphic origins. The primary allanites also show variable Fe^{3+}/Fe_{tot} ratios, which seem to range from 35 to 45 wt% Fe^{3+} (of total Fe amount) whereas the oxidated domains reach up to 55 % Fe^{3+} (Fig. 5). While the I-type granite allanite typically coexists with magnetite the reduced allanite may not coexist with any Fe-Ti oxide. Reduced allanites (0–30 % Fe^{3+}) are found rarely in peraluminous granitoids where they may coexist with ilmenite or without Fe-Ti oxide.

In terms of REE content, allanite is closely related to monazite, another important accessory REE carrier. In most granitoids these minerals show dichotomy (antagonism) allanite occurring in metaluminous varieties (I-type) and monazite in more peraluminous varieties (S-type). However, magmatic differentiates of REE-rich, S-type granites can begin with allanite precipitation which soon turns to monazite after decreasing Ca activity in the melt due to massive crystallisation of plagioclase (Broska et al. 2000). In metaluminous more mafic hornblende bearing granitoids where monazite is absent, the REE are distributed between titanite and allanite. The thermodynamic stability of allanite is not known, however the Mg species (dissakisite) (Janots et al. 2007) in the system SiO_2–Al_2O_3–FeO–Fe_2O_3–MgO–CaO–Na_2O–K_2O–P_2O_5–La_2O_3–CO_2–H_2O is stable in a window between 2–15 kb and 300–550 °C. Beyond these limits, monazite is stable. However, with increasing Ca content, the stability field of dissakisite shifts towards higher T (up to 800 °C at 15 kb and >3 mol.% Ca). The stability of metamorphic allanite increases with higher Ca and Al contents in parental rock, and with metamorphic degree (Wing et al. 2003). In granulite facies metamorphism (700–760 °C, 500 MPa) granitoid allanite breaks down to monazite and thorite (Bingen et al. 1996).

The REE patterns of allanite are strongly fractionated (La/Yb = 250–2,340) with pronounced negative Eu anomaly (Bea 1996). The preferential gain of LREE is expressed by K_d (= REE crystal/REE glass) decreased by two orders of magnitude between La and Yb (Gieré and Sorensen 2004).

Magmatic allanite is commonly mantled by late- to postmagmatic epidote. Allanite is usually zoned with outward growth of Ca, Th and decreasing REE: they decrease mainly due to fractionation of allanite itself (Gromet and Silver 1983; Chesner and Ettlinger 1989; Gieré and Sorensen 2004). However, the decrease of REE in allanite is accompanied by increase in Th following the vector (2REE) $Ca_{-1}Th_{-1}$. Therefore, the allanite zoning is not due to simple increase in epidote end-member. The latter grows as a new, Th-poor epitaxial phase (Gromet and Silver 1983). The REE fractionation is not simple: Ca, La fractionate more effectively than Nd, and heavy REE and Y may increase in a manner so the REE patterns crossed over more evolved allanite zones. Beard et al. (2006) have shown that this is due to increasing K_d values of Nd, Sm and Y during the course of allanite crystallization. This is explained by modification of allanite structure changing with REE contents. However, the zoning in crystals may also be interpreted in terms of melt structure. The decrease of the REE from core to rims indicates a decrease of K_d due to increasing ratio of non-bridging oxygens (NBO) to tetrahedral cations (T) thus allowing more large cations to remain in the melt. This is mostly by the increasing role of F at the end stages of solidification (Sawka et al. 1984).

Except the REE, allanite structure prefers also Th and U (other minor elements are Zr, P, Ba, Cr). Characteristic concentrations are 2–3 wt% ThO_2 and 0.1 wt% UO_2. Based on these elements allanite is suitable for chemical dating by ion microprobe although its compositional variability (matrix effects) limits the accuracy of the results (e.g. Catlos et al. 2000). Poitrasson (2002) also showed that due to significant common lead contents allanite, in contrast to monazite, is not suitable for microprobe chemical dating.

2.7 Monazite

Monazite [monoclinic $(Ce,La,Th)PO_4$] is, along with allanite and titanite, the most important REE repository in granitoid and felsic volcanic rocks. It may have a suffix according to the dominant RE element [e.g. monazite-(Ce)]. Monazite is also a common metamorphic mineral in quartzofeldspathic rocks in a wide range of P–T conditions. Besides the REE, dominated mostly by Ce and La, Th and Y are important elements accompanied by minor Si and Ca, Through substitutions: $REE^{3+} + P^{5+} = Th^{4+} + Si^{4+}$ and $2REE = Th^{4+} + Ca^{2+}$.

Monazite changes to huttonite [monoclinic $(Th,U,Pb)SiO_4$] and cheralite [monoclicnic $Ca(Th,U,Pb)(PO_4)_2$, older name brabantite]. Natural granite monazites always show variable contents of both end members, typically 0.1–8, up to 15 mol.% huttonite and 2–20 mol.% cheralite (Fig. 7). However, both end members are also found: cheralite is found in specialised Li–F granites (Förster, 1998) wheras huttonite exsolves as a secondary phase in huttonite-rich monazite (e.g. Broska et al. 2000). The highest Th contents are found in monazites from specialised Li–F granites where ThO_2 reaches 20–25 wt% (40 mol.% cheralite). However, granulite-grade metamorphic monazites, show up to 35 wt% ThO_2

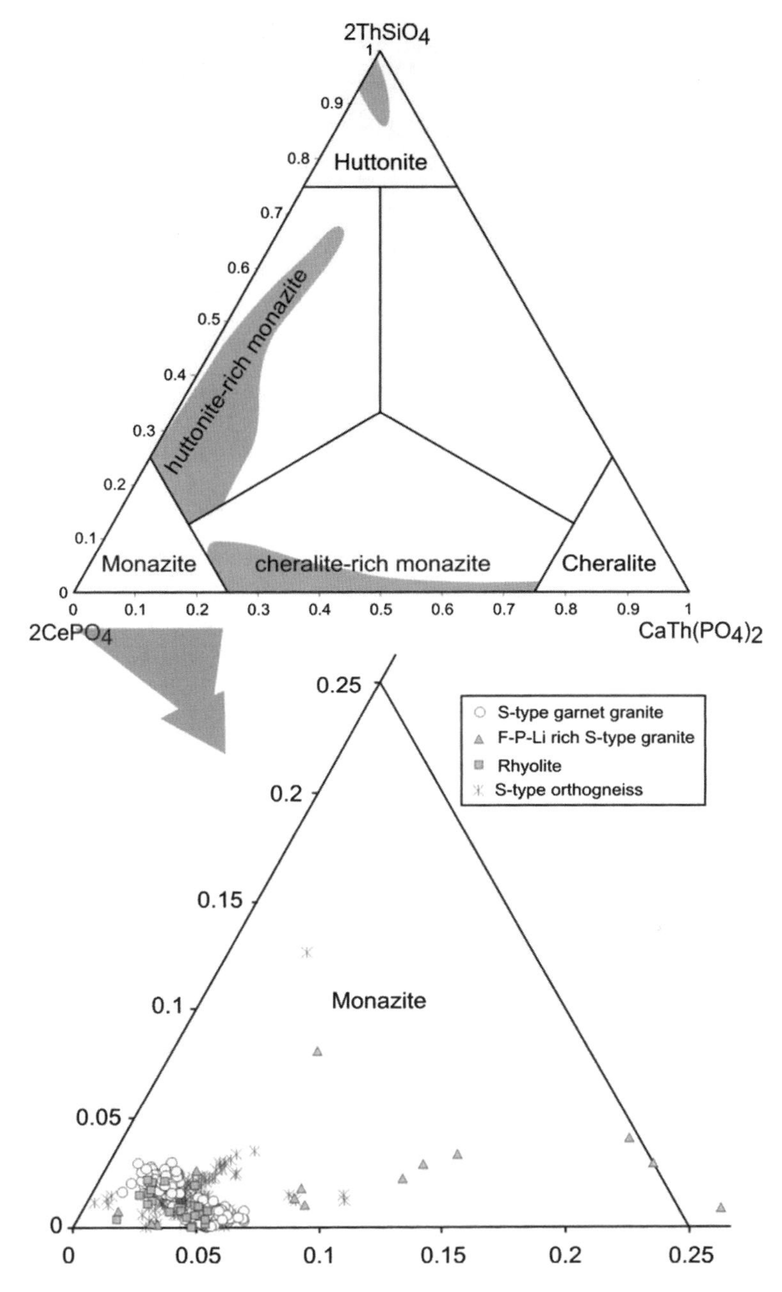

Fig. 7 Monazites from various felsic rocks (Western Carpathians) in the system $2CaPO_4$–$2ThSiO_4$–$CaTh(PO_4)_2$. The fields of huttonite—huttonitic monazite, and cheralite series are according to compilation of Förster and Harlov (1999). Following their procedure (REE + Y) are included in monazite and (U + Pb) are included in cheralite, which results in higher mol.% of monazite and lower mol.% of cheralite as compared to calculation involving xenotime. All monazites fall to the monazite field. Three points at the boundary with cheralite series are Th-rich monazites from Permian Li-F-P specialised microgranite

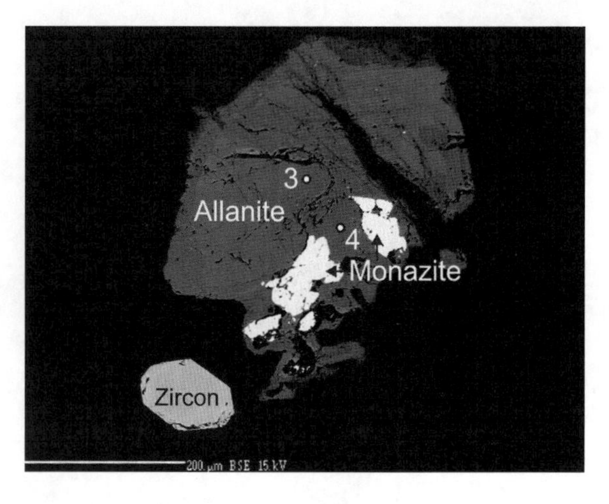

Fig. 8 Allanite enclosing an earlier monazite from a reduced, peraluminous biotite tonalite (Western Carpathians). Numbers refer to points in Fig. 5

(Förster and Harlov 1999). Monazite is also excellent monitor of the presence of S-rich brines in magmatic and metamorphic rocks because in such environment monazite is also enriched in S (Ondrejka et al. 2007).

Monazite usually prevails in peraluminous and more leucocratic granitoids, while in metaluminous (mafic) types it is substituted by allanite. Although typical for the S-type granites, it is present in late leucocratic differentiates of the I-type only, commonly in dikes. Such tendency was also confirmed experimentally (Broska et al. 2000). Nevertheless, both LREE phases may show more complicated relations apparently controlled by the Ca activity, for example monazite may enclose allanite and vice versa (Fig. 8). It seems possible to inter-relate monazite stability with the Ca activity, REE and water contents (the latter determining stability of plagioclase). In peraluminous granites with moderate LREE contents the temperature of monazite saturation (see below) is suppressed, plagioclase precipitates early lowering the Ca activity of the melt—monazite becomes stable. In contrast, in Ca-bearing hornblende-biotite tonalites and granodiorites with high REE contents allanite becomes stable prior to plagioclase crystallization due to higher water content inferred in such magmas (Petrík 1999; Broska et al. 2000). Alternatively, the allanite → monazite or monazite → allanite (Fig. 8) relationship may be interpreted in *P–T–X* space by a change of *P–T* conditions (in contrast to the above explanation in terms of *X* parameter, i.e. melt composition). The positive slope of dissakisite stability field (Janots et al. 2007) implies that the replacement of dissakisite by monazite would occur along decreasing *T* or *P* paths. The dissakisite field is shifted to much higher temperatures with increasing Ca content which may also hold true for allanite.

In felsic magmas monazite is responsible for LREE fractionation due to its high LREE partitioning coefficients (Miller and Mittlefeldt 1982; Michael 1988; Bea 1996) which produce characteristic low REE, flat patterns with pronounced negative Eu anomalies.

Monazite stability may be expressed by saturation of the LREE in the melt. Montel (1993) derived an equation applicable to Ca, Fe and Mg-poor compositions, which produces the REE concentrations in equilibrium with monazite, i.e. its solubility. It depends on temperature *T*, water content and bulk composition (Si, Al, alkalies). Reversely, the *T* calculated from coexisting melt (approximated by whole rock REEs) indicates the last equilibration of monazite with melt. The dependence of known saturation temperature on water may be used for estimating the primary water content in granite melt.

Kinetics of the monazite dissolution can be estimated from equations of Rapp and Watson (1986), which show that monazite dissolves slowly and the dissolution rate is strongly dependent on the water content of the melt and crystal size, for example at 1 wt% of H_2O in the melt and 890 °C, a 5 μm grain dissolves in 100 Ka, but 10 μm grain dissolves in 1 Ma.

Being phosphate, monazite stability is related to apatite and xenotime. Xenotime component in monazite commonly ranges between 1and 7 mol.%. When buffered by the presence of xenotime, Y in monazite correlates with *T* and may be used as thermometer (Heinrich et al. 1997; Pyle et al. 2001). When coexisting with garnet Y is distributed between Y-component of garnet ($Y_3Al_2Al_3O_{12}$) and xenotime (YPO_4) component in monazite. The equilibrium constant K_{Eq} of the reaction strongly depends on *T* and the assemblage can be used as an empirical geothermometer for garnet- and monazite-bearing granites (Pyle et al. 2001). Solubility of apatite in peraluminous melt is by greater than monazite an order of magnitude. Therefore, a REE enriched apatite dissolves incongruently crystallizing monazite (Wolf and London 1995). Another origin of monazite, by fluid-induced metasomatism (Cl-brines) of a REE-rich apatite, was suggested by Harlov and Förster (2003). Monazite commonly breaks down to apatite and allanite (REE-epidote) forming coronas under high- to medium-grade P-T conditions (Finger et al. 1998). In hypergene conditions with available fluoride, carbonate bearing fluids monazite alters to rhabdophane [(LREE,Ca,Th)PO_4.H_2O] (e.g. Berger et al. 2008).

The high Th and U contents of monazite make it an important phase for isotopic (Th, U)/Pb dating, moreover, the lack of common lead in its structure, enables chemical dating by means of microprobe (Parrish 1990; Suzuki and Adachi 1991; Montel et al. 1996). The low solubility of monazite in melt implies preservation of inherited monazite cores (e.g. Harrison et al. 1995) or whole grains in granite magma (Petrík and Konečný 2009). The inherited monazite may suffer from interaction with late-magmatic fluids and change its composition considerably: large inherited monazites from a garnet granite show younger Y- and U-rich rims which are metasomatically changed probably by dissolution-reprecipitation mechanism (Petrík and Konečný 2009).

2.8 Xenotime-(Y)

Xenotime-(Y) (tetragonal YPO_4) is a more rare accessory mineral usually occurring in felsic and peraluminous granites with S-type affinity. The heavy REE content of it compensate the LREE (along with zircon) bound in monazite of garnet free granites. Xenotime crystals commonly show complex zoning pattern suggesting crystallisation from melt and display a distinct euhedral core. In general, U, Th and Ca contents decrease from core to rim, whereas Y and HREE increase. A positive correlation between (U + Th + Si) and (REE + P) in atoms per formula unit indicates a coupled substitution corresponding to (U + Th)SiREE$_{-1}$P$_{-1}$. The coffinite [U(SiO_4)1−x $(OH)_4$x] or thorite ($ThSiO_4$) components in xenotime are analogous to the huttonite component in monazite solid solutions but coffinite prevails (Spear and Pyle 2002). The positive correlation between (U + Th + Ca) and 2REE in xenotime indicates the presence of a cheralite-like molecule—a constituent typical of monazite (Spear and Pyle 2002). Xenotime in granitic rock is commonly derived by replacement of zircon by postmagmatic fluids that produce low-Y zircon and xenotime (Pan 1997).

2.9 Tourmaline

The general formula of tourmaline supergroup may be written as $XY_3Z_6(T_6O_{18})(BO_3)_3V_3W$ where $X = Na^+$, Ca^{2+}, K^+, vac.; $Y = Fe^{2+}$, Mg^{2+}, Mn^{2+}, Al^{3+}, Li^+, Fe^{3+}, Cr^{3+}; $Z = Al^{3+}$, Fe^{3+}, Mg^{2+}, Cr^{3+}; $T = Si^{4+}$, Al^{3+}, B^{3+}; $B = B^{3+}$; $V = OH^{1-}$, O^{2-} a $W = OH^{1-}$, F^{1-}, O^{2-} (Henry et al. 2011). Three main groups of tourmaline are determined by X position. The Na predominates in X—site in the most common alkali tourmalines (typically schorl-dravite solid-solutions), vacancies in vacant tourmaline leading to foitite—magnesiofoitite and Ca (calcic tourmaline) are the rarest. Although schorl commonly crystallizes with K-feldspar, K^+ due to large ionic radius is incompatible in tourmaline structure and Na prevails. Deficit of alkalies in X-site is charge-compensated also by substitution of Al for divalent cations such as Mg and Fe^{2+} in Y-site. The heterovalent substitutions commonly control the tourmaline composition. According to the W—site hydroxy-, fluor- and oxy-tourmalines are distinguished, which are known to occure mainly in evolved igneous rocks such as pegmatites.

Tourmaline, as the most significant mineral of B, is common and locally abundant in the crustal igneous rocks, particularly in evolved granite and associated hydrothermal activity (Anovitz and Grew 1996). In closed magmatic system in granites without rapid changes in composition the magmatic tourmaline shows simple and minor chemical variation and precipitates as schorl (Henry and Guidotti 1985). Typically, this phenomenon is also observed in SW England (Cornwall), where magmatic tourmaline is homogeneous with high Fe/Mg, high F, and high Al in the Y-site (London and Manning 1995). Iron in granitic tourmaline is mostly divalent. Tourmaline in granites is later than biotite, and precipitates by reaction with biotite and B-bearing silicate melt in low Ti abundance (Shearer et al. 1987; Scaillet et al. 1995; Nabelek et al. 1992). The enrichment of F, B and H_2O in granitic magmas with tourmaline is explained by fractional crystallization of initially B- and F-enriched parental melts (Thomas et al. 2005). Boron in the felsic melt significantly changes rheological character such as density or viscosity and consenquently influences the primary nature of the melt by decreasing the liquidus and solidus temperatures, which are result of melt depolymerization (Pichavant 1981). The commonly observed tourmalinization (B metasomatism) is consistent with the preferential partitioning of B into aqueous vapour of magmatic–hydrothermal systems (Schatz et al. 2004).

Tourmaline in granitic rocks precipitates in several stages. In the Bohemian massif leucocratic granite shows the presence of two distinct textural types: (1) euhedral disseminated tourmaline crystallizing during early magmatic stages, and (2) interstitial nodular from the stage of transitional late- to subsolidus crystallization (Buriánek and Novák 2007). The tourmaline nodules (up to several cm in diameter) is commonly suggested to have formed as magmatic with later penetration of hydrothermal fluids (e.g. Samson and Sinclair 1992; Buriánek and Novák 2007; Trumbull et al. 2008). In granites of Moslavačka Gora (Croatia) the formation of tourmaline nodules has been interpreted as a result of crystallisation in the vapour bubbles which migrate through the granite body during ascent of fluids in decompression regime (Balen and Broska 2011). The tourmaline in such environment has oval 3D clusters of nodules either separated or arranged in chains (Balen and Broska 2011). A similar origin of tourmaline nodules was also suggested in the Western Carpathians (Kubiš and Broska 2010) (Fig. 9).

Of great importance is B isotopic system, which shows a wide range in $\delta^{11}B$ values in tourmalines from granites, pegmatites, and associated hydrothermal veins (−29.9 per mil to +8.6 per mil) which discriminate various sources and geotectonic settings (Jiang and Palmer 1998). The B isotopic composition of granitic tourmalines varies between −15 per mil and 5 per mil. But tourmalines from granite-related veins at granite-wall rock contacts or within adjacent country rocks show generally higher $\delta^{11}B$ values with a maximum at −10 per mil to −5 per mil. During magma degassing $\delta^{11}B$ is preferentially partitioned into the vapour phase, leading to lower $\delta^{11}B$ in magmatic tourmalines relative to the exsolved hydrothermal fluids. Degassing also leads to a decrease in $\delta^{11}B$ from early to late magmatic tourmalines (Jiang and Palmer 1998).

3 Granitoid Accessory Minerals: Western Carpathians Case Study

Crystalo-chemical properties of accessory minerals together with zircon morphology (and whole rock chemistry) allowed discrimination of granitic rocks in the Western Carpathians into S-, specialised S, I- and A-type granitoid suites. A typical paragenesis of accessory minerals from granites, helped to recognise comagmatic relationships among granitoids of Western Carpathians, which are attributed to orogenic I-and S-type granitoids of Variscan in age with peak felsic magmatism occurred at ca 350 Ma (e.g. Finger et al. 2003, Broska et al. 2013).

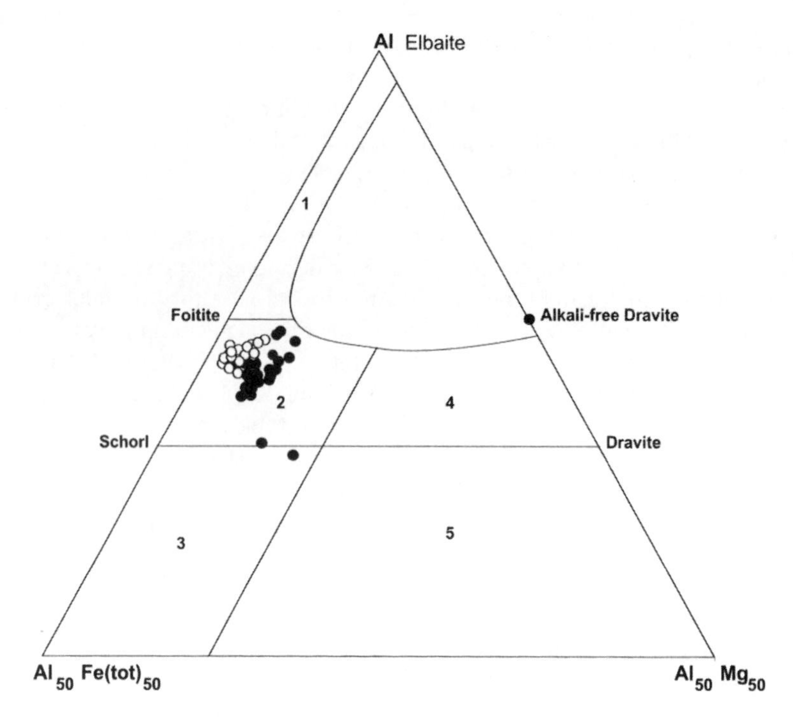

Fig. 9 The Al-Fe-Mg diagram of Henry and Guidotti (1985) is used for determination of environment of tourmaline growth. Two generations of tourmaline are shown as an example from the Western Carpathian specialised granites: *black symbols*—older phase schorl, *open symbols*—younger more alkali deficient. Explanations: (*1*) Li-rich granites; (*2*) Li-poor granites and aplites; (*3*) Fe^{3+}-rich quartz-tourmaline rocks (hydrothermally altered granites); (*4*) metapelites and metapsamites coexisting with Al-saturating phase; (*5*) metapelites and metapsamites not coexisting with an Al-saturating phase

The I-type metaluminous or subaluminous granites (ASI up to 1.05) originated in a more oxidised environment compared to those in S-type granites reflecting environment of different sources, show a more significant input from mantle and presumably have subduction-related geotectonic origin. The assemblages of I- and S-type granitoids in this sense are not just products of contrasting protoliths, an igneous and sedimentary for I-type and S-type respectively, but can also be explained as having variable mantle input shown by $^{143}Nd/^{144}Nd$ ratios (Kohút et al. 1999). The Permian S-type granites, rich in volatiles may contain primary tourmaline, fluorite or topaz. In this special type of mineralization, the most evolved granite bodies also contain cassiterite, wolframite, Nb-Ta oxides, U-Th-phases, especially in altered apical parts (Uher and Broska 1996). These granites have been named as specialised S-type granites (Ss-type granite). Anorogenic Permian A-type granites are also present within the Western Carpathians, which were primarily recognised based on zircon morphology. Most important, typomorphic accessory minerals used for the characterization and discrimination of Western Carpathian granite suites are briefly described below. The granite discrimination is based on zircon morphology and primary magmatic accessory mineral paragenesis, whole rock trace element and isotopic composition.

3.1 Zircon

Precipitation of zircon in calk-alkaline melts of Western Carpathian granitoids covers early to late stages. Metamictic zircon represents typical of late stage differentiated products largely influenced by fluid regime. Compatible behaviour of Zr in bulk rock during fractionation results in contineous decrease in granite melt along with quantity of zircons in rock. The highest concentrations are observed in I-type tonalites which also have low content of inherited cores in comparison to S-type granites. The zircon Zr/Hf_{at} ratio from S-type granites shows typical average of ca 35, similar to as noted in aluminious leucogranites (Pupin 2000) with increased content of Hf in late differentiates, especially in pegmatites. Zircon from barren pegmatites posses average 3 wt% of HfO_2 wheras beryl-columbite pegmatite zircon contains 7 wt% of HfO_2 with very low Zr/Hf ratios (locally up to 5) via possible heterovalent substitutions driven by fluids $HfZr_{-1}$, $Al^{3+}P^{5+}Si^{4+}_{-2}$ and $(Y, HREE, Fe)^{3+}P^{5+}(Zr, Hf)^{4+}_{-1}Si^{4+}_{-1}$ (Uher and Černý 1998). Isotopic composition of O in zircons indicating crustal sources varies from 6.8 to 8.56 ‰ SMOW (unpublished data). The zircons have been used for numerous conventional datings including the sensitive high resolution in microprobe (SHRIMP) method to determine the evolution of igneous magmatism in the Western Carpathians (e.g. Kohút et al. 2009 and references therein).

Assesment of zircon morphology provides a strong basis to define granite affiliation and ascertaining their comagmatic nature. The zircon morphology of Variscan I-type granitoids shows prevalence of zircon subtypes S_{12}, S_{13} and S_{25}, and granitoids with zircon subtypes with low L and S indexes (L_1, L_2, S_2, S_3, S_8) are typical of peraluminous, or S-type granites with I.T below 350. In both the suites, zircon morphology during differentiation terminates by G_1 type and morphology of zircons with I.T below 200 is typical of aplite dikes and pegmatites. The Permian specialised orogenic S-type granites with elevated fluid contents, rich in B and F, contains zircons with characteristic features of S_8 subtype. Zircons showing high I.A index (>600) due to high contents of subtypes P_1, P_2, P_3 and subtype D defined anorogenic A-type nature of granites. In Permian anorogenic A-type granites P_1 zircons determine a group of subsolvus granites, but D type indicates hypersolvus granites generated from hot and dry melt, which directly points to post-orogenic, rift-related granite of the Western Carpathians (Uher and Broska 1996). According to Pupin (1980) classification, the zircon typology helped the recognition of the A-type granites in the Western Carpatians (Uher and Puskharev 1994; Uher and Broska 1996) which was further substantiated based on bulk rock geochemistry (Fig. 10).

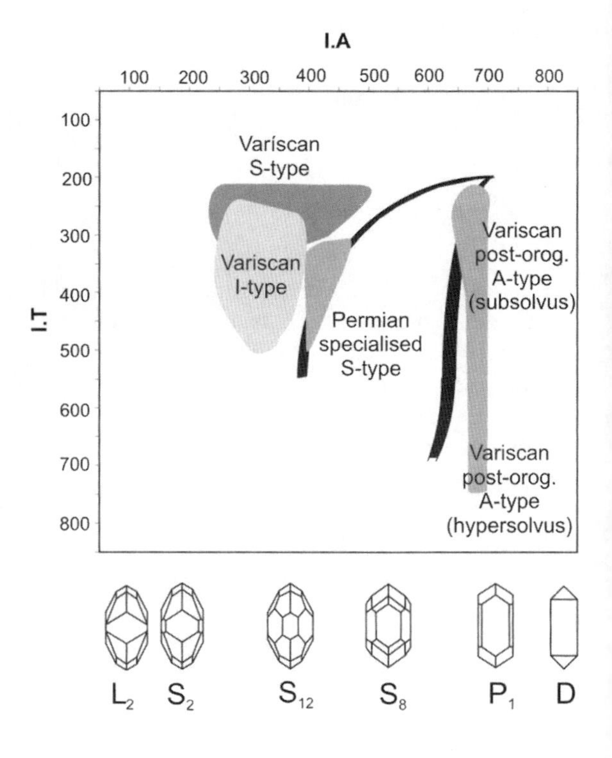

Fig. 10 Zircon typological diagram (Pupin 1980), used to characterise zircon morphology of main granitoid suites from the Western Carpathians. Typical zircon morphology for the granitoid suites is also shown: L_2 and S_2 Variscan S-type granite; S_{12} Variscan I-type tonalite; S_8 Permian specialised S-type granite; P_1 and *D* Permian A-type granite (subsolvus/hypersolvus)

3.2 Apatite, Monazite and Xenotime Distribution

Apatite is abundant in early magmatic granite differentiates with I-type affinity and locally it may reach to high concentrations. Monazite-(Ce) is typically found in S-type granites but it is also present in late-magmatic I-type differentiates, where xenotime is rare or absent. Primary xenotime is more typical of S-type granites but it is also common as a secondary phase formed from apatite and zircon.

3.2.1 Apatite

This typical accessory mineral of I-type granitoids in the Western Carpathians is usually colourless or slightly yellow due to iron oxides, but locally occurs also as dark pigmented variety (black or dusky apatite). Apatite crystallizes both as an early- or a late-magmatic phases. Commonly zoned early-magmatic apatite (Fig. 12a) is associated with zircon and other accessory phases poikilitically enclosed in biotite where they are trapped during its growth. In contrast, the late apatite is found associated with late titanite and magnetite (Fig. 12b). This paragenesis reflects the interaction between Ca^{2+} released from anorthite and the products liberated from titanomagnetite, biotite and an earlier apatite (Broska et al. 2007). Late apatite occurs in higher quantity in S-type granites due to higher P

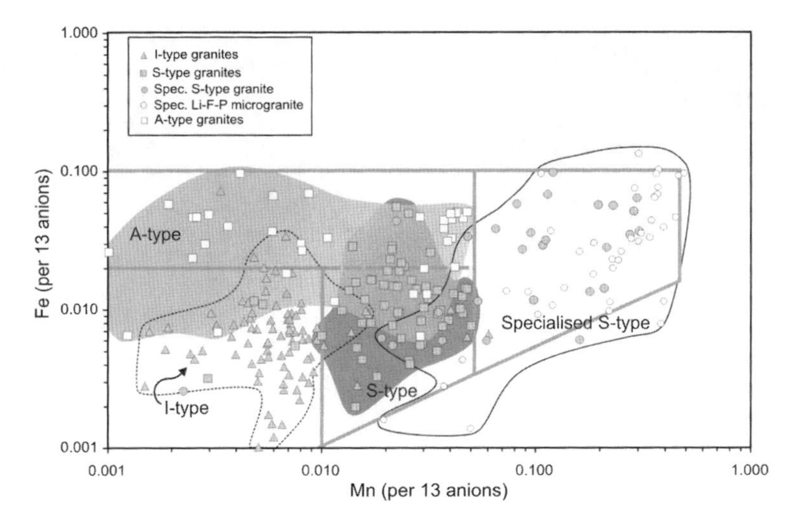

Fig. 11 Fe versus Mn (pfu) in apatites from S-, I-, A- and specialised S-type granitoids (Western Carpathians). The fields cover most of the compositional range of individual rock types. Tentative discrimination boundaries (*thick lines*) are also suggested for the apatites. The well defined field for S-type granite shows a significant overlap with A-type granite apatites

solubility in peraluminous melts (cf. Fig. 2). But locally, due to increase of alkaline character, e.g. in rock of syenitic composition, the apatite concentration may increase to rock-forming concentrations (Fig. 12c). For the recognition of apatite affiliation to S, I or A-type granites Mn and Fe are of primary importance (Fig. 11). The apatite from the S-type granites has significantly higher Mn content than apatite from the I-type granites. Apatite from specialized S-type granites has the highest MnO up to 6 wt%. Moreover, the apatite from A-type granites is enriched in Fe and HREE, reflecting their higher bulk concentration. Increased sulphur contents distinguish the apatites from I-type granites. Apatite from topaz-bearing Li-F-P microgranite may have very high SrO up to 15 wt% indicating a late enrichment by Sr-rich fluids (Petrík et al. 2011). Enrichment in S is indicator of I-type granite as well Sr, which generaly correlates with the Sr in bulk rock.

Besides primary apatite, there are also clusters of small secondary apatite grains distributed in alkali feldspar grains of evolved specialised granites (Fig. 12d). This apatite with a distinct composition is considered to have formed post-magmatically from P-rich alkali feldspars during decreasing temperature by reaction with F and B bearing fluid-rich, which are also enriched in alkaline and alkaline earth metals. It occurs in highly peraluminous granites, rich in volatiles, typically in post-orogenic Permian S-type granites.

3.2.2 Monazite

Monazite characterizes peraluminous, mostly S-type granites where it occurs in the form of typical prismatic and tabular crystals of yellow, yellow-green or

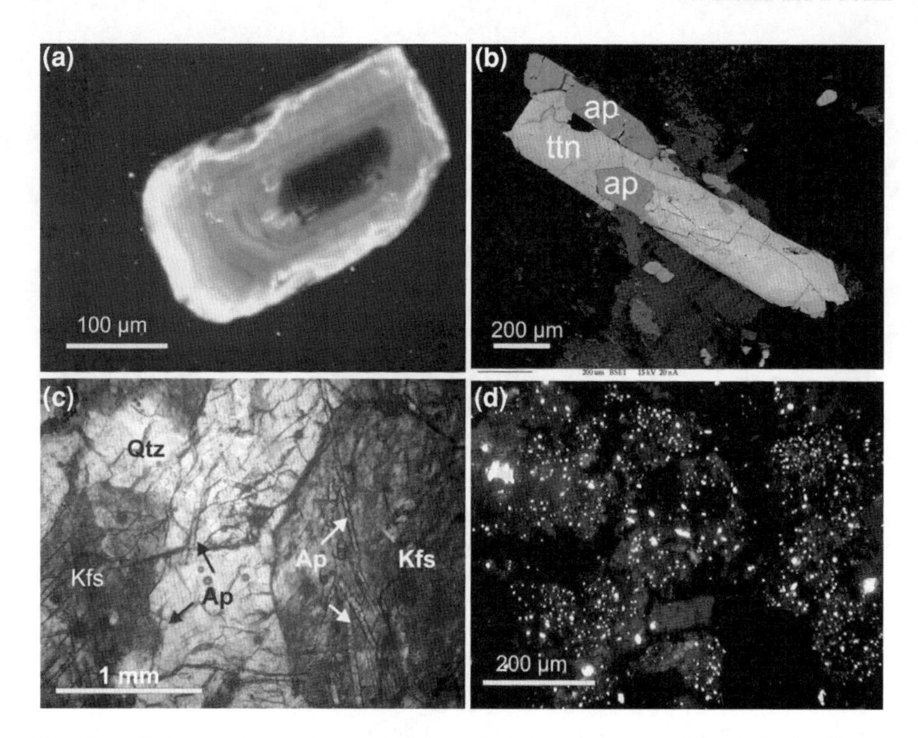

Fig. 12 **a** CL image of an early zoned apatite, which is commonly enclosed in biotite; **b** BSE image of late apatite (*ap*) associated with titanite (*ttn*), crossed nicols; **c** abundant apatite hosted in rock-forming minerals; **d** CL image of subsolidus apatite (numerous tiny bright dots hosted in albites) formed by leaching of Ca and P from albite

orange colour. Monazite is rarely present in I-type granite suite, but can be found as primary phase only in late differentiates (dikes). Its composition is quite uniform but some compositional variations can be observed. The huttonite component in monazite-(Ce) is related to temperature; higher content typically characterises the monazites of higher temperature origin (Broska et al. 2000). In the most evolved S-type granites, monazite has an increased cheralite component.

Monazites in S-type granite show lower content of the cheralite component compared to those found in most evolved specialized S-type granite: e.g. usually <10 mol.% compared to >20–30 cheralite mol.% from the specialised granites, Slovak Ore Mts (Fig. 7). Besides huttonite and cheralite, monazite and xenotime end members may reach up to a considerable proportion in monazite. Petrík et al. (2006) reported monazite coexisting with xenotime in orthogneisses containing 2–7 mol.% xenotime. Monazite from I-type granitoids, as similar to apatite, is enriched in sulphur. Arsenic in monazite has a very small concentration (0.1–0.2 wt%) and can form transitional to monazite and gasparite ($REEPO_4$ and $REEAsO_4$) solid solutions (Ondrejka et al. 2007).

3.2.3 Xenotime

Xenotime-(Y) is generally considered a late magmatic phase on the basis of magmatic structural features developed in equilibrium with monazite but its abundance is very variable and no significant correlation between Y-HREE and xenotime-(Y) concentrations in bulk rock was noted, which indicates the presence of a secondary xenotime.

In many Western Carpathian granitoids, xenotime-(Y) grains are compositionally zoned with a slight core-to-rim increase of Y and decrease in Th + U. The mole fraction of YPO_4 in S-type granitic rocks ranges between 0.84 and 0.75 with an average of 0.78. As in the case of monazite, the coffinite and cheralite substitutions are also characteristic in xenotime. The heavy REE, principally Gd, Dy, Er and Yb, reach a mole fraction of 10–20 mol.%. The actinide elements, Th and U, vary in concentration with U always enriched relative to Th.

Monazite and xenotime coexisting in equilibrium give a temperature of <650 °C (Broska et al. 2005; Petrík et al. 2006) using the monazite-xenotime thermometer of Heinrich et al. (1997) which suggested a late- to post- magmatic origin. Partial replacement of igneous zircon followed by post-magmatic re-crystallization resulted in production of low-Y zircon and secondary xenotime. A similar origin for xenotime is also known from the Li-F specialised granites of the Gemeric unit (Dlhá dolina valley), which are enriched in volatiles such as B, F, LIL and HFS elements (Rb, Nb, Ta, W) and from the Tatra Mts. (Michalik et al. 2000). New xenotime was probably precipitated following the leaching of P and Y from apatite, which contains ca 2 wt% of Y_2O_3 (Broska et al. 2004). Similar leaching effects have been described in experimental studies conducted by Harlov and Förster (2003).

3.2.4 Fe–Ti Oxides

Magnetite is present in I-type granites as individual grains or clusters, as well as inclusions in other accessory and rock-forming minerals. The crystals are mostly euhedral, occasionally in octahedral form. The content of magnetite in bulk rock is variable, usually reaches ca hundreds g/t (calculated from heavy fraction) but it can even achieve several thousands g/t in some parts of the granitoid body, usually in the areas with local occurrences of microgranular mafic enclaves. The increase of magnetite concentration is typical also of some late magmatic leucocratic dikes. The increased magnetic susceptibility in the vicinity of microgranular mafic enclaves is an indication of mixing event in granite magma chamber with more mafic magma and a late oxidation. Fe–Ti oxide crystalisation is the result of such mixing, which has increased the volume of femic components (Broska and Petrik 2011). The presence of abundant magnetite, microgranular enclaves in I-type granitoids of the Western Carpathians supports their origin by magma mixing (Kumar 1995) similarly as proposed by Castro et al. (1991) for Iberian granites.

In general, the Fe–Ti oxides have formed in the Western Carpatians granitoids in three main stages: (1) orthomagmatic (pre-mixing) (2) late-magmatic (post-mixing) and (3) hydrothermal. The orthomagmatic stage produced primary titanomagnetite, which possibly witnesses the incorporation of mafic magma into felsic chamber. The titanomagnetite soon broke down to ilmenite and magnetite composite crystals at 750–650 °C by inter-oxide re-equilibration. The finest measurable spindles from Western Carpathian granites give ca. 670–650 °C using the above calibration, which means that exsolution ceases near the solidus (for granites cooled in mid-crustal depths).Titanium was removed by the exsolution/oxidation reaction to ilmenite and rutile. Mineral equilibria indicate that a late magmatic magnetite was stoichiometrically close to pure end-member, which indeed crystallised in granitoid magma at an increased f_{O2} as a consequence of oxidation including earlier titanomagnetite, biotite, and anorthite in post-mixing late- to post-magmatic stages. If all the phases in reactions (5, 6) are in equilibrium, the invariant point at 590–600 °C indicates that characteristic assemblage of the I-type tonalites, responsible for their increased magnetic susceptibility, is result of the late- to post-magmatic processes and original magma may have been more reduced similar to the S-type magma. The evolution of Fe–Ti oxides terminates by oxidation to hematite (±rutile) along with other low-temperature subsolidus alteration products of the host rock.

3.2.5 Titanite

Titanite is an abundant accessory phase of I-type granitoids of the Western Carpathians. It is usually in a paragenesis with stoichiometrically pure magnetite, quartz and locally with amphibole. Calculations of mineral equilibria in the system K_2O–CaO–FeO–Al_2O_3–TiO_2–SiO_2–H_2O–O_2 (KCFATSHO) indicate that titanite forms in these granitoids as a consequence of the reaction involving Ti–magnetite, biotite, and anorthite in a fluid-rich environment under oxidizing conditions during late stages of granite crystallisation. Evidence for this reaction can be found in relics of Ti-rich magnetite and biotite inclusions in titanite in the Tribeč Mts. and Slovak Ore Mts. Titanite crystalised in the environment with increased Ca/Al ratio is commonly enriched in TiO_2 (min 0.4 wt%). Once titanite precipitates it takes more than 70 wt% Ti of the bulk rock. According to Mössbauer spectroscopy concentration of ferrous iron in titanite is significant; locally almost half of iron is in divalent form. Substitutions in titanite are heterovalent, the site X (Ca) contains trivalent elements and Fe^{2+}, Y (Ti) contains mainly Al^{3+} and Fe^{3+}. Secondary titanite is common in biotite and on the crystal faces of primary titanite. Titanite needles were also found in sagenitic biotite where they are normally considered as rutile. Locally, the titanite alteration process has lead to the complete replacement of titanite by ilmenite, quartz, REE-bearing epidote, and allanite preserves the characteristic euhedral diamond shape of the former titanite. Completely altered titanite grains become optically opaque due to the presence of ilmenite. The formation of REE-bearing epidote and allanite is probably due to the release of the REE from titanite. This can be explained by model hydration reactions such as titanite + anorthite + annite + H_2O = ilmenite + clinozoisite + muscovite + quartz.

Calculation of mineral equilibria involving magmatic precursor minerals such as plagioclase and biotite indicate that the formation of ilmenite from titanite requires an influx of H_2O and/or an increase in f_{O2}. The breakdown of titanite to ilmenite was found in some tonalites from the southwest area of the Tribeč Mts. as a result of decreasing f_{O2} during subsolidus cooling.

3.2.6 Tourmaline

Tourmaline is a widespread accessory mineral in Permian granites of Gemeric and Veporic units. Four types of tourmaline can be described based on petrographic observations and mineral composition of composite Betliar granite, a typical tourmaline-rich granite system. The earliest one is represented typically by schorl with high molar Fe/(Fe + Mg) ratio of 0.7–1.0 and low X-site vacancy (**X**vac= 0.1–0.5 apfu). This type is a euhedral, irregularly disseminated tourmaline that could crystallized directly from granitic magma.

The second type, the big nodular tourmaline forming clusters, formed near solidus or at early subsolidus temperatures. These tourmalines have molar Fe/(Fe + Mg) ratios of 0.8–1.0 and X-site vacancy, **X**vac of 0.29–0.40 apfu. The origin of tourmaline nodules is suggested similar to those found in Croatian granites: due to degassing of granite melt to vapour bubbles.

The third type of tourmaline represented by schorl occurs in quartz–tourmaline veins cross-cutting the medium-grained and fine-grained equigranular granites. Tourmaline crystals from the quartz–tourmaline veins show a wide compositional variations between schorl–dravite solid solution with molar Fe/(Fe + Mg) = 0.44–0.93 and Xvac = 0.07–0.43 apfu.

The fourth tourmaline type is represented by hydrothermal schorl to dravite occurring in all granite types where it forms very thin veins or irregular grain clusters, filling cracks and fractures in pre-existing tourmaline crystals. This tourmaline type shows lower molar Fe/(Fe +Mg) ratio (0.45–0.65) and its mineral chemistry is mainly controlled by substitutions represented by exchange vectors of $FeMg_{-1}$ and $XvacYAlXNa_{-1}Y(Fe,Mg)_{-1}$ increasing Al in the structure (Fig. 13).

The multistage tourmaline origin was shown also in the Hnilec area. The magmatic tourmaline is rich on Fe, Al and Mn with boron isotopic composition which varies from 10.3 to −15.4 ‰. The second tourmaline type forms veins and is Mg-enriched with $\delta^{11}B$ value of −16.0 ‰ to −17.1 ‰. These trends reflect a changing fluid source from a dominant magmatic–hydrothermal fluid derived from the granites to a late-stage metamorphic fluid derived from regional metamorphism (chlorite and biotite zone) of metapelites. The significantly higher Fe^{3+} in the late stage than the magmatic stage tourmalines reflects changing redox conditions towards a more oxidising environment (Jiang et al. 2008).

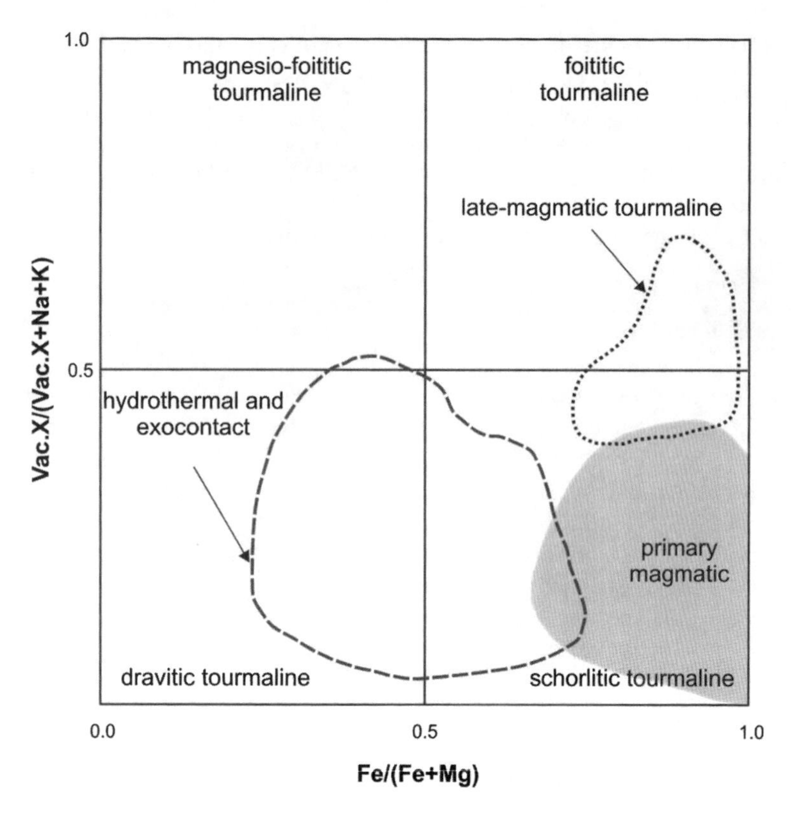

Fig. 13 Tourmaline compositional variation: schorl prevails in specialised granite and compositionally up to younger foitite. Dravite-schorlitic torumalines are product of late-, to subsolidus crystallisation

3.3 Other Significant Accessory Mineral Phases

Other accessory minerals may occur in any particular granite types. The most significant is *garnet,* which characterizes especially peraluminous garnet bearing leucogranites and aplites, as products of biotite dehydration melting where the garnet is incongruent breakdown product of biotite. The relatively MnO rich zoned (6–10 wt%) garnet occurring in the Western Carpathian S-type leucogranites and pegmatites is of magmatic origin. More rarely, garnet is found in association with biotite (Petrík and Konečný 2009). The garnet-bearing peraluminous granites also contain accessory *sillimanite* (fibrolite), considered part of peritectic assemblage, originated by breakdown reaction of biotite and/or muscovite. It is usually overgrown by muscovite or late biotite indicating post-magmatic retrogression by reaction with K-feldspar and H_2O. I-type tonalites commonly contain *sulphides*, typically *pyrite*, rarely *chalcopyrite* indicating increased fugacity of S. The sulphides are almost always strongly retrogressed to *goethite*. In the Western Carpathians amphibole (*Mg-hornblende*) is an accessory phase occurring in I-type tonalites with microgranular enclaves. When coexists with K-feldspar, titanite and magnetite e.g. in Tribeč I-type tonalite, it indicates pressure of ca. 420 MPa (Anderson and Smith 1995). The specialised S-type granites contain characteristic oxides of high field strength elements, *cassiterite, wolframite, columbite, Nb-rutile* and *uraninite*, which are formed in high P microgranites (cf. Fig. 2) accompanied by rare phosphates, *arrojadite, lacroixite* and *viitaniemite* (Petrík et al. 2011).

3.4 Behaviour of Accessory Minerals Superimposed by Metamorphism and Alteration

In polymetamorphic terrains the accessory mineral assemblages may exprience several metamorphic episodes. Typical alterations are found mainly in phosphates, which are discussed in detail. Apatite dissolves easily in felsic fluids, and hence both monazite-(Ce) and xenotime-(Y) are unstable during fluid-activated overprinting. The REE accessories (monazite and allanite) show an especially complex behaviour replacing each other. It has been shown elsewhere (Broska and Siman 1998; Finger et al. 1998) that monazite is not a stable phase in metagranitoids and breaks down to apatite and allanite, or REE-rich epidote. All types of replacement (monazite → allanite, allanite → monazite) are increasingly noted during both progressive and regressive metamorphism (Michalik and Skublicki 1998; Majka and Budzyň 2006; Finger and Krenn 2007; Budzyň et al. 2010; Ondrejka et al. 2012). In hydrothermal conditions allanite may alter to form secondary fluorocarbonates (Buda and Nagy 1995)

In S-type granites, low temperature alteration of monazite leads to the formation of apatite enriched in the britholite component, while in hypergene conditions monazite hydrates to rhabdophane (Nagy et al. 2002). Medium grade metamorphism results in the formation of apatite and LREE-enriched epidote (partly allanite) as a corona enclosing monazite-(Ce). Xenotime-(Y) shows a similar metamorphic pattern, but with different REE distributions in the metamorphic products. At greenschist/amphibolite facies, rims of secondary Y-rich apatite and Y-rich epidote form around xenotime-(Y). In low-Ca granites, however, apatite is missing in the assemblage because xenotime-(Y) breaks down directly to Y-enriched epidote. The fluids responsible for the breakdown of monazite and xenotime contain elements released from altered anorthite (Ca) and biotite (Si, Al and F).

4 Typology of Granites from Accessory Paragenesis

Paragenesis of typomorphic accessory minerals together with zircon morphology can be used to discriminate the granitic rocks into S-, specialised S, I- and A-type granitoids as has been discussed in the case of Western Carpathians. The following magmatic assemblages of accessory minerals were found to determine the granite typology:

- I-type granite: zircon of high S subtypes, apatite, allanite, magnetite, titanite, ± amphibole, pyrite
- S-type granite: zircon with low L and S subtypes, monazite, xenotime, garnet, sillimanite
- Specialised S type: zircon with prevalence S_8 subtypes, tourmaline, monazite, xenotime, garnet, in evolved granites Nb–Ta–W–Sn oxides, topaz and fluorite.
- A-type: zircon P_1 (subsolvus) and D subtype (hypersolvus), monazite, xenotime, garnet ± allanite.

Inter-relating crystallo-chemical properties of accessory minerals, their paragenesis, zircon morphology, along with whole rock chemistry have allowed the discrimination of granitic rocks in the Western Carpathians. Apatite is abundant in early magmatic granite differentiates of I-type affinity where locally it may reach to a high concentration. Monazite-(Ce) is typical of S-type granites but it is found also in late-magmatic I-type differentiates, whereas xenotime is rare or absent. Primary xenotime is more typical of S-type granites but is common mostly as a secondary phase formed from apatite and zircon. LREE distribution in the silicic melt is controlled mainly by distribution of REE minerals such as allanite, titanite, monazite and apatite. The presence or absence of accessory monazite, allanite and apatite depends on P, LREE and Ca concentrations, and is controlled by the process of magmatic differentiation. In this regard the most important is alkalinity and the oxidation-reduction stage of the silicic melt indicated by Fe–Ti oxides, especially magnetite and ilmenite (e.g. higher f_{O2} and Ca activity lead to the crystallization of apatite and allanite, in contrast, monazite indicates lower f_{O2} and Ca activity).

The I-type metaluminous or subaluminous granites (ASI 0.9–1.05) show features of late oxidation, which appears to be responsible for their characteristic mineralogical behaviour, and reflects involvement of different source rocks that are richer in H_2O, and presumably have a different geotectonic position during their evolution (cf. Broska et al. 2013). The allanite, magnetite and titanite (+oxidated Mg-biotite) as well apatite, form the typical paragenesis of I-type granitoids. In contrast, monazite, garnet and xenotime are typical of S-type granites. The assemblages of I- and S-type granitoids are not just products of contrasting protoliths (an igneous and sedimentary source are required for I-type and S-type respectively), a variable amount of mantle input is presumably nedded for I-type granitoids based on their $^{143}Nd/^{144}Nd$ ratios (Kohút et al. 1999), or high amounts of magnetite ultimately resulted from assimilation of microgranular mafic enclaves. The Permian S-type granites rich in volatiles may contain primary tourmaline, topaz or fluorite. In this special type of mineralization, the most evolved part of granite bodies crystallise cassiterite, wolframite, Nb–Ta oxides, and especially in altered parts (Uher and Broska 1996), whereas P-rich melt crystallised aluminophosphates.

5 Conclusions

Accessory minerals bear vital information on age, primary conditions or possible late- to post-magmatic alterations of the parental granitoid rocks. By their compositions and internal structure they record character of granitic protoliths, geochemical and petrological evolution of the magma, reworked under the influence of repeated burials, thus revealing signature of later orogens. Temperature, pressure, oxygen fugacity, fluid compositions may all be retrieved from the study of proper accessory assemblages. Zircon and monazite may record both the old history and evidence of new magma evolution. Apatite discriminates well among various granite types. Magnetite, allanite, titanite record the $f_{O2} - T$ evolution whereas xenotime offers another thermometer. Tourmaline may witness the volatile activity in the melt. Thus, the data on accessory minerals are essential for the recognition of geotectonic settings of granitoids.

Acknowledgments Grant agency Vega (project 2/0060/10 and 2/0159/13) and APVV (project 080-11) are thanked for financial support. Santosh Kumar is thanked for thoughtful review and thorough editorial work. Figures 2, 4, 5, 6, 8 and 11 are published with kind permission of the Veda Publishing House (Broska et al. 2012)

References

Andersen DJ, Lindsley DH (1988) Internally consistent solution models for Fe-Mg-Mn-Ti oxides: Fe-Ti oxides. Am Miner 73:714–726

Anderson JL, Smith DR (1995) The effects of temperature and f_{O2} on Al-hornblende barometer. Am Miner 80:549–559

Anovitz LM, Grew ES (1996) Mineralogy, petrology and geochemistry of boron: an introduction. In: Grew ES, Anovitz LM (eds) Boron: mineralogy, petrology, and geochemistry. Reviews in mineralogy, vol 33, pp 1–40

Bačík P, Uher P (2010) Dissakisite-(La), mukhinite, and clinozoisite: (V, Cr, REE)-rich members of the epidote group in amphibole - Pyrite pyrrhotite metabasic rocks from Pezinok, Rybníček mine, Western Carpathians, Slovakia. Can Miner 48:523–536

Balen D, Broska I (2011) Tourmaline nodules: products of devolatilization within the final evolutionary stage of granitic melt? vol 350, Geological Society, Special Publications, London, pp 53–68

Bea F (1996) Residence of REE, Y, Th and U in granites and crustal protoliths: implications for the chemistry of crustal melts. J Petrol 37:521–532

Bea F, Fershtater G, Corretgé LG (1992) The geochemistry of phosphorus in granite and the effect of aluminium. Lithos 29:43–45

Beard JS, Sorensen SS, Gieré R (2006) REE zoning in allanite related to changing partition coefficients during crystallization: implications for REE behaviour in an epidote bearing tonalite. Miner Mag 70:419–435

Belousova EA, Walters S, Griffin WL, O'Reilly SY (2001) Trace-element signatures of apatites in granitoids from the Mt Isa Inlier, northwestern Queensland. Aust J Earth Sci 48:603–619

Belousova EA, Griffin WL, O'Reilly SY, Fisher NI (2002) Igneous zircon: trace element composition as an indicator of source rock type. Contrib Miner Pet 143:602–622

Belousova EA, Griffin WL, O'Reilly SY (2006) Zircon crystal morphology, trace element signatures and Hf isotope composition as a tool for petrogenetic modelling: examples from eastern Australian granitoids. J Petrol 47:329–353

Belousova EA, Kostitsyn YA, Griffin WL, Begg GC, O'Reilly SY, Pearson NJ (2010) The growth of the continental crust: constraints from zircon Hf-isotope data. Lithos 119:457–466

Benisek A, Finger F (1993) Factors controlling the development of prism faces in granite zircons: a microprobe study. Contrib Miner Pet 114:441–451

Berger A, Gnos E, Janots E, Fernandez A, Giese J (2008) Formation and composition of rhabdophane, bstnäsite and hydrated thorium minerals during alteration: implications for geochronology and low-temperature processes. Chem Geol 254:238–248

Bingen B, Demaiffe D, Hertogen J (1996) Redistribution of rare earth elements, thorium and uranium over accessory minerals in the course of amphibolite to granulite facies metamorphism: the role of apatite and monazite in orthogneisses from southwestern Norway. Geochim Cosmochim Ac 60:1341–1354

Bónová K, Broska I, Petrík I (2010) Biotite from Čierna Hora Mountains granitoids (Western Carpathians, Slovakia) and estimation of water contents in granitoid melts. Geol Carpath 61:3–17

Borodina NS, Fershtater GB, Votyakov SL (1999) The oxidation ratio of iron in coexisting biotite and hornblende from granitic and metamorphic rocks: the role of P, T, and f(O2). Can Miner 37:1423–1429

Broska I, Uher P (1991) Regional typology of zircon and its relationship to allanite/monazite antagonism (on an example of Hercynian granitoids of Western Carpathians). Geol Carpath 42:271–277

Broska I, Siman P (1998) The breakdown of monazite in the West-Carpathian Veporic orthogneisses and Tatric granites. Geol Carpath 49:161–167

Broska I, Petrík I, Wiliams T (2000) Co-existing monazite-(Ce) and allanite-(Ce) from S-type granitoids (on the example of the Tribeč Mts. Western Carpathians). Am Miner 85:22–32

Broska I, Uher P (2001) Whole-rock chemistry and genetic typology of the West-Carpathian Variscan granites. Geol Carpath 52(2):79–90

Broska I, Williams CT, Uher P, Konečný P, Leichmann J (2004) The geochemistry of phosphorus in different granite suites of the Western Carpathians, Slovakia: the role of apatite and P-bearing feldspar. Chem Geol 205:1–15

Broska I, Williams CT, Janák M, Nagy G (2005) Alteration and breakdown of xenotime-(Y) and monazite-(Ce) in granitic rocks of the Western Carpathians, Slovakia. Lithos 82:71–83

Broska I, Harlov D, Tropper P, Siman P (2007) Formation of magmatic titanite and titanite—ilmenite phase relations during granite alteration in the Tribeč Mountains, Western Carpathians, Slovakia). Lithos 95:58–71

Broska I, Petrík I (2011) Accessory Fe-Ti oxides in the West-Carpathian I-type granitoids: witnesses of the granite mixing and late oxidation processes. Miner Pet 102:87–97

Broska I, Petrík I, Uher P (2012) Accessory minerals of the Carpathian granitic rocks. Veda Publishiong House, Slovak Academy of Sciences, Bratislava, p 235

Broska I, Petrík I, Shlevin ZB, Majka M, Bezák V (2013) Devonian/Mississipian I-type granitoids in the Western Carpathians: a subduction-related hybrid magmatism. Lithos 162–163:27–36

Buddington AF, Lindsley DH (1964) Iron-titanium oxide minerals and synthetic equivalents. J Petrol 5:310–357

Buda G, Nagy G (1995) Some REE-bearing accessory minerals in two types of Variscan granitoids, Hungary. Geol Carpath 46(2):67–78

Budzyń B, Hetherington CJ, Williams ML, Jercinovic MJ, Michalik M (2010) Fluid -mineral interactions and constraints on monazite alteration during metamorphism. Mineral Magaz 74:659–681

Buriánek D, Novák M (2007) Compositional evolution and substitutions in disseminated and nodular tourmaline from leucocratic granites: examples from the Bohemian Massif, Czech Republic. Lithos 95:148–164

Carswell DA, Wilson RN, Zhai M (1996) Ultra-high pressure aluminous titanites in carbonate-bearing eclogites at Shuanghe in Dabieshan, central China. Mineral Mag 60:461–471
Castro A, Moreno-Ventas I, De la Rosa JD (1991) H-type (hybrid) granitoids: a proposed revision of the granite-type classification and nomenclature. Earth Sci Rev 31:237–253
Catlos EJ, Sorensen SS, Harrison TM (2000) Th-Pb ion-microprobe dating of allanite. Am Mineral 85:548–633
Cavosie AJ, Kita NT, Valley JW (2009) Primitive oxygen-isotope ratio recorded in magmatic zircon from the Mid-Atlantic Ridge. Am Mineral 94:926–934
Chappell BW, White AJR (1998) Development of P-rich granites by sequential restite fractionation and fractional crystallization: the Koetong Suite in the Lachlan fold belt. Acta Univ Car Geol 42:23–27
Cherniak DJ, Watson EB (2000) Pb diffusion in zircon. Chem Geol 172:5–24
Chesner CA, Ettlinger AD (1989) Composition of volcanic allanite from the Toba tuffs, Sumatra, Indonesia. Am Mineral 74:750–758
Corfu F, Hanchar JM, Hoskin PWO, Kinny P (2003) In: Hanchar JM and Hoskin PWO (eds) Reviews in mineralogy and geochemistry, v 53, pp 468–500
Czamanske GK, Mihálik P (1972) Oxidation during magmatic differentiation, Finnmarka Complex, Oslo area, Norway: part I, the opaque oxides. J Petrol 13:493–509
Czamanske GK, Wones DR (1973) Oxidation during magmatic differentiation, Finnmarka Complex, Oslo area, Norway: part 2, the mafic silicates. J Petrol 14:349–380
Davidson C, Rosenberg C, Schmidt SM (1996) Synmagmatic folding of the base of the Bergell pluton, Central Alps. Tectonophysics 265:213–238
Enami M, Suzuki K, Liou JG, Bird DK (1993) Al-Fe^{3+} and F-OH substitutions in titanite and constraints on their P-T dependence. Eur J Mineral 5:219–231
Ewart A, Hildreth W, Carmichael ISE (1975) Quaternary acid magma in New Zealand. Contrib Miner Petrol 51:1–27
Fielding PE (1970) The distribution of uranium, rare earth and color centers in a crystal of natural zircon. Am Mineral 55:429–440
Finch RJ, Hanchar JM (2003) Structure and chemistry of zircon and zircon-group minerals. In: Hanchar JM, Hoskin PWO (eds) Reviews in mineralogy and geochemistry, vol 53, pp 1–25
Finger F, Broska I, Roberts M, Schermeier A (1998) Replacement of primary monazite by apatite-allanite-epidote coronas in an amphibolite facies granite gneiss from the eastern Alps. Am Mineral 83:248–258
Finger F, Broska I, Haunschmid B, Hraško Ľ, Kohút M, Krenn E, Petrík I, Riegler G, Uher P (2003) Electron microprobe dating of monazites from Western Carpathian basement granitoids: plutonic evidence for an important Permian rifting event subsequent to Variscan crustal anatexis. Int J Earth Sci (Geol. Rundsch.) 92:86–98
Finger F, Krenn E (2007) Three metamorphic monazite generations in a high-pressure rock from the Bohemian Massif and the potentially important role of apatite in stimulating polyphase monazite growth along a P-T loop. Lithos 95:115–125
Fowler A, Prokoph A, Stern R, Dupuis C (2002) Organization of oscillatory zoning in zircon: analysis, scaling, geochemistry, and model of a zircon from Kipawa, Quebec, Canada. Geochim Cosmochim Ac 66:311–328
Förster H-J (1998) The chemical composition of REE-Y-Th-U-rich accessory minerals in peraluminous granites of the Erzebirge—Fichtelgebirge region, Germany, part I: the monazite-(Ce)—brabantite solid solution series. Am Mineral 83:259–272
Förster H-J, Harlov DE (1999) Monazite-(Ce)—huttonite solid solutions in granulite-facies metabasites from the Ivre-Verbano zone, Italy. Mineral Mag 63:587–594
Franz G, Spear FS (1985) Aluminous titanite (sphene) from the eclogite zone, south central Tauern window, Austria. Chem Geol 50:33–46
Franz G, Liebscher A (2004) Physical and chemical properties of the epidote minerals—an introduction. In: Liebscher A, Franz G (eds) Epidotes. Reviews in mineralogy and geochemistry, vol 56. Mineralogical Society of America, pp 1–82

Frýda J, Breiter K (1995) Alkali feldspars as a main phosphorus reservoir in rare-metal granites: three examples from the Bohemian Massif (Czech Republic). Terra Nova 7:315–320

Garver JI, Kamp PJJ (2002) Integration of zircon color and zircon fission-track zonation patterns in orogenic belts: application to the Southern Alps, New Zealand. Tectonophysics 349:203–219

Geisler T, Pidgeon RT, Kurtz R, Bronswijk W, Schleiche H (2003) Experimental hydrothermal alteration of partially metamict zircon. Am Mineral 88:1496–1513

Ghiorso MS, Sack RO (1991) Fe-Ti oxide geothermobarometry: thermodynamic formulation and the estimation of intensive variables in silicic magmas. Contrib Miner Petrol 108:485–510

Ghiorso MS, Evans BW (2009) Thermodynamics of rhombohedral oxide solid solutions and a revision of the Fe-Ti two oxide geothermometer and oxygen-barometer. Am J Sci 308:957–1039

Gieré R, Sorensen SS (2004) Allanite and other REE-rich epidote minerals. In: Liebscher A, Franz G (eds) Epidotes. Reviews in mineralogy and geochemistry, vol 56. Mineralogical Society of America, pp 431–493

Green TH, Watson EB (1982) Crystallization of apatite in natural magmas under high pressure, hydrous conditions, with particular references to "orogenic" rock series. Contrib Mineral Petrol 79:96–105

Grew ES, Essene EJ, Peacor DR, Su Sh-Ch, Asami M (1991) Dissakisite-(Ce), a new member of the epidote group and the Mg analogue of allanite-(Ce), from Antarctica. Am Mineral 76:1990–1997

Grey IE, Reid AF (1975) The structure of pseudorutile and its role in the natural alteration of ilmenite. Am Mineral 60:898–906

Griffin WL, Wang X, Jackson SE, Pearson NJ, O'Reilly SY, Xu X, Zhou X (2002) Zircon chemitry and magma genesis, SE China: in situ analysis of Hf isotopes, Pingtan and Tonglu igneous complex. Lithos 61:237–269

Gromet LP, Silver LT (1983) Rare earth element distributions among minerals in a granodiorite and their petrogenetic implications. Geochim Cosmochim Acta 47:925–939

Halden NM, Hawthorne FC (1993) The fractal geometry of oscilatory zoning in crystals: application to zircon. Am Mineral 78:1113–1116

Hanchar JM, Finch RJ, Hoskin PWO, Watson B, Cherniak DJ, Mariano AN (2001) Rare earth elements in synthetic zircon: part 1. Synthesis, and rare earth element and phosphorus doping. Am Mineral 86:667–680

Harlov DE, Förster H-J (2003) Fluid-induced nucleation of (Y-REE)-phosphate minerals within apatite: nature and experiment part II. Fluorapatite. Am Mineral 88:1209–1229

Harlov D, Tropper P, Seifert W, Nijland T, Förster H-J (2006) Formation of Al-rich titanite ($CaTiSiO_4O$-$CaAlSiO_4OH$) reaction rims on ilmenite in metamorphic rocks as a function of fH_2O and fO_2. Lithos 88:72–84

Harrison TM, Watson EB (1983) Kinetics of zircon dissolution and zirconium diffusion in granitic melts of variable water content. Contrib Mineral Petrol 84:66–72

Harrison TM, Watson EB (1984) The behaviour of apatite during crystal anatexis: equilibrium and kinetic considerations. Geochim Cosmochim Ac 48:1467–1477

Harrison TM, McKeegan KD, LeFort P (1995) Detection of inherited monazite in the Manaslu leucogranite by 208Pb/232Th ion microprobe dating: crystallization age and tectonic implications. Earth Planet Sci Lett 133:271–282

Heaman LM, Bowins R, Crocket J (1990) The chemical composition of igneous zircon suites: implications for geochemical tracer studies. Geochim Cosmochim Acta 54:1597–1607

Heinrich W, Andrehs G, Franz G (1997) Monazite-xenotime miscibility gap thermometry. I. An empirical calibration. J Metamorph Geol 15:3–16

Henry DJ, Guidotti CV (1985) Tourmaline as a petrogenetic indicator mineral: an example from the staurolite-grade metapelites of NW Maine. Am Mineral 70:1–15

Henry DJ, Novák M, Hawtorne FC, Ertl A, Dutrow BL, Uher P, Pezzotta F (2011) Nomenclature of tourmaline—supergroup minerals. Am Mineral 96:895–913

Holland TJB, Powell R (1998) An internally consistent thermodynamic data set for phases of petrological interest. J Metamorph Geol 16:309–343
Holtstam D, Andersson UB, Mansfeld J (2003) Feriallanite-(Ce) from the Bastnäs deposit, Västmanland, Sweden. Can Mineral 41:1233–1240
Hoskin PWO (2000) Patterns of chaos: fractal statistics and the oscillatory chemistry of zircon. Geochim Cosmochim Ac 64:1905–1923
Hoskin PWO, Black LP (2000) Metamorphic zircon formation by solid-state recrystallization of protolith Igneous zircon. J Metamorph Geol 18:423–439
Hoskin PWO, Schaltegger U (2003) In: Hanchar JM, Hoskin PWO (eds) Reviews in mineralogy and geochemistry, vol 53, pp 27–62
Hughes JM, Rakovan J (2002) The crystal structure of apatite, $Ca_5(PO_4)_3(F,OH,Cl)$. In: Reviews of mineralogy and geochemistry, vol. 48, pp 1–12
Ishihara S (1977) The magnetite series and ilmenite-series granitic rocks. Min Geol 27:293–305
Janots E, Brunet F, Goffé B, Poinssot C, Burchard M, Cemič L (2007) Thermochemistry of monazite-(La) and dissakisite-(La): Implications for monazite and allanite stability in metapelites. Contrib Miner Petrol 154:1–14
Jiang SY, Palmer MR (1998) Boron isotope systematics of tourmaline from granites and pegmatites; a synthesis. Eur J Mineral 10:1253–1265
Jiang SI, Radvanec M, Nakamura E, Palmer M, Kobavashi K, Zhao HX, Zhao KD (2008) Chemical and boron isotopic variations of tourmaline in the Hnilec granite-related hydrothermal system, Slovakia: constraints on magmatic and metamorphic fluid evolution. Lithos 106:1–11
Kempe U, Gruner T, Renno AD, Wolf D, René M (2004) Discussion on Wang et al.(2000) 'Chemistry of Hf-rich zircons from the Laoshan I- and A-type granites, Eastern China', Mineral Mag 64:867–877. Mineral Mag 68:669–675
Klemm DD, Henckel J, Dehm R, Von Gruenewaldt G (1985) The geochemistry of titanomagnetite in magnetite layers and their host rocks of the Eastern Bushveld complex. Econ Geol 80:1075–1088
Kohút M, Putiš M, Ondrejka M, Sergeev S, Larionov A, Paderin I (1999) Sr and Nd isotope geochemistry of Hercynian granitic rocks from the Western Carpathians—implications for granite genesis and crustal evolution. Geol Carpath 50:477–487
Kohút M, Uher P, Putiš M, Ondrejka M, Sergeev S, Larionov A, Paderin I (2009) SHRIMP U-Th-Pb zircon dating of the granitoid massifs in the Malé Karpaty Mountains (Western Carpathians): evidence of the Meso-Hercynian successive S- to I-type granitic magmatism. Geol Carpath 60:345–350
Kubiš M, Broska I (2010) The granite system near Betliar village (Gemeric Superunit, Western Carpathians): evolution of a composite silicic reservoir. J Geosci 55(2010):131–148
Kumar S (1995) Microstructural evidence of magma quenching inferred from enclaves hosted in the Hodruša granodiorites, Western Carpathians. Geol Carpath 46:379–382
Kumar S (2010) Magnetite and ilmenite series granitoids of Ladakh batholith, Northwest Indian Himalaya: implications on redox conditions of subducted zone magmatism. Curr Sci 99:1260–1264
Liou JG (1973) Synthesis and stability relations of epidote $Ca_2Al_2FeSi_3O_{12}(OH)$. J Petrol 14:381–413
London D, Černý P, Loomis JL, Pan JJ (1990) Phosphorus in alkali feldspars of rare-element granitic pegmatites. Canad Mineral 28:771–786
London D (1992) Phosphorus in S-type magmas: the P_2O_5 content of feldspars from peraluminious granites, pegmatites and rhyolites. Am Mineral 77:126–145
London D, Manning DAC (1995) Chemical variation and significance of tourmaline from Southwest England. Econ Geol 90:495–519
London D, Wolf MB, Morgan GB, Garrido MG (1999) Experimental silicate-phosphate equilibria in peraluminous granitic magmas, with a case study of the Alburquerque batholith at Tres Arroyos, Badajoz, Spain. J Petrol 40:215–240

Majka J, Budzyń B (2006) Monazite breakdown in metapelites from Wedel Jarlsberg land, Svalbard—preliminary report. Mineralogia Polonica 37:61–69

McNear E, Vincent MG, Parthe E (1976) The crystal structure of vuagnatite, $CaAl(OH)SiO_4$. Am Mineral 61:831–838

Michael PJ (1988) Partition coefficients for rare earth elements in mafic minerals of high silica rhyolites: the importance of accessory mineral inclusions. Geochim Cosmochim Ac 52:275–282

Michalik J, Skublicki Ł (1998) Breakdown of monazite during alterations of the High Tatra granitoids. Polskie Towarzystwo Mineralogiczne—Prace Specjalne 11:145–147 (in Polish)

Michalik M, Popczyk R, Kusiak M, Paszkowski M (2000) Xenotime zircon intergrowths in the Western Tatra leucogranites. Polskie Towarzystwo Mineralogiczne-Prace Specjalne. Mineral Soc Poland-Special Papers 17:249–251

Miller CF, Mittlefeldt DW (1982) Depletion of rare-earth elements in felsic magmas. Geology 10:129–133

Miller CF, McDowell SM, Mapes RW (2003) Hot and cold granites? Implications of zircon saturation temperatures and preservation of inheritance. Geology 31:529–532

Mojzsis SJ, Harrison TM, Pidgeon RT (2001) Oxygen-isotope evidence from ancient zircons for liquid water at the Earth's surface 4300 Myr ago. Nature 409:178–181

Montel J-M (1993) A model for monazite/melt equilibrium and application to the generation of granitic magmas. Chem Geol 110:127–146

Montel J-M, Foret S, Veschambre M, Nichollet Ch, Provost A (1996) Electron microprobe dating of monazite. Chem Geol 131:37–53

Nabelek P, Russ-Nabelek C, Denison JR (1992) The generation and crystallization of Proterozoic Harney Peak leucogranite, Black Hills, South Dakota, USA: petrologic and geochemical constrains. Contrib Mineral Petrol 110:173–191

Nagy G, Draganits E, Demény A, Pantó Gy, Árkai P (2002) Genesis and transformations of monazite, florncite and rhabdophane during medium grade metamorphism: examples from the Sopron Hills, Eastern Alps. Chem Geol 191:25–46

Nakada S (1991) Magmatic processes in titanite-bearing dacites, central Andes of Chile and Bolivia. Am Mineral 76:548–560

Nasdala L, Hanchar JM, Kronz A, Whitehouse MJ (2005) Long-term stability of alpha particle damage in natural zircon. Chem Geol 220:83–103

Nasdala L, Kronz A, Wirth R, Váczi T, Pérez-Soba C, Willner A, Kennedy AK (2009) Alteration of radiation-damaged zircon and the related phenomenon of deficient electron microprobe totals. Geochim Cosmochim Ac 73:1637–1650

Noyes HJ, Wones DR, Frey A (1983) A tale of two pluons: petrographic and mineralogical constraints on the petrogenesis of the Red Lake and Eagle Peak plutons, central Sierra Nevada. J Geol 91:353–379

Ondrejka M, Uher P, Pršek J, Ozdín D (2007) Arsenian monazite-(Ce) and xenotime-(Y), REE arsenates and carbonates from the Tisovec-Rejkovo rhyolite, Western Carpathians, Slovakia: composition and substitutions in the (REE, Y)XO_4 system (X = P, As, Si, Nb, S). Lithos 95:116–129

Ondrejka M, Uher P, Putiš M, Broska I, Bačík P, Konečný P, Schmiedt I (2012) Two-stage breakdown of monazite by post-magmatic and metamorphic fluids: an example from the Veporic orthogneiss, Western Carpathians, Slovakia. Lithos 142–143:245–255

Pan Y (1997) Zircon- and monazite-forming metamorphic reactions at Manitouwadge, Ontario. Can Mineral 35:105–118

Pan Y, Fleet ME (2002) Compositions of the apatite-group minerals: substitution mechanism and controlling factor. Rev Mineral Geochem 48:13–49

Parrish RR (1990) U-Pb dating of monazite and its application to geological problems. Can J Earth Sci 27:1431–1450

Pasero M, Kampf AR, Ferraris C, Pekov IV, Rakovan J, White TJ (2010) Nomenclature of the apatite supergroup minerals. Eur J Mineral 22:163–179

Petrík I (1999) Allanite vs. monazite in granitoid magmas: the role of water and REE content. Berichte der Deutschen mineralogischen Gesselschaft. Beihefte zum. Eur J Mineral 11:176

Petrík I, Broska I (1994) Petrology of two granite types from the Tribeč Mountains, Western Carpathians: an example of allanite (+ magnetite) versus monazite dichotomy. Geol J 29:59–78

Petrík I, Broska I, Lipka J, Siman P (1995) Granitoid allanite-(Ce): Substitution relations, redox conditions and REE distributions (on example of I-type granitoid, Western Carpathians, Slovakia). Geol Carpath 46:79–94

Petrík I, Konečný P, Kováčik M, Holický I (2006) Electron microprobe dating of monazite from the Nizke Tatry Mountains orthogneisses (Western Carpathians, Slovakia). Geol Carpath 57:227–242

Petrík I, Konečný P (2009) Metasomatic emplacement of inherited metamorphic monazite in a biotite-garnet granite from the Nízke Tatry Mountains, Western Carpathians, Slovakia: chemical dating and evidence for disequilibrium melting. Am Mineral 94:957–974

Petrík I, Kubiš M, Konečný P, Broska I, Malachovský P (2011) Rare phosphates from the Surovec topaz – Li-mica microgranite, Gemeric unit, Western Carpathians, Slovakia: the role of the F/H_2O in the melt. Can Mineral 49:521–540

Piccoli P, Candela P (1994) Apatite in felsic rocks: a model for the estimation of initial halogen concentrations in the Bishop Tuff (Long Valley) and Tuolumne Intrusive Suite (Sierra Nevada Batholith) magmas. Am J Sci 294:92–135

Pichavant M (1981) An experimental study of the effect of boron on a water saturated haplogranite at 1 kbar vapour pressure. Geological applications. Contrib Miner Petrol 76:430–439

Pichavant M, Montel JM, Richard LR (1992) Apatite solubility in peraluminous liquids: experimental data and an extension of the Harrison-Watson model. Geochim Cosmochim Ac 56:3855–3861

Poitrasson F (2002) In situ investigations of allanite hydrothermal alteration: examples from calc-alkaline and anorogenic granites of Corsica. Contrib Mineral Petrol 142:485–500

Pupin JP (1980) Zircon and granite petrology. Contrib Miner Petrol 73:207–220

Pupin JP (2000) Granite genesis related to geodynamics from Hf-Y in zircon. Trans R Soc Edinb Earth Sci 91:245–256

Pyle JM, Spear FS, Rudnick RL, McDonough WF (2001) Monazite-xenotime-garnet equilibrium in metapelites and a new monazite-garnet thermometer. J Petrol 42:2083–2107

Rapp RP, Watson EB (1986) Monazite solubility and dissolution kinetics: implications for the thorium and light rare earth chemistry of felsic magmas. Contrib Miner Petrol 94:304–316

Ribbe PH (ed) (1980) Titanite. Orthosilicates. Reviews in mineralogy, vol 5. Mineralogical Society of America, pp 137–155

Ripp GC, Karmanov NS, Kanakin SV, Doroshkevich AG (2002) Allanites of Wetsern Transbaikalia. Proc RMS (Zapisky vsesoyuznogo mineralogicheskogo obshchestva) 4:92–106 (in Russian)

Robinson DM, Miller CF (1999) Record of magma chamber processes preserved in accessory mineral assemblages, Aztec Wash pluton, Nevada. Am Mineral 84:1346–1353

Sakoma EM, Martin RF (2002) Oxidation-induced postmagmatic modifications of primary ilmenite, NYG-related aplite, Tibchi complex, Kalato, Nigeria. Mineral Mag 66:591–604

Samson IM, Sinclair WD (1992) Magmatic hydrothermal fluids and the origin of quartz-tourmaline orbicles in the Seagull Batholith, Yukon Territory. Can Mineral 30:937–954

Sato M, Wright TL (1966) Oxygen fugacity directly measured in volcanic gases. Science 153:1103–1105

Sauerzapf U, Lattard D, Burchard M, Engelmann R (2008) The titanomagnetite-ilmenite equilibrium: new experimental data and thermo-oxybarometric application to the crystallization of basic to intermediate rocks. J Petrol 49:1161–1185

Sawka WN, Chappell BW, Norrish K (1984) Light-rare-earth element zoning in sphene and allanite during granitoid fractionation. Geology 12:131–134

Scaillet B, Pichavant M, Roux J (1995) Experimental crystallization of leucogranite magmas. J Petrol 36:663–705

Sha LK, Chappell BW (2000) Apatite chemical composition, determined by electron microprobe and laser-ablation inductively coupled plasma mass spectrometry, as a probe into granite petrogenesis. Geochim Cosmochim Acta 63:3861–3881

Schatz OJ, Dolejš D, Stix J, Williams-Jones AE, Layne GD (2004) Partitioning of boron among melt, brine and vapor in the system haplogranite–H2O–NaCl at 800°C and 100 MPa. Chem Geol 210:135–147

Shearer CK, Papike JJ, Laul JC (1987) Mineralogical and chemical evolution of a rare-element granite–pegmatite system: Harney Peak Granite, Black Hills, South Dakota. Geochim Cosmochim Ac 51:473–486

Schmidt MW, Thompson AB (1996) Epidote in calc-alkaline magmas: an experimental study of stability, phase relationships, and the role of epidote in magmatic evolution. Am Mineral 81:462–474

Schmidt MW, Poli S (2004) Magmatic epidote. In: Liebscher A, Franz G (eds) Epidotes. Reviews in mineralogy and geochemistry, vol 56. Mineralogical Society of America, pp 399–430

Simpson DR (1977) Aluminium phosphate variants of feldspar. Am Mineral 62:351–355

Spear FS, Pyle JM (2002) Phosphates in metamorphic rocks. In: Kohn ML, Rakovan J, Hughes JM (eds) Phosphates, Reviews in mineralogy, vol 48. Mineralogical Society of America, pp 293–335

Suzuki K, Adachi M (1991) Precambrian provenance and Silurian metamorphism of the Tsubonosawa paragneiss in the South Kitakami terrane, Northeast Japan, revealed by the chemical Th-U-total Pb isochron ages of monazite, zircon, and xenotime. Geochem J 25:357–376

Thomas R, Förster H-J, Rickers K, Webster JD (2005) Formation of extremely F-rich hydrous melt fractions and hydrothermal fluids during differentiation of highly evolve tin-granite magmas: a melt/fluid-inclusion study. Contrib Miner Petrol 148:582–601

Tropper P, Manning CE, Essene EJ (2002) The substitution of Al and F in titanite at high pressure and temperature: experimental constraints on phase relations and solid solution properties. J Petrol 43:1787–1814

Trumbull RB, Krienitz MS, Gottesmann B, Wiedenbeck M (2008) Chemical and boron-isotope variations in tourmalines from a S-type granite and its source rocks: the Erongo granite and tourmalinites in the Damara Belt, Namibia. Contrib Mineral Petrol 155:1–18

Turner MB, Cronin SJ, Stewart RB, Bebbington M, Smith IEM (2008) Using titanomagnetite texture to elucidate volcanic eruption histories. Geology 36:31–34

Uher P, Puskharev Y (1994) Granitic pebbles of the Cretaceous flysh of the Pieniny Klippen belt, Western Carpathians: U/Pb zircon ages. Geol Carpath 45:375–378

Uher P, Broska I (1996) Post-orogenic Permian granitic rocks in the Western Carpathian-Pannonian area: geochemistry, mineralogy and evolution. Geol Carpath 47:311–321

Uher P, Černý P (1998) Zircon in Hercynian granitic pegmatites of the Western Carpathians, Slovakia. Geol Carpath 49:261–270

Valley JW, Chiarenzelli JR, McLelland JM (1994) Oxygen isotope geochemistry of zircon. Earth Planet Sci Lett 126:187–206

Vavra G (1990) On the kinematics of zircon growth and its petrogenetic significance: a cathodoluminescence study. Contrib Miner Petrol 106:90–99

Vavra G (1994) Systematics of internal zircon morphology in major Variscan granitoid types. Contrib Mineral Petrol 117:331–344

Vavra G, Gebauer D, Schmidt R, Compston W (1996) Multiple zircon growth and recrystallization during polyphase late carboniferous to triassic metamorphism in granulites of the Ivrea Zone (Southern Alps): an ion microprobe (SHRIMP) study. Contrib Miner Petrol 122:337–358

Watson EB, Capobianco CJ (1981) Phosphorus and rare earth elements in felsic magmas. An assesment of the role of apatite. Geochim Cosmochim Ac 45:2349–2358

Watson EB, Harrison TM (1983) Zircon saturation revised: temperature and composition effects in a variety of crustal magma types. Earth Planet Sci Lett 64:295–304

Whalen JB, Chappell BW (1988) Opaque mineralogy and mafic mineral chemistry of I- and Stype granites of the Lachlan fold belt, southeast Australia. Am Mineral 73:281–296

White T, Ferraris C, Kim J, Madhavi S (2005) Apatite—an adaptive framework structure. In: Ferraris G, Merlino S (eds) Reviews in mineralogy and Geochemistry, vol 57, pp 307–401

Wing BA, Ferry JM, Harrison TM (2003) Prograde destruction and formation of monazite and allanite during contact and regional metamorphism of pelites: petrology and geochronology. Contrib Mineral Petrol 145:228–250

Wolf MB, London D (1994) Apatite dissolution into peraluminous haplogranitic melts: an experimental study of solubilities and mechanism. Geochim Cosmochim Ac 58:4127–4145

Wolf MB, London D (1995) Incongruent dissolution of REE- and Sr-rich apatite in peraluminous granitic liquids: differential apatite, monazite, and xenotime solubilities during anatexis. Am Mineral 80:765–775

Wones DR (1989) Significance of the assemblage titanite+magnetite+quartz in granitic rocks. Am Mineral 74:744–749

Wones DR, Eugster HP (1965) Stability of biotite: experiment, theory and application. Am Mineral 50:1228–1272

Zen E-an, Hammarstrom JM (1984) Magmatic epidote and its petrological significance. Geology 12:515–518

Self-Similar Pattern of Crystal Growth from Heterogeneous Magmas: 3D Depiction of LA-ICP-MS Data

Ewa Słaby, Michał Śmigielski, Andrzej Domonik and Luiza Galbarczyk-Gasiorowska

Abstract Crystals grown from mixed magmas are characterized by extreme geochemical heterogeneity. The system is self-similar which is reflected in a complex pattern of element distribution in the crystal. New tools are required to show the complexity. 3D depiction (digital concentration-distribution models DC-DMs) combined with fractal statistics is an ideal tool for the identification and description of any subsequent change occurring due to the chaotic processes. LA-ICP-MS analysis gives simultaneous information on the concentration of many elements from the same analysed crystal volume. Thus the data collected are an ideal basis for the calculation of both DC-DMs and fractals. Simultaneous information retrieved by LA-ICP -MS on both compatible and incompatible elements and further data processing allow the determination of the process dynamics in terms of element behavior: antipersistent/persistent, being incorporated according to Henry's Law or beyond it. The multi-method approach can be used for any system showing geochemical variability.

E. Słaby (✉)
Institute of Geological Sciences, Polish Academy of Sciences,
Research Centre in Warsaw, Warsaw, Poland
e-mail: e.slaby@twarda.pan.pl

M. Śmigielski
Department of Geology, Pope John Paul II State School of Higher Education in Biala Podlaska, Biała Podlaska, Poland

A. Domonik
Institute of Hydrogeology and Engineering Geology, University of Warsaw,
Warsaw, Poland

L. Galbarczyk-Gasiorowska
Institute of Geochemistry, Mineralogy and Petrology, University of Warsaw,
Warsaw, Poland

S. Kumar and R. N. Singh (eds.), *Modelling of Magmatic and Allied Processes*,
Society of Earth Scientists Series, DOI: 10.1007/978-3-319-06471-0_7,

1 Introduction

Magma differentiation in an open system is a complex phenomenon. Mixing occurring between two melts showing different compositional and rheological features is triggered by chaotical, mechanical stretching and folding followed by chemical exchange (Barbarin and Didier 1992; Hallot et al. 1994, 1996; Perugini and Poli 2004; Perugini et al. 2002, 2003). Well-mixed and poorly-mixed domains can appear simultaneously close to each other and induce extreme geochemical heterogeneity. The system is self-similar, e.g. an arbitrarily chosen part of the system is similar to itself. The whole has the same pattern as one or more of the parts. This also means that the parts of the system show statistical self-similarity. Self-similarity refers to a fractal. Fractal statistics is the best method to precisely describe the magma mixing-mingling process (Perugini et al. 2003, 2005). As mentioned, the system provides extreme geochemical heterogeneities in the magma volume where crystallization proceeds. It is larger on the micro- than macro- scale, thus it is better reflected in the complexity of a single crystal growth morphology and composition than in the whole rock composition (Perugini et al. 2002; Pietranik and Waight 2008; Pietranik and Koepke 2009; Słaby et al. 2007a, b, 2008). A crystal migrating across a heterogenous environment composed of variably mixed magma domains registers all the details of the change of element concentration along the migration path. Experiments on element mobility within magmas show, that due to different diffusivity of elements, chemical exchange between magma domains progresses differently for individual elements (Perugini et al. 2006, 2008). These differences are time-dependent and will go to full completion with magma blending and homogenisation. However, the migrating crystal is fed by the local environment before blending occurs and its composition preserves all the geochemical complexity of the magma domains. Special tools are needed to retrieve this complexity.

The data presented in this chapter have been retrieved mainly from feldspar crystals collected from the Karkonosze pluton, an igneous body of well-studied mixed origin (Słaby and Martin 2008). The feldspars have been the subject of many detailed investigations (Słaby et al. 2002, 2007a, b, 2008; Słaby and Götze 2004). Many crystals were analyzed along several transects, in constant steps, from margin to margin with the use of LA-ICP-MS. This technique gives simultaneous information on the concentration of many elements from the same analysed crystal volume. The data are ideal basis for solving many problems concerning the complicated process of crystal formation in a dynamic, open system. A new multi-method approach towards the interpretation and 3D depiction of LA-ICP-MS data has allowed us insights into the mechanism of crystal growth in such a system.

The 3D quantification of objects, rocks and minerals, is a matter of increasing interest in geosciences (Jerram and Davidson 2007; Jerram and Higgins 2007). It is usually applied to an analysis of crystal-size and shape distribution (Bozhilov et al. 2003; Gualda 2006; Mock and Jerram 2005 to mention only few). The new approach presented in this paper is directed towards integrating textural and geochemical features observed within a single crystal.

2 Growth Morphology Versus Composition: 3D Depiction of Geochemical Data

The typical growth morphology resulting from crystal migration across poorly- and well- mixed magma domains is observed as zones. Such zoned crystals have been investigated many times with the use of geochemical data collected mostly along single linear traverses (Gagnevin et al. 2005a, b; Ginibre et al. 2002, 2004, 2007; Pietranik and Koepke 2009; Pietranik and Waight 2008; Słaby et al. 2007a, b, 2008, to mention only a few). The depiction of the obtained data consists of a diagram including two variables: the coordinates of the points along the traverse and the corresponding element concentrations. Such a diagram can show a completely random cross-section which does not truly reflect the complexity of the growth process. Multi-dimensional models can be better tools (Słaby et al. 2008, 2011, 2012; Śmigielski et al. 2012) and the ideal data for such models can be collected using LA-ICP-MS.

Early attempts to use LA-ICP-MS data for raster digital distribution models was shown in Słaby et al. (2008) and Woodhead et al. (2008). The data for the 3D surface depiction of the spatial element distribution have been collected at every individual spot (Fig. 1). Operation conditions and data acquisition parameters are given in Słaby et al. (2008). About 70–100 laser impulses were used at one spot. Each pulse (60–120 μm in diameter) gives data collected from a 5 μm thick layer of the investigated crystal. A single measurement is completed in 0.943 s. The next, ablating the same amount of crystal, does not mix with the previous one as the aerosol is removed by the carrier gas within 1 s. Thus the number of laser impulses reflects the duration and at the same time the depth of each LA-ICP-MS analysis, composed of 70–120 individual measurements.

An important question is whether data from each measurement can be used as indicating local heterogeneity in the crystal composition, which in turn reflects the compositional heterogeneity of chaotically advected magma domains to the growing crystal surface. As the whole system is self-affine, fractal statistic should verify the data so obtained. Another verification is an error estimation on a single measurement. Such an estimation is given for Ba by Słaby et al. (2008). At the limit of detection of ca 1 ppm for ^{137}Ba, each ppm corresponds to a signal of ca 100 cps. An analysis lasting 20 ms gives 10,000 counts with a counting noise of 100. This contributes to a relative standard deviation of 1 % from counting statistics alone. Thus, data on trace elements incorporated into the crystal in high concentrations can be used for the depiction presented below and total error is mostly dependend on the quality of external standard.

The data for the raster digital obeyed the number of spots along the analysed traverse considered as the first dimension—X, the timing of the analysis or the depth of the ablation and the concentration of an element corresponding to single ablation within an individual spot considered, respectively, as second—Y and third—Z dimension (Fig. 2). The base prepared in this way has been converted into a grid of nodes for digital concentration-distribution models (DC-DM). Thus

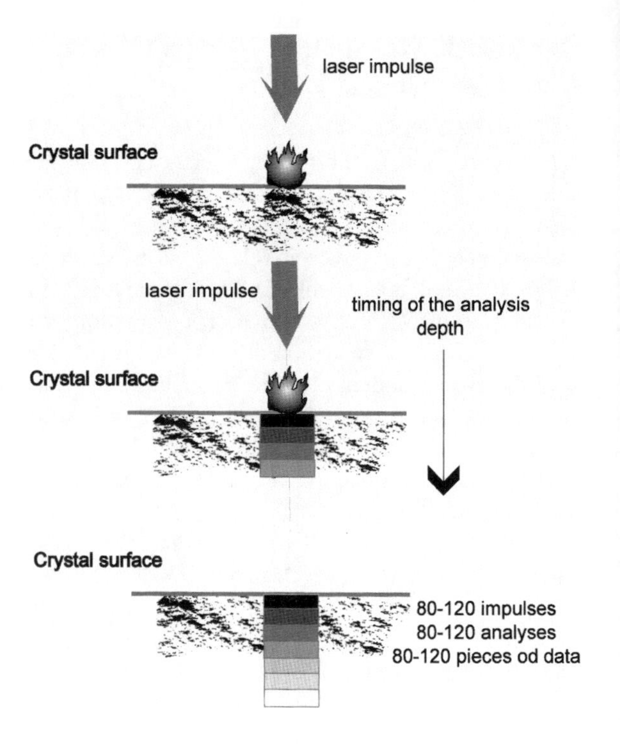

Fig. 1 Schematic diagram outlining the concept of LA-ICP-MS pieces of data collection for spatial depiction

two of the parameters X and Y determined a node position within the grid, located on the investigated crystal transect surface, the third one a concentration of an element for each marked node.

Three methods of interpolation have been used: Inverse Distance to Power (IDP), Kriging and Natural Neighbor (Śmigielski et al. 2012). The IDP interpolation method, called also Shepard's method (Shepard 1968), weights surrounding points (grid nodes) to that estimated according to their Inverse Distance with a user-specified power; power 1.0 means linear interpolation, higher power polygonal estimation. The main disadvantage of the method is that the local extremes are located at grid points, which causes the spatial depiction to have a poor shape (Fig. 3a–b). Kriging assigns weights to each point (grid node) in order to give the smallest possible error of assessment on average (Fig. 3c–d). It uses a measure of the variance in the data as a function of distance e.g. based on variograms. The Natural Neighbor interpolation (Sibson 1981) is based on Voronoi cells of a discrete set of spatial points. In contrast to IDP, it brings a relatively smooth approximation. In this method grid node value can be found by estimating which Voronoi cell from the data set will intersect the Voronoi cell of the interpolated (inserted) point. It means the node value is determined from the weighted average of the neighbor data points (Fig. 3e–f).

From the three interpolations employed, Kriging and Natural Neighbor seem to most reflect most accurately the details of crystal compositional heterogeneity (Fig. 3c, e). Thus, these methods were usually applied to the data. The depiction of

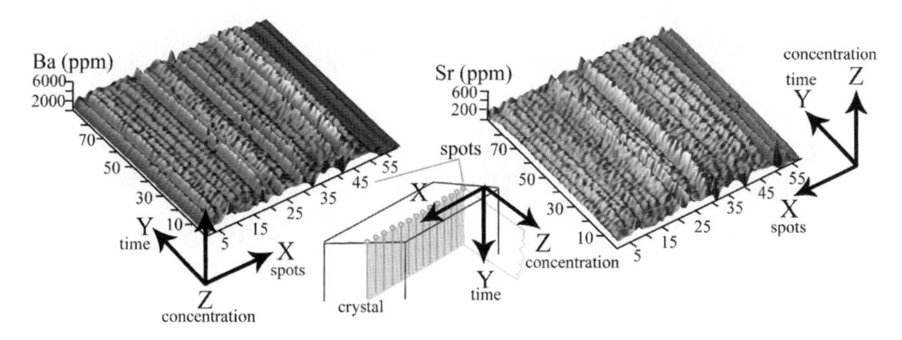

Fig. 2 Digital concentration-distribution models (DC-DM). Barium and strontium distribution within the same inward transect of the feldspar. The schematic crystal between the DC-DMs illustrates the way in which data have been collected

the data as a contour map and a 3D surface model of element distribution involved the use of Surfer 8.0 (Golden Software), classical isoline maps and 3D models merged with shaded-relief images (Yoeli 1965). Depiction of the models was preceded by a spline smooth procedure (Fig. 4) without any recalculation of the model. To further improve the resolution of the data depiction and to best display the spatial distribution of element concentration, the vertical scale on the plots was arbitrarily chosen. A detailed description of all the steps in model preparation can be found in Śmigielski et al. (2012).

Comparing models shown on Figs. 2 and 3 one can easily conclude, that although both crystals were formed in a mixed magma, they present different growth morphologies and compositions. The Ba and Sr spatial distributions in Fig. 2 point to a regular change in element concentration in magma domains advected to the crystal's growth surface. The domains demonstrate contrasting compositions, one being Ba- and Sr rich, the other poor in those elements. Consequently the zones reflecting the magma composition, are either enriched or depleted in both elements. The zones are sharply separated but the variability of the element concentration within a single zone is negligible. It seems that such a steady-state crystal migration between magma domains occurred during a relatively long time. Note that the zonal patterns are not entirely compatible for both elements. Whereas Ba shows enriched zones close to spots 20, 42 and 47, the maximum concentrations for Sr occur close to spots 30, 25 and 22. In parental magmas Ba and Sr are strongly correlated. The lack of a perfect correlation revealed by DC-DM can be used to estimate element mobility during migration across magma domains in relation to other elements.

In turn the Rb DC-DMs of the second crystal (Fig. 3) point to a more dynamic and chaotic process. The differences in element concentration between zones are significantly larger. The distribution of an element within a single zone is very irregular and variable. Even where the pattern is irregular, the growth morphology of the crystal can still be recognised as zoned. Such a pattern is less visible for zones grown from intensively stirred magmas (Figs. 5 and 6). The chaotic,

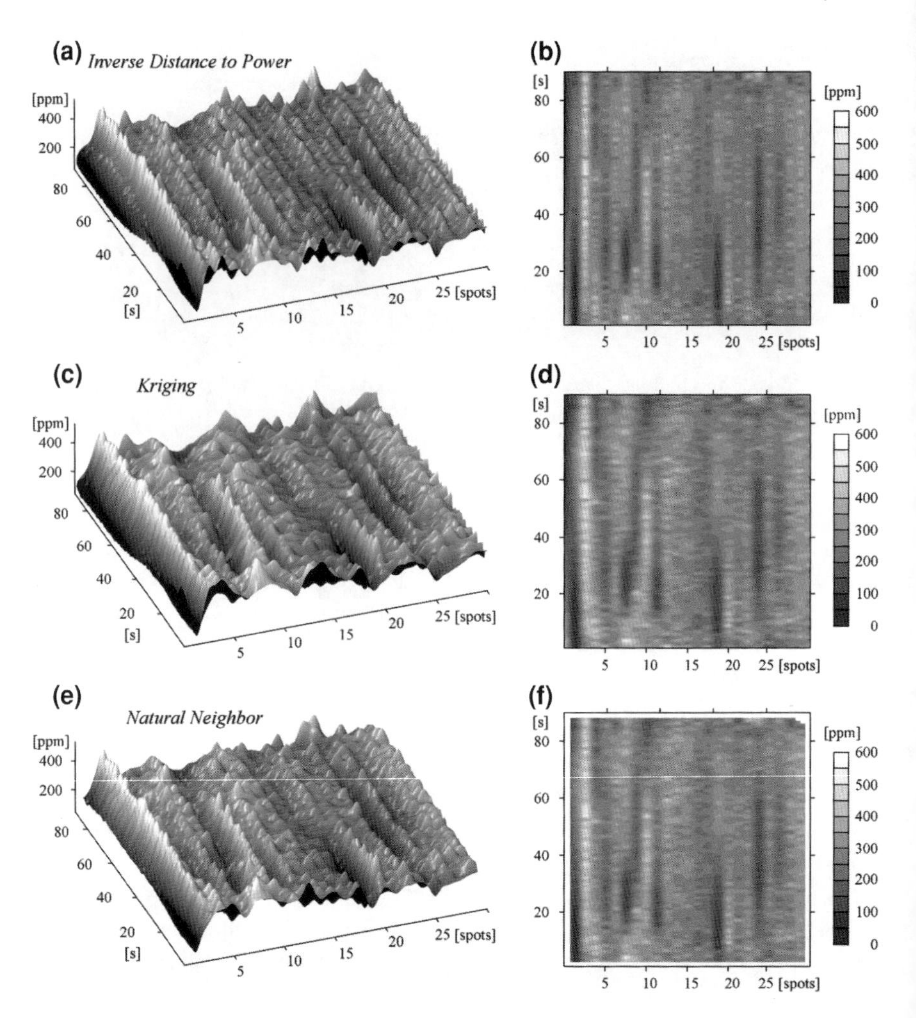

Fig. 3 Comparison of interpolation methods **a-b** Inverse Distance to a Power, **c-d** Kriging, **e-f** Natural Neighbor used for the Rb spatial distribution model. Note that DC-DM constructed with IDP, Kriging and Natural Neighbor demonstrate progressive image smoothing

irregular pattern of Sr distribution visible in the upper part of DC-DM passes into a regular zoned type (Fig. 5a, b). Barium distribution within a crystal grown from intensively stirred magmas splits into numerous domains, chaotically distributed and variably enriched or impoverished in the element. This deterministic chaos is reflected not only in trace element scattering, but also in an increased density of structural defects and in the variability of structural ordering between domains (Słaby et al. 2008).

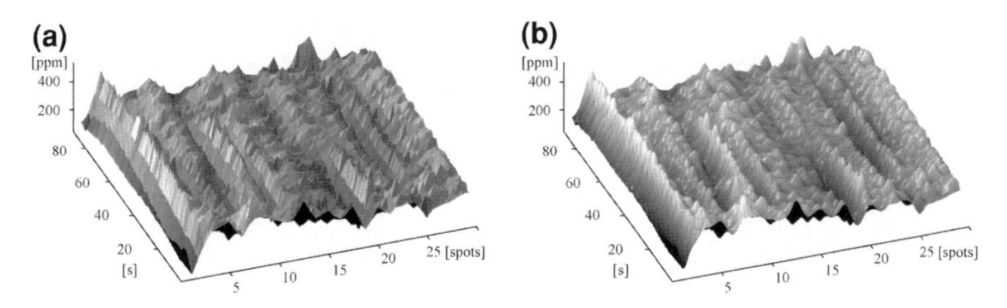

Fig. 4 DC-DMs before (**a**) and after (**b**) the spline smooth procedure. The models of rubidium distribution are shown in basic grid resolutions 49 × 90 spline smoothed to 891 × 481

The element distribution maps and 3D surface models point to a more or less dynamic change of element concentration during incorporation (Fig. 5a–b). Thus, it is valuable to introduce a tool giving information about the dynamics. An analysis of local gradient, e.g. maps of the direction of maximum gradient and the maximum gradient value in element concentration, can provide such information (Fig. 5c–d). The direction of the maximum gradient is an azimuth of a maximum gradient line (e.g. local greatest rate of change in element concentration; Fig. 5d), whereas the maximum gradient value is the dip angle of that line (Fig. 5c). Both the values for a grid node are properties of a tangent plane to a point on an interpolated surface, which quantifies the variability of the element concentration. These derivative maps, especially the map of the direction of the maximum gradient (Fig. 5d), display a better type of growth morphology, paying more attention to areas of larger regularity-irregularity in zoned growth than is visible from spatial concentration distribution maps. The other map, of maximum gradient value, shows places where the process was most dynamic and changeable. On the basis of spatial distribution map (Fig. 5a, b), one can conclude that Sr incorporation is characterized by relatively low variability. Looking at the derivative map (Fig. 5c, d), it is clear that the dynamics of Sr incorporation changed from chaotic to almost steady-state, going back finally to more chaotic.

3 Comparison of Element Behavior

During magma mixing, elements show considerable differences in behavior, including element mobility (Perugini et al. 2006, 2008). The chemical exchange between magma domains proceeds with different speed for different elements (Perugini et al. 2006, 2008) Simultaneous measurements of their concentrations by LA-ICP-MS and further data processing allows us to track such differences (Słaby et al. 2011).

The elements are partitioned into the crystal with different intensity, such that the absolute concentrations are not the best input for the tracking. Normalization is required. The proposed normalization is based on a “cut off value”, a value

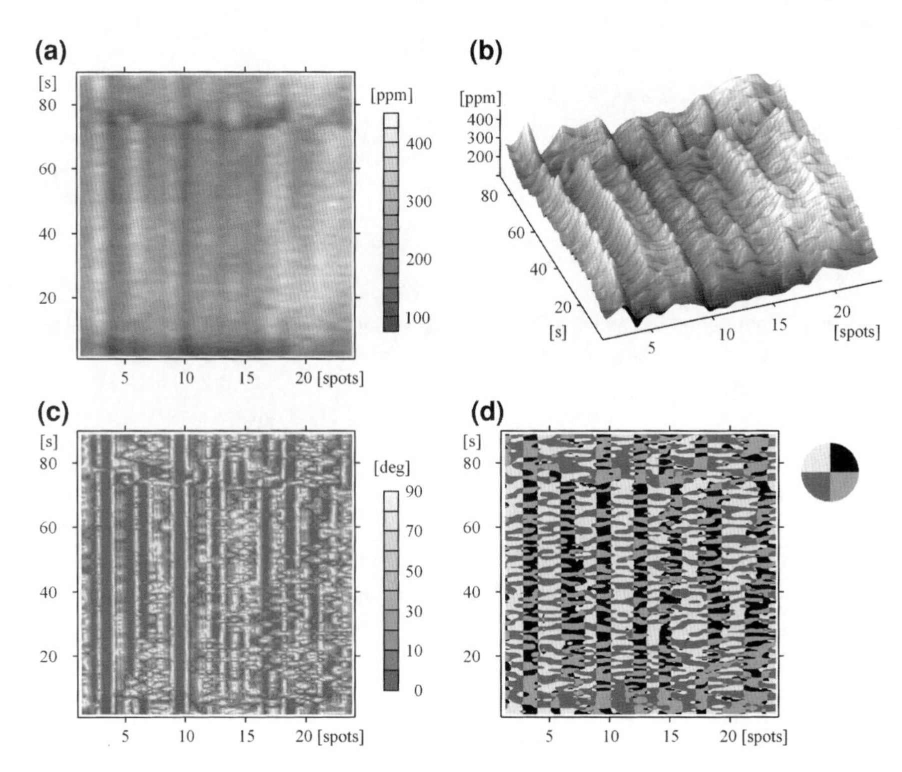

Fig. 5 Strontium DC-DMs (**a**, **b**) as well as maps of the maximum element gradient value (**c**) and the direction of the element maximum gradient (**d**—quarters of the circle show the maximum decrease azimuth). The upper part of the DC-DMs reveals a chaotic pattern of element distribution, which passes into a regular one towards the crystal surface. Regularity and irregularity in the Sr distribution are better expressed as the map of maximum gradient value, where the increasing and changeable degree of gradient value (slope) marks the areas of most intensive change in element concentration. In turn, the map of the direction of maximum gradient splits the whole pattern into three areas, the uppermost and lowest parts point to a more-, and the middle part to a less-, chaotic growth process

separating the local minimum from the neighboring local maximum along each analyzed traverse (Fig. 6). Usually it is the average value of an element over the total area analyzed (Słaby et al. 2011). The "cut off value" rearranges the whole population into a binary system, with two subsets of points (concentration values), one below ("reduced" concentration) and one above ("high" concentration). Such a normalization can be used to determine the degree of spatial accordance/non-accordance in the behaviour of two elements during incorporation into the crystal. Areas where both elements show similar behavior, e.g. their concentrations within an area are reduced or high, can be marked as accordance area. Areas, where a "high" value in the concentration of one element is spatially linked to a "reduced" value of the second will be non-accordance areas. On the basis of the spatial

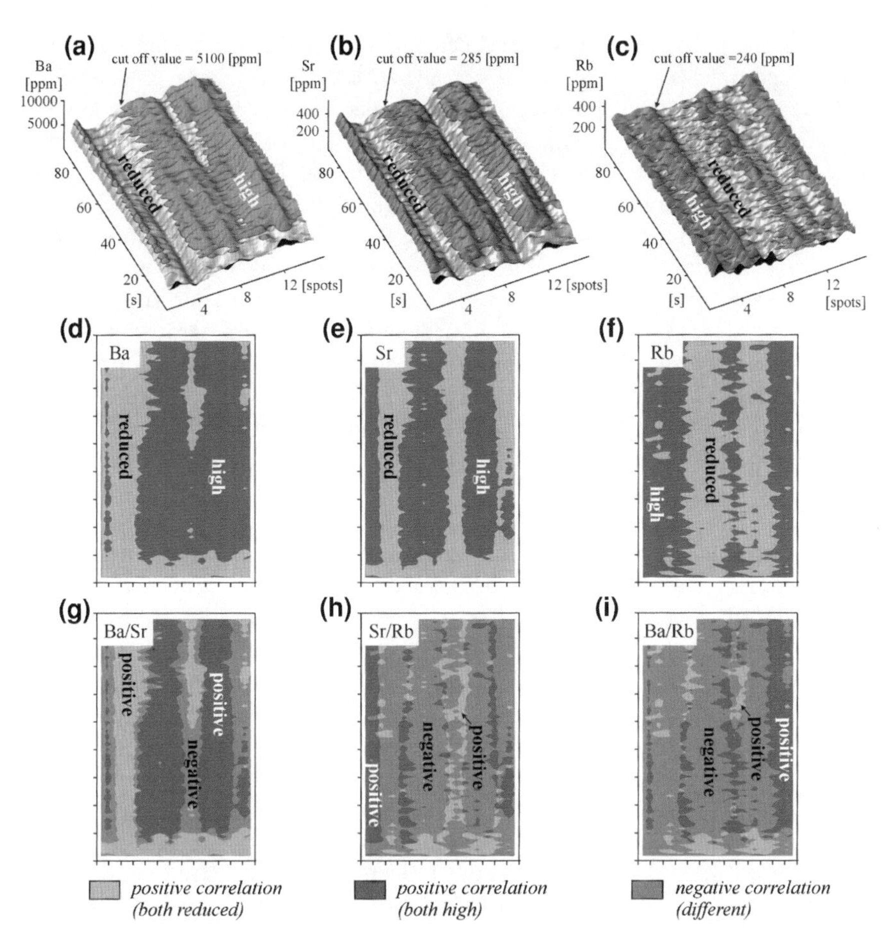

Fig. 6 Comparison of the behavior of two elements during incorporation into the crystal. An average concentration of an element, the cutoff value (**a-c**), sets apart areas of "high" and "reduced" content along the whole surface (**d-f**). DC-DMs of Ba, Sr and Rb (**a-c**) and derivative maps (**g-i**) demonstrating the fields of mutually similar and non-similar element behavior.

distribution (again a binary system) of 'positively' and 'negatively' correlated values, an output concentration map can be constructed (Fig. 6). The map will show areas of mutually accordant and non-accordant behavior.

4 Fractal Statistics

The mutual relationship between elements can possibly be described without any use of fractals (see subsection "Comparison of element behaviour"). Digital models illustrating different element mobility during spreading across magma domains and its effect on crystal geochemistry, do not use fractal statistics. On the other hand, magma mixing can be considered as a deterministic process and be defined by deterministic mathematical formulae. Thus the advantage of using fractals arises from its sensitivity to all subtleties of a stochastic process, better detected by fractal than classical statistical methods. Therefore, fractal statistics can better facilitate the detailed determination of the evolution of the system in time e.g. progressive element incorporation from mixed magmas into the crystal and a quantitative evaluation in term of its dynamics (Domonik et al. 2010).

A self-similarity parameter, the Hurst exponent, can be used to show the long-range dependence (memory effect) of element behavior during growth proceeding from chaotically mixed magma domains (Domonik et al. 2010). As shown above, chaos is reflected in crystal geochemistry. The Hurst exponent was used to describe the pattern of chaos in minerals by Hoskin (2000). It is closely related to the self-affinity concept. It describes processes undergoing scaling, e.g. processes related to affine transformations. The self-similarity, a key feature of fractals preferably used for fractal analysis, is a particular case of self-affinity. The relation between the Hurst exponent (H) and the fractal dimension (D) is described with the simple equation $D = 2 - H$. The exponent itself was calculated by a rescaled range analysis method (R/S) (Peters 1994). The relationship between R/S and H can be expressed as follows: $R/S = an^H$. The Hurst exponent can be estimated using the following regression: $\mathrm{Log(R/S)} = \mathrm{Log}(a) + H(\mathrm{Log}(n))$ and by plotting it, where $Y = \log R/S$ and $X = \log n$ and the exponent H is the slope of the regression line.

The expected values of H lie between 0 and 1. For $H = 0.5$, element behavior shows a random walk and the process produces an uncorrelated white noise. For values greater or less than 0.5, the system shows non-linear dynamics. $H < 0.5$ represents anti-persistent (more chaotic and appropriate for mixing of magmas) behavior, whereas $H > 0.5$ corresponds to increasing persistence (less chaotic).

Crystals grown in mixed magmas show different degree of homogenization of their geochemical composition (Figs. 3, 5 and 6). Those grown in more chaotic, active regions of mingled magmas are strongly zoned. Simultaneously the 3D depiction of element distribution shows a particularly complicated pattern of compositionally heterogeneous domains within each zone, being products of intensive, chaotic magma stirring (Fig. 7, crystal III). The Hurst exponent for such a crystal domain usually ranges from almost zero to 0.5 (for Karkonosze feldspars, $H = 0.06-0.47$), reflecting intensive chemical mixing and the underlying strong non-linear dynamics of the system (Fig. 7, diagrams on the right). Feldspars grown from homogenized mixed magmas are almost homogeneous. The DC-DM of such a crystal (Fig. 7, II) illustrates the "steady-state" migration between mixed magma domains, where the process progresses and tends to completion. Despite homogenization, the fractal statistics reveal that trace elements were incorporated chaotically into the growing crystals. The anti-persistent, chaotic behavior of elements during growth of these feldspars is preserved. For comparison a sector of crystal grown from end-member magma is presented in Fig. 7, I. A relatively small variation in trace element contents can be still recognized due to DC-DM depiction.

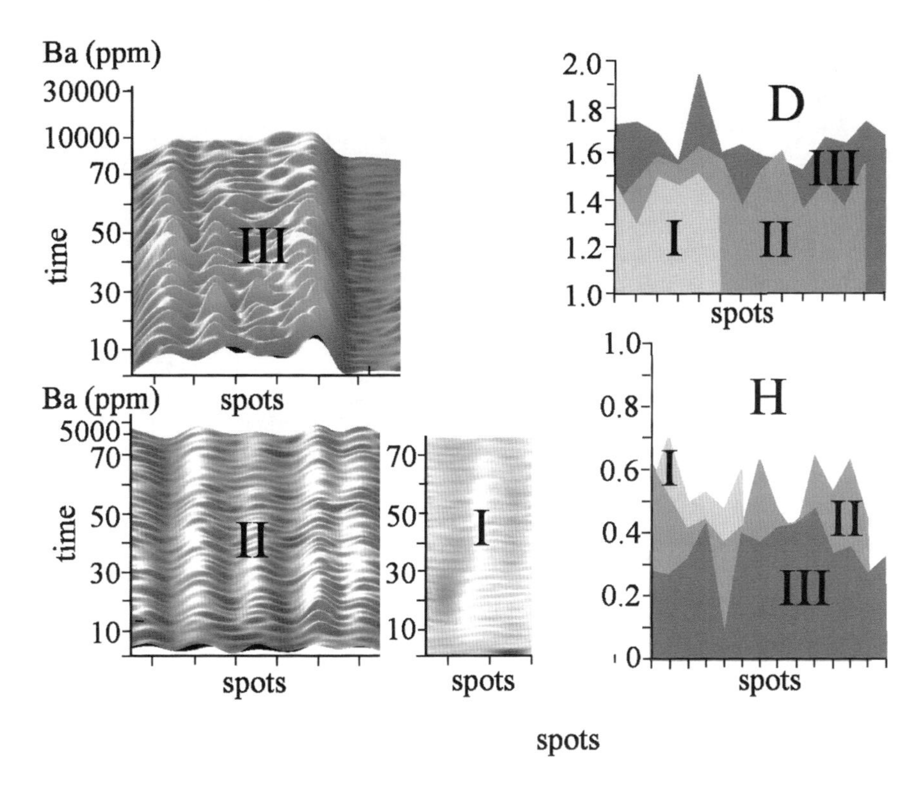

Fig. 7 The relationship between fractal dimension (*D*)—Hurst exponent value (*H*) and DC-DMs of crystal sectors grown from variably mixed magmas. The biggest differences in fractal dimension (lowest *H* values) are observed for crystal sectors grown from intensively stirred magmas (*III*). With a decrease of the process dynamics *D* decreases too. The element incorporation shows a more persistent tendency

Fractal statistics point to persistent element behavior during this phase of crystallization. It appears that 3D depiction and analysis combined with fractal statistics is an ideal tool for the identification of the growth mechanism and any subsequent changes occurring due to chaotic processes in on open system.

5 Equilibrium: Non-Equilibrium Processes

An important problem in the investigation of crystal compositional heterogeneity is the determination of whether crystal growth proceeded close to or far from surface equilibrium. As mentioned earlier, LA-ICP-MS analysis is an ideal tool for collecting data to solve such a problem, since it gives simultaneous information on the concentration of many elements from the same crystal area. The relationship between all the analysed elements can be easily retrieved and considered as

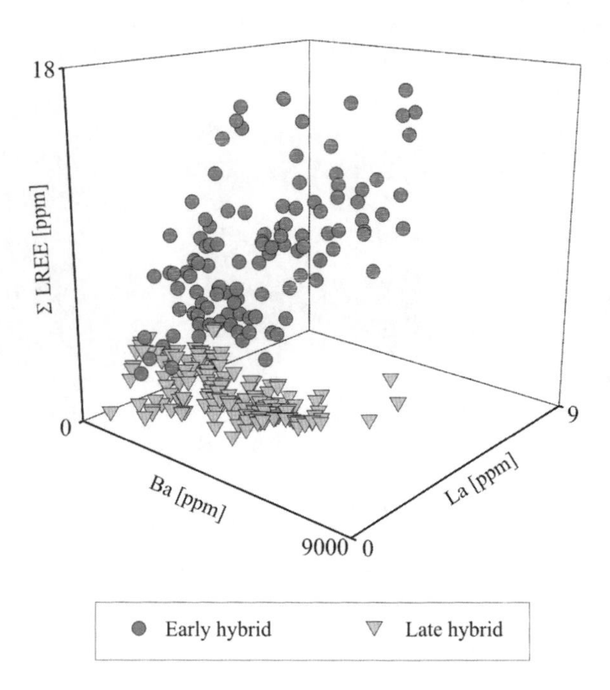

Fig. 8 3D depiction of the relationship between compatible and incompatible elements in crystals grown in early and late hybrids. La ICP MS data collected from alkali feldspar megacrysts.

non-random. On the basis of this relationship it is possible to determine whether the mechanism of element (compatible and incompatible) incorporation is similar or not, e.g. whether it obeyed or disobeyed Henry's Law (Blundy and Wood 1994; Morgan and London 2003; Słaby et al. 2007b, Tilley 1987). Crystallization in a dynamic system can promote conditions of growth far from the attainment of crystal–melt equilibrium. Some investigations have shown that during magma mixing some compatible elements may be supplied to the surface of growing crystals under near-equilibrium (Słaby et al. 2008) or far-from-equilibrium (Gagnevin et al. 2005a) conditions. Increased compositional zonation with different compatible element concentrations can thus be accomplished by changeable undercooling during mixing, without any significant chemical exchange.

Whereas compatible elements can be incorporated in excess into the growing crystal, incompatible elements can continue to show equilibrium behavior. Morgan and London (2003) showed that the incorporation of incompatible elements is not affected much, even by non-equilibrium crystal growth. Consequently, the simulation of the crystallization process in a dynamic system can be done more precisely by simultaneously considering the behavior of both compatible and incompatible elements. Simultaneously collected LA-ICP -MS data on both element populations are well suited to this purpose. Their Cartesian plots or 3D data depiction are again very helpful (Fig. 8). Crystals grown during magma mixing can be formed due to the various types of hybridization occurring at different stages of progressive magma differentiation (Barbarin 2005). As an example, Fig. 8 presents data from feldspars grown in early and late hybrids. They follow two different but consistent trends. The trend for crystals formed in early hybrids shows a distinctly positive correlation, whereas that for late hybrids shows an increasing Ba content and constant LREE concentration. In turn, the whole rock compositions, for early and late hybrids, demonstrate a similar positive correlation between LREE and Ba resulting from mixing between mantle-derived components (Ba-LREE-rich) and crustal (Ba-LREE poor) (Słaby et al. 2007b; Słaby and Martin 2008). The plot provides important information on equilibrium-disequilibrium crystallization conditions during both stages of hybridization. Whereas during the early stage the incorporation of compatible elements conforms more closely to surface equilibrium, during the late stage it does not.

Compatible elements are frequently used in models of feldspar crystallization. Here the LA-ICP-MS data demonstrate that the information on compatible elements from the early stages of hybridization can be used for such models, but not from the later stages. Such a model using LA-ICP -MS data on associated minerals, was published by Słaby et al. (2007b).

6 Conclusions

The proposed new multi-method approach to LA-ICP-MS data processing and depiction gives new insights into the mechanism of crystal growth in geochemically heterogeneous environments. One such environment is a system of interacting magmas; however, any other process leading to compositionally heterogeneous crystal formation can be considered. The 3D depiction combined with fractal analysis allows us to separate crystal sectors grown under different dynamic conditions. Due to different element behavior, the more or less persistent/antipersistent, successive stages of a single process can also be distinguished. Similarly, the methods of data processing and depiction can be successfully applied to the separation of overlapping effects caused by different processes, e.g. magmatic and post-magmatic, if element activities and the mechanism of their incorporation are different. In addition, simultaneous data processing of compatible and incompatible elements may help to determine whether the process being investigated proceeded close to or far from equilibrium.

Acknowledgments The work has been funded by NCN grant 2011/01/B/ST10/04541. We are very grateful to S. Kumar for his invitation to provide our contribution to this book. We greatly appreciate peer reviews by R. MacDonald and A. Pietranik. We extend our sincere appreciation to Ch. Gallacher and R. MacDonald, who in addition corrected grammar and style.

References

Barbarin B (2005) Mafic magmatic enclaves and mafic rocks associated with some granitoids of the central Sierra Nevada batholith, California: nature, origin, and relations with the hosts. Lithos 80:155–177

Barbarin B, Didier J (1992) Genesis and evolution of mafic microgranular enclaves through various types of interaction between coexisting felsic and mafic magmas. Trans R Soc Edinb: Earth Sci 83:145–153

Bozhilov KN, Green HW II, Dobrzhinetskaya LF (2003) Quantitative 3D measurement of ilmenite abundance in Alpe Arami olivine by confocal microscopy: confirmation of high-pressure origin. Am Mineral 88:596–603

Blundy J, Wood B (1994) Prediction of crystal-melt partition coefficients from elastic moduli. Nature 372:452–454

Domonik A, Słaby E, Śmigielski M (2010) Hurst exponent as the tool for description of the magma field heterogeneity reflected in the geochemistry of growing crystal. Acta Geol Pol 60:437–443

Gagnevin D, Daly JS, Poli G, Morgan D (2005a) Microchemical and Sr isotopic investigation of zoned K-feldspar megacrysts: insights into the petrogenesis of a granitic system and disequilibrium crystal growth. J Petrol 46:1689–1724

Gagnevin D, Daly JS, Waight T, Morgan D, Poli G (2005b) Pb isotopic zoning of K-feldspar megacrysts determined by laser ablation multiple-collector ICP-MS: insights into granite petrogenesis. Geochim Cosmochim Acta 69:1899–1915

Ginibre C, Wörner G, Kronz A (2002) Minor-and trace-element zoning in plagioclase: implications for magma chamber processes at Parinacota volcano, northern Chile. Contrib Mineral Petrol 143:300–315

Ginibre C, Wörner G, Kronz A (2004) Structure and dynamics of the Laacher See magma chamber (Eifel, Germany) from major and trace element zoning in sanidine: a cathodoluminescence and electron microprobe study. J Petrol 45:2197–2223

Ginibre C, Wörner G, Kronz A (2007) Crystal zoning as an archive for magma evolution. Elements 3:261–266

Gualda GAR (2006) Crystal size distributions derived from 3D datasets: sample size versus uncertainties. J Petrol 47:1245–1254

Hallot E, Auvray B, de Bremond d'Ars J, Davy P, Martin H (1994) New injection experiments in non-newtonian fluids. Terra Nova 6:274–281

Hallot E, Davy P, de Bremond d'Ars J, Auvray B, Martin H, van Damme H (1996) Non-Newtonian effects during injection in partially crystallised magmas. J Volcanol Geoth Res 71:31–44

Hoskin PWO (2000) Patterns of chaos: fractal statistics and the oscillatory chemistry of zircon. Geochim Cosmochim Acta 64:1905–1923

Jerram DA, Davidson JP (2007) Frontiers in textural and microgeochemical analysis. Elements 3:235–238

Jerram DA, Higgins MD (2007) 3D analysis of rock textures: quantifying igneous microtextures. Elements 3:239–245

Mock A, Jerram DA (2005) Crystal size distributions (CSD) in three dimensions: Insight from 3D reconstruction of a highly porphyritic rhyolite. J Petrol 46:1525–1541

Morgan GB, London D (2003) Trace-element partitioning at conditions far from equilibrium: Ba and Cs distributions between alkali feldspar and undercooled hydrous granitic liquid at 200 MPa. Contrib Mineral Petrol 144:722–738

Perugini D, De Campos CP, Dingwell DB, Petrelli M, Poli D (2008) Trace element mobility during magma mixing: preliminary experimental results. Chem Geol 256:146–157

Perugini D, Petrelli M, Poli G (2006) Diffusive fractionation of trace elements by chaotic mixing of magmas. Earth Planet Sci Lett 243:669–680

Perugini D, Poli G (2004) Analysis and numerical simulation of chaotic advection and chemical diffusion during magma mixing: petrological implications. Lithos 78:43–66
Perugini D, Poli G, Gatta GD (2002) Analysis and simulation of magma mixing processes in 3D. Lithos 65:313–330
Perugini D, Poli G, Mazzuoli R (2003) Chaotic advection, fractals and diffusion during mixing of magmas: evidence from lava flows. J Volcanol Geoth Res 124:255–279
Perugini D, Poli G, Valentini L (2005) Strange attractors in plagioclase oscillatory zoning: petrological implications. Contrib Mineral Petrol 149:482–497
Peters EE (1994) Fractal market analysis: applying Chaos theory to investment and economics. Wiley, New York
Pietranik A, Koepke J (2009) Interactions between dioritic and granodioritic magmas in mingling zones: plagioclase record of mixing, mingling and subsolidus interactions in the Gęsiniec Intrusion, NE Bohemian Massif, SW Poland. Contrib Mineral Petrol 158:17–36
Pietranik A, Waight T (2008) Processes and sources during late Variscan Dioritic-Tonalitic magmatism: insights from Plagioclase chemistry (Gęsiniec Intrusion, NE Bohemian Massif, Poland). J Petrol 49:1619–1645
Shepard D (1968) A two-dimensional interpolation function for irregularly-spaced data. In: Proceedings of the 1968 ACM national conference 517–524
Sibson R (1981) A brief description of natural Neighbor interpolation. In: Barnett V (ed) Interpreting multivariate data. Wiley, New York
Słaby E, Galbarczyk-Gąsiorowska L, Baszkiewicz A (2002) Mantled alkali-feldspar megacrysts from the marginal part of the Karkonosze granitoid massif (SW Poland). Acta Geol Pol 52:501–519
Słaby E, Galbarczyk-Gąsiorowska L, Seltmann R, Műller A (2007a) Alkali feldspar megacryst growth: geochemical modelling. Mineral Petrol 68:1–29
Słaby E, Seltmann R, Kober B, Műller A, Galbarczyk-Gąsiorowska L, Jeffries T (2007b) LREE distribution patterns in zoned alkali feldspar megacrysts—implication for parental melt composition. Mineral Mag 71:193–217
Słaby E, Götze J, Wörner G, Simon K, Wrzalik R, Śmigielski M (2008) K-feldspar phenocrysts in microgranular magmatic enclaves: A cathodoluminescence and geochemical study of crystal growth as a marker of magma mingling dynamics. Lithos 105:85–97
Słaby E, Śmigielski M, Domonik A, Simon K, Kronz A (2011) Chaotic three-dimensional distribution of Ba, Rb and Sr in feldspar megacrysts grown in an open magmatic system. Contrib Mineral Petrol 162:909–927
Słaby E, Martin H, Hamada M, Śmigielski M, Domonik A, Götze J, Hoefs J, Hałas S, Simon K, Devidal J-L, Moyen J-F, Jayananda M (2012) Evidence in Archaean alkali-feldspar megacrysts for high-temperature interaction with mantle fluids. J Petrol 53:67–98
Słaby E, Götze J (2004) Feldspar crystallization under magma-mixing conditions shown by cathodoluminescence and geochemical modelling—a case study from the Karkonosze pluton (SW Poland). Mineral Mag 64:541–557
Słaby E, Martin H (2008) Mafic and felsic magma interactions in granites: the Hercynian Karkonosze pluton (Sudetes, Bohemian Massif). J Petrol 49:353–391
Śmigielski M, Słaby E, Domonik A (2012) Digital concentration distribution models—tools for a description of the heterogeneity of the magmatic field as reflected in the geochemistry of a growing crystal. Acta Geol Pol 62:129–141
Tilley RJD (1987) Defect crystal chemistry. Blackie, London
Woodhead I, Hellstrom J, Paton C, Hergt J, Greig A, Maas R (2008) A guide to deep profiling and imaging application of LA ICP MS. In P.Sylvester (Ed.) Laser Ablation-ICP-MS in Earth Sciences; Current practices and outstanding issues. Mineralogical Association of Canada. Short Course Series, Vol. 40, 135–146
Yoeli P (1965) Analytical hill shading. Surveying and Mapp 25:573–579

Microanalytical Characterization and Application in Magmatic Rocks

Naresh C. Pant

Abstract Chemical characterization of magmatic rocks is a primary requirement in interpreting their evolutionary history. Magmatic rocks are nearly always heterogeneous and thus require micro-scale chemical characterization. Two of the main micro-characterization techniques are Scanning Electron Microscopy (SEM) and Electron Probe Micro Analysis (EPMA). Both are near surface characterization techniques and utilize the effects of interaction of an electron beam with the targeted sample. Back Scattered Electrons (BSE) represent an atomic number dependent elastic scattering effect which provides high resolution petrographic information of heterogeneities while generation of characteristic X-rays from inner shell energy-level transitions of different atoms, a type of inelastic scattering effect, provides quantitative chemical characterization at micron scale. Both are highly useful for describing magmatic rocks as well as inferring the operative magmatic processes.

1 Introduction

Chemical characterization of magmatic rocks is an important parameter in understanding and modelling magmatic processes. Bulk rock characterization is commonly carried out to describe the attributes of parent magma. In many ways it is an artificial homogenization of an essentially heterogeneous product. The evolution of a magmatic rock is a complex process and often the bulk characterization attributes oversimplify this process. Magmatic rocks generally form by partial melting of silicates mainly in the mantle. The crystallization of these melts is often in the magma reservoirs within the crust. The phenocrysts form, enlarge

N. C. Pant (✉)
Department of Geology, University of Delhi, Delhi, India
e-mail: pantnc@gmail.com

S. Kumar and R. N. Singh (eds.), *Modelling of Magmatic and Allied Processes*,
Society of Earth Scientists Series, DOI: 10.1007/978-3-319-06471-0_8,

and develop chemical (and resultant optical) zoning as a consequence of crystallization conditions which, in turn, are a function of pressure, temperature, volatile content, successive magma fluxing, assimilation of host rocks and even the tectonic events. Thus, the complex zoning patterns of phenocrysts contain the records of these processes. A magmatic process can be considered in terms of a set of intensive and extensive variables. Some of the intensive variables can be listed as (Blundy and Cashman 2008):

1. Temperature and pressure of magma.
2. The concentration, distribution and migration of volatile species.
3. The extent, distribution and timing of crystallization and degassing.
4. Experimental simulation of sub-volcanic systems.

Working out these requires micron-scale chemical characterization of heterogeneities e.g. the temperature and pressure from the chemical analysis of co-existing phases or from glass, volatile species from the microanalysis of glass, combined texture, mineral chemistry and in situ geochronology for extent distribution and timing of crystallization and analysis of reactants and products for experimental simulation studies. It may be remembered that the classical concept of a simple gravity induced sinking and floating of phenocrysts is inadequate in view of well-documented magma recharge events in many magma chambers.

Development and use of microbeam techniques since ~1960 has been responsible for most of such high-resolution studies of the magmatic rocks. Analysis of both major and trace elements in combination with high resolution elemental mapping has led to the characterization and quantification of magmatic processes to a great extent. The present article details the basics of one of the microbeam technique, namely Electron Probe Micro Analysis (EPMA) in the context of major and trace element quantification and elemental mapping and discusses its application in deciphering some of the magmatic processes.

2 Scanning Electron Microscope and Electron Probe Micro Analysis

In both, SEM and EPMA, the probe used is an electron beam whose interaction with the target matter (rocks and minerals in the case of magmatic rocks) produces several effects such as;

- Secondary electrons (SE)
- Backscattered electrons (BSE)
- Cathodoluminescence (CL)
- Continuum X-ray radiation (bremsstrahlung)
- Characteristic X-ray radiation
- Phonons (heat).

Though the two techniques are considered surface characterization techniques, the beam-matter interaction occurs in a volume which depends on the extent of penetration of electron beam. The penetration depth is dependent upon interaction or scattering of electron in a particular cross sectional area which is generally of the order of 10^{-4} pm^2. The scattering is either elastic (change of trajectory but kinetic energy and velocity nearly unchanged; e.g. BSE) or inelastic (no change of trajectory possible but loss of energy; e.g. SE, bremsstrahlung) and on account of presence of large number of atoms even in a small cross sectional area, the probability of scattering is unity and an electron may be scattered many times. The depth up to which an electron will travel in the matter is dependent on its energy (or accelerating voltage used for generating the beam), the atomic number (Z) or the average atomic number of the matter, the magnitude of the current and the angle of incidence of the electron beam. Kanaya and Okayama (1972) provided a theoretical expression for the range (r) up to which an electron can travel before coming to rest.

$$r(\mu m) = \frac{2.76 \times 10^{-2} A E_0^{1.67}}{\rho Z^{0.89}} \tag{1}$$

where, A = average atomic weight (g/mol); E_0 = accelerating voltage (keV); Z = average atomic number; ρ = density (g/cm^3).

For a vertically incident electron beam the depth of penetration (x) is generally around 1–5 μm for geological material and it can be expressed in terms of the following empirical expression (Potts 1987; Potts et al. 1995):

$$x(\mu m) = \frac{0.1 E_0^{1.5}}{\rho} \tag{2}$$

In context of the magmatic rocks, backscattering of electrons is of great use as a petrological tool as backscattering coefficient (η = proportion of incident electrons to the backscattered electrons) is a function of atomic number of the matter. In case of minerals or compounds an average atomic number can be computed using the following equation;

$$\mathrm{Av} \cdot \mathrm{Z} = \sum \mathrm{ZiCi} \tag{3}$$

where Zi is the atomic number of element i and Ci is the mass concentration of element i.

The back scattered electron (BSE) images are very useful in bringing out not only the inter-grain compositional contrast but also the intra-grain compositional variation as illustrated in Figs. 1 and 2. In Fig. 1, monazite of two stage growth is observed where large part of the nearly homogeneous metamorphic monazite grew around a preexisting euhedral monazite of possibly magmatic parentage. There was another overgrowth represented by a thin, bright and incomplete rim and the contrasting grey tone of the rim is on account of higher Th content. In Fig. 2,

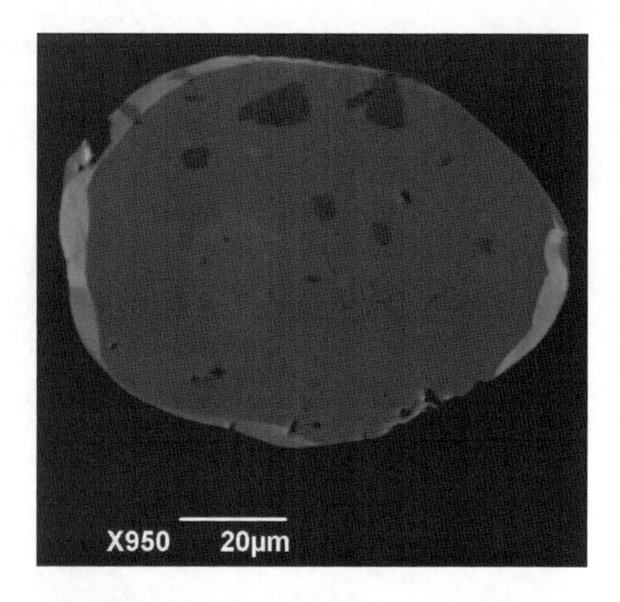

Fig. 1 BSE image of monazite. Note a euhedral magmatic monazite in the core and two overgrowth, possibly metamorphic, the last being in form of a thin Th-rich rim

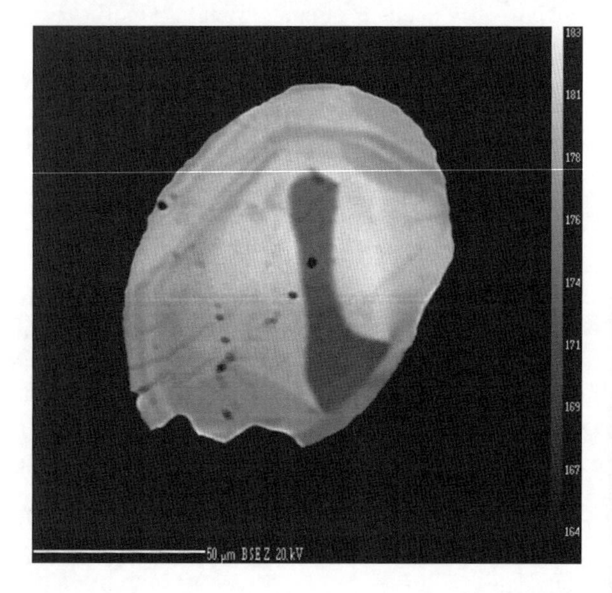

Fig. 2 BSE image of oscillatory zoned magmatic monazite over an anhedral, Th-poor monazite in the core (Bhandari et al. 2011)

oscillatory zoned magmatic growth of monazite has taken around anhedral, low Th pre-existing core.

The most important result of the electron beam-matter interaction in context of chemical characterization of matter is in the generation of characteristic X-rays as well as continuum X-ray radiation from a volume of matter. The X-rays produced are in the wave length range of 0.1–100 Å out of which longer wave length ($\lambda \sim$ 50–100 Å) are known as 'soft' X-rays and the shorter wave length are called as 'hard' X-rays as λ is inversely proportional to the energy of the electromagnetic spectrum as illustrated below.

The relationship between the energy of any part of electromagnetic spectrum can be expressed in form of Eq. 4

$$E = h\nu \tag{4}$$

where h is the Planck's constant and ν is the frequency which for electromagnetic radiations can be related to wavelength by

$$\nu = c/\lambda \tag{5}$$

where c = speed of light and λ is the wavelength in m.

After substituting Eq. 5 in Eq. 4 we get

$$E = hc/\lambda \tag{6}$$

And after substituting h = 6.6260×10^{-34} J s and c = 2.9978×10^{8} m/s we get.

$$E\,(\text{in joules}) = 1.98636 \times 10^{-25}/\lambda \tag{7}$$

And in units of nanometers and electron volts (1 m = 10^9 nm and 1 eV = 1.6021×10^{-19} J) Eq. 7 is reduced to Duane-Hunt equation of relationship between energy and wavelength

$$E\,(\text{in eV}) = 1239.84/\lambda\,(\text{in nm}) \tag{8}$$

Characteristic X-rays can either be measured as energies or can be detected in terms of their wavelengths.

A small fraction of the incident electron beam leads to ejection of inner-shell electrons from their orbital, thus, ionizing it for a small time period before the vacancy is filled by an outer-shell electron. This requires the energy of the beam electrons to be higher than the energy of the electron to be ejected and the threshold above which such displacement is possible is called as critical ionization potential (E_c) or excitation potential. Substitution of an L-shell electron in K-shell produces Kα family of X-rays, M-shell to K-shell transition leads to Kβ X-rays, M-shell to L-shell transition generates Lα X-rays and N-shell to M-shell transition produces Mα family of X-rays. All the transitions and resultant X-rays are shown in Fig. 3. The Ec is always higher than the energy of the X-ray it produces. The Ec associated with Kα family of X-rays for an element is much higher than for L or M-series X-rays (e.g. Ec for Fe-Kα is 7.11 keV; whereas, Ec for Fe-Lα is only 0.71 keV). Generally K-family X-rays are used for lower ($<\sim30$) atomic number elements, L- for intermediate ($\sim30-70$) and M- for higher (>70) atomic number elements.

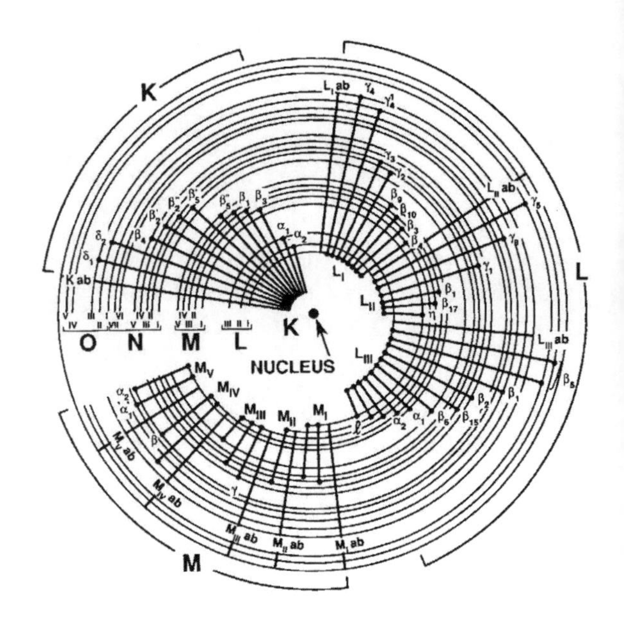

Fig. 3 Diagram illustrating energy level transitions and resultant characteristic X-rays. (Originally Woldseth 1973, reprinted in Goldstein et al. 1981, p. 125)

Measurement of any characteristic X-ray of an element indicates the quantum of that element but this measurement also includes the continuum or bremsstrahlung X-rays and thus for proper quantification we need to understand and quantify these X-rays.

Electrons of incident electron beam interact with the electromagnetic field of nuclei of atoms in the interaction volume and are slowed to different degrees from 0 to complete stoppage. This results in X-rays which have continuous range for all the wavelengths (or energies) with the highest energy (and thus the shortest wavelength) equivalent to the accelerating voltage of the incident beam which in terms of Eq. 6 can be expressed as;

$$\text{Limiting}\ \lambda = hc/E_0 \quad (9)$$

This constitutes the background X-ray counts and its intensity is required to be subtracted from the characteristic X-ray measurements for accurate chemical estimation. The maximum intensity of the continuum occurs at ~ 1.5 times the limiting λ value (Eq. 9). Dependence of the intensity of continuum X-rays on a specific energy, E, can be described by Kramer's equation;

$$I_c = i_b Z(E_0 - E)/E \quad (10)$$

where i_b is the beam current, Z is the atomic number and E_0 is the accelerating voltage. I_c will be 0 at E projected to exponentially increase to ∞ at $E = 0$ though in the observed spectrum, the longer wavelength (low E), being the 'soft' X-rays are absorbed and actually the continuum or 'background' decreases towards lower energy and, thus, behaves in a non-linear manner. In EPMA continuum X-rays are measured on both sides of the concerned characteristic peak and the background subtracted considering a linear behavior. The proportion of continuum X-rays can be expressed by the following equation.

$$p = \left(1.1 \times 10^{-6}\right)\overline{Z}E_o \tag{11}$$

where p represents the proportion of continuum X-rays, bar Z the average atomic number and E_o is the acceleration voltage. Since it represents a very small fraction ($\sim$0.05 %) the non-linearity of the background is insignificant for major element quantification but it assumes significance in trace element measurements.

The relationship between the wavelength of the characteristic X-ray (λ) and atomic number (Z) was empirically established by Moseley in 1914 which can be expresses as;

$$\lambda = K/(Z - \sigma)^2 \tag{12}$$

where K and σ are constants for that characteristic X-ray. The latter (σ) is known as shielding constant and is approximately 1 for K lines and 7.4 for more shielded L-lines.

Overvoltage, which is the ratio between accelerating voltage, E_o, and critical ionization potential, E_c, for an X-ray series controls the capture cross section for each shell. The probability of ionization is a function of target versus ionization cross section generally expressed as Q and the rate of ionization along with the probability of electron transition controls the intensity of X-ray lines (I) and it can be expressed in form of following equation:

$$I = Ci_b(E_o - E_c)^p \tag{13}$$

where i_b is beam current, E_o is accelerating voltage, E_c is critical ionization potential for X-ray under consideration and C is a constant. Value of $p = 1.7$ for $E_o < 1.7E_c$. As is evident from Eq. 13, an increase in beam intensity is possible by either increasing the beam current or by optimizing the overvoltage.

To avoid effects of peak shifts in EPMA analyses use of oxide standards for analysis of magmatic material is made as these are made up of oxide minerals.

3 Characteristic X-ray Intensities, Corrections and Quantification

Characteristic X-ray intensities (generally in units of counts per second or cps) need to be detected, measured and quantified. In situ non-destructive detection and measurement of geological material on micron scale is carried out using electron microprobe. X-rays emerging at some specific angle (known as take-off angle) are diffracted through appropriate crystals (artificial and of specific d-spacing) following Bragg's law ($n\lambda = 2d \sin \theta$) to a gas-proportional flow counter (calibration gas is P-10 which is a mixture of 90 % Argon and 10 % methane) type detector which has a central metal wire acting as anode with a positive voltage (bias) ranging from 1,200–2,000 V. The specimen, the diffracting crystal and the receiving screen of the detector are on the periphery of a circle known as Rowland Circle so that all three components are in focus throughout. The diffracting crystal is curved to fit either in Johann symmetry (curved crystal with the curvature to match diameter of the Rowland Circle) or in Johanson symmetry (the inner surface of the curved crystal ground to the radius of the Rowland Circle) for precise in-focus measurement of X-rays. Since, $\sin \theta$ varies from 0.2 to 0.8, and λ varies over a wide range from hundreds to fractions of an Å it is not possible to diffract all the X-rays using one crystal and, therefore, several crystals are normally required for analysis of geological material.

Quantification requires measurement of characteristic X-rays at its wavelength position and measurement of continuum X-rays (background) at a suitable distance from peak position and than background X-ray counts require to be modeled to that below the peak position for deduction from the peak counts. X-ray counts at peak or background positions can also be influenced by interference from other peaks which may be present in the matrix of analyzed volume of mineral. Such influences, if present, are required to be corrected from the raw X-ray counts. A detailed treatment of corrections is not within the scope of this chapter but the principal aspects are given in the following lines.

Prior to measurement of composition of an element, the characteristic X-rays for that element are measured on standard. The standard for EPMA analysis is a grain/piece of homogeneous composition which has been measured by an independent accurate determination. For geological material, commonly, natural mineral standards are used. Since characteristic X-rays are counted in the same instrument under same operating conditions on both the standard and the unknown, the ratio of two expressed as 'K' ratio, it is assumed that many physical parameters which may be necessary to compute errors in a rigorous physical model, tend to cancel out each other.

$$K_i = I^{un} / I^{sd} \tag{14}$$

where i refers to the element being measured and I^{un} and I^{sd} refers to X-ray counts of unknown and standard respectively. Following this Castaing in his Ph.D. thesis in 1957 proposed the following equation for the first approximation of composition of element I in the unknown.

$$C^{un} = K_i C_i^{sd} \tag{15}$$

Correction to this first approximation of the composition of unknown are required as the matrix (other elements present in association with the unknown element being measured) affects the quantum of characteristic X-rays of the concerned element. The effect can be assessed in terms of a sum of effects induced by the atomic number of the matrix elements (Z), absorption of characteristic X-rays of i by these elements (A) and amount of fluorescence (F) effect of the matrix on i. It also implies that an incomplete definition of the matrix will also lead to errors in quantification. Atomic number correction resolvable in terms of mutually opposite stopping power correction and back scatter correction, absorption correction (using mass absorption coefficients) and fluorescence correction (F) in terms of Eq. 16 are automatically incorporated in modern electron microprobes.

$$F = \frac{1}{1 + \sum \frac{I_f}{I_p}} \tag{16}$$

where I_f/I_p is the ratio of X-ray intensity from fluorescence to the X-ray intensity from inner shell ionization.

4 Application of Microanalysis in Magmatic Rocks

4.1 Geothermobarometry

Large magmatic bodies, especially plutons, have a complex crystallization history. Their evolution in Pressure-Temperature-Time (P-T-t) space is necessary to document to understand the magmatic processes responsible for their generation. Thus, estimation of P-T of crystallization assumes significance. Microanalysis is invaluable is this endeavor as illustrated in some applications below.

In many plutonic rocks conventionally pressure (and temperature) estimates of emplacements were recovered from the contact metamorphic assemblages. However, several new formulations, which utilize lower detection limits now achievable using EPMA, can be used in plutonic rocks. The Ti-in quartz thermometer is an illustration in this context for which following experimental relation was worked out by Wark and Watson (2006).

$$\log \mathrm{Ti(ppm)qtz} = 5.69 - [3,765/\mathrm{T(K)}] \tag{17}$$

and with ~20 ppm as the lower detection limits the temperature of the magma can be worked out with reasonable accuracy if there is an accurate and precise determination of trace elements. In a manner similar to above, Ti in zircon has also been used as a sensitive temperature indicator (Watson et al. 2006) by following empirical-cum-experimental relation.

$$\log \mathrm{Ti(ppm)zircon} = 6.01 - [5080/\mathrm{T(K)}] \tag{18}$$

In above formulations, the low concentration of Ti can be measured accurately using an ICP-MS but the advantage of EPMA is having a measurement radius of one tenth to one twentieth of the minimum area analyzed by ICP-MS.

Use of Al-content of hornblende, estimated through electron microprobe, was one of the early geobarometer to demonstrate the depth of emplacement of post-orogenic granites and for estimation of rate of their uplift (Zen and Hammarstorm 1984; Hammarstorm and Zen 1986; Zen 1989). Considering Al to be sum of $Al^{iv} + Al^{vi}$ on 13(O) basis, Anderson and Smith (1995) recalibrated the barometer including the effects of temperature in form of Eq. 19.

$$P\,(\text{kbar}) = 476\text{Al} - 3.1 - \left[\frac{(T^{o}C - 675)}{85}\right] \times [0.530\text{Al} - 0.005924(T^{o}C - 675)] \tag{19}$$

The above has been used successfully in conjunction with hornblende-plagioclase thermometer (Blundy and Holland 1990; Holland and Blundy 1994) for estimating P-T of emplacements of several Palaeozoic plutons (Anderson 1996).

In the garnet bearing per aluminous granites and considering that the grossular content of the garnet increases on the expanse of anorthite content of plagioclases in granite, pressure estimates can be arrived at using the formulation of Wu et al. (2004a, b).

In the continental margin batholiths the magnetite and ilmenite series classification of granites (Ishihara 1977) which explores granites in terms of oxidation-reduction context is of special use (Frost et al. 2001). The reaction quartz + spinel = ilmenite + fayalite was used in QUILF program (Anderson et al. 1993) to provide temperature estimates using oxygen fugacity-temperature relations. This was expanded to include sphene (Xirouchakis et al. 2001) and has been applied to several magmatic plutons.

In basic and ultrabasic rocks two pyroxene thermometry (Wells 1977; Brey and Kohler 1990) and Al-in pyroxene thermometry (Pattison et al. 2003) is commonly applied using EPMA analysis and provides reliable results.

Details of the geothermobarometry of the plutonic and volcanic rocks have been reviewed by Anderson et al. (2008) and Putirka (2008).

4.2 Time Scales of Magmatic Processes

Microanalysis using EPMA can be of use in putting constraints on time scales of magmatic processes especially in bodies older than ~ 100 Ma using chemical dating technique. A protocol for chemical dating tested using externally dated standards has been described earlier (Pant et al. 2009) and a brief theoretical background is given below.

The chemical dating of a mineral assumes that (i) the radioactive decay of an element took place in a closed system and (ii) that all Pb is radiogenic. Thus, in case of monazite the decay of Th and U can be considered to occur directly to the various isotopes of Pb.

In a mineral which contains measurable amount of U and Th (radioactive parent element) and Pb (daughter element), the total Pb is equal to the sum of three isotopes of Pb i.e. ^{208}Pb, ^{207}Pb and ^{206}Pb resulting from decay of ^{232}Th, ^{235}U and ^{238}U respectively. The time t required for formation of certain number of Pb isotopes is related to the current abundance of Th and U radiogenic nuclides present in a radioactive mineral by the equations:

$$^{208}\mathrm{Pb} = ^{232}\mathrm{Th}\{\exp(\lambda_{232}t) - 1\} \tag{20}$$

$$^{207}\mathrm{Pb} = ^{235}\mathrm{Th}\{\exp(\lambda_{235}t) - 1\} \tag{21}$$

$$^{206}\mathrm{Pb} = ^{238}\mathrm{U}\{\exp(\lambda_{238}t) - 1\} \tag{22}$$

where λ is the decay constants for each parent; $\lambda_{232} = 4.9475 \times 10^{-11}$/year, $\lambda_{235} = 9.8485 \times 10^{-10}$/year and $\lambda_{238} = 1.551255 \times 10^{-10}$/year (Steiger and Jäger 1977).

Assuming initial Pb to be negligible in monazite (Parrish 1990) and $^{238}U/^{235}U = 137.88$, the Eqs. 1–3 can be combined to derive the age of the mineral, t required to accumulate Pb (in wt%) present in the mineral by the equation (Suzuki and Kato 2008):

$$\begin{aligned}\mathrm{PbO}/W_{\mathrm{Pb}} = {} & \mathrm{ThO_2}/W_{\mathrm{Th}}\{\exp(\lambda_{232}t) - 1\} + (\mathrm{UO_2}/W_{\mathrm{U}}) \times [\{\exp(\lambda_{235}t) \\ & + 137.8\exp(\lambda_{238}t)\}/138.88 - 1]\end{aligned} \tag{23}$$

where W refers to gram molecular weights of each oxides (for Th-rich minerals W_{Pb} is 224, the daughter product being ^{208}Pb while for U-rich minerals it is 222 as the daughter product is ^{206}Pb). The above equation has to be solved iteratively for t.

The accuracy of chemical dating greatly depends upon the detection limit of elements which are crucial in age estimation. Detection limit refers to X-ray counts of an element close to the estimated background counts. If X-ray counts at peak position of an element is C_{pk} and counts at background is C_{bk} than detection limit will refer to the lowest C_{pk} (which is above C_{bk}) at a certain confidence level. For major elements the counts are reported at 1σ (or 0.683 probability) which is not sufficient for low concentrations. Corrected counts and their standard deviations can be written as;

$$C_{corr} = C_{pk} - C_{bk} \tag{24}$$

$$\sigma_{pk} = \sqrt{C_{pk}} \quad \text{and} \quad \sigma_{bk} = \sqrt{C_{bk}} \tag{25}$$

Error associated with the corrected counts are:

$$\sigma_{corr} = \sqrt{\sigma_{pk}^2 + \sigma_{pk}^2} \tag{26}$$

or

$$\sigma_{corr} = \sqrt{C_{pk} + C_{bk}} \tag{27}$$

For detection an element will have to have X-ray counts at peak position higher by some factor (e.g. m) from the background.

$$\text{Det. Limit}\, C_{pk} - C_{bk} > m\sigma_{corr} \tag{28}$$

This factor m is 2.326 for 99.73 % confidence level (3σ) and 1.96 for 95 % confidence level for large data set. X-ray counts are directly proportional to rate and time of counts. The detection limit can nearly be halved by increasing the counting time by a factor of four. However, there are other issues including damage to the sample which may affect the counts and thus the time can not be indiscriminately increased.

4.3 Reconstruction of Magmatic System Processes

Melt Inclusions: Melt inclusions are small quantities of melt included in crystals and separated from main magma. These are especially important in basaltic melt as these indicate the conditions of mantle from where this magma originate and provide important information on magma generation (e.g. mantle heterogeneities), evolution, transport etc. Since melt inclusions are trapped at pressures significantly higher than the erupting magma, these provide information on high P-volatile composition of the magma (e.g. Blundy and Cashman 2008). On the basis of melt compositions indicated by the melt inclusions, diversity of basaltic magma appears to be much more than seen in the erupted volcanic rocks (Sobolev 1996; Slater et al. 2001). EPMA is the most common technique to determine the chemical composition of the melt inclusions (Nielsen et al. 1998; Davis et al. 2003; Straub and Layne 2003). A review of time-scales of magmatic processes which largely utilizes the chemical data of micro-domains and utilizes the heterogeneities of the magmatic rocks is given by Costa et al. (2008).

Behaviour of magma in the context of open system process requires mineral data on micron scale and zoning of magmatic minerals characterized using EPMA provides a critical input (Jerram and Davidson 2007; Streck 2008). Since melt inclusions preserve relatively closed system attributes of magma, the major preservation of open system processes (e.g. magma-mixing and crustal contamination) is in the zoning of minerals as time sequenced record of melt chemistry

(Ginibre et al. 2002, 2007). Contrasting mineral assemblages, textures and major, trace and isotope chemistry assists in documenting the open system behaviour (e.g. magma mixing) of magmatic systems.

Acknowledgments Amitava Kundu and Sonalika Joshi, erstwhile colleagues from Geological Survey of India are acknowledged for numerous discussions and clarifications of many ideas. This contribution was possible on account of an invitation from Professor Santosh Kumar.

References

Andersen DJ, Lindsley DH, Davidson PM (1993) qUILF: A Pascal program to assess equilibria among Fe-Mg-Mn-Ti oxides, pyroxenes, olivine, and quartz. Comput Geosci 19:1333–1350
Anderson JL (1996) Status of thermobarometry in granitic batholiths. Trans R Soc Edinburgh 87:125–138
Anderson JL, Smith DR (1995) The effect of temperature and oxygen fugacity on Al-in-hornblende barometry. Am Mineral 80:549–559
Anderson JL, Barth AP, Wooden JL, Mazdab F (2008) Thermometers and thermobarometers in granitic systems. Rev Miner Geochem 69:121–142
Bhandari A, Pant NC, Bhowmik SK (2011) 1.6 Ga ultrahigh-temperature granulite metamorphism in the central Indian tectonic zone: insights from metamorphic reaction history, geothermobarometry and monazite chemical ages. Geol J 46:198–216. doi:10.1002/gj.1221
Blundy J, Cashman K (2008) Petrologic reconstruction of magmatic system variables and processes. Rev Mineral Geochem 69:179–239
Blundy JD, Holland TJB (1990) Calcic amphibole equilibria and a new amphibole-plagioclase geothermometer. Contrib Mineral Petrol 104:208–224
Brey GP, Kohler T (1990) Geothermobarometry in four-phase lherzolites II. New thermobarometers and practical assessment of existing thermobarometers. J Petrol 31:1353–1378
Costa F, Dohmen R, Chakraborty S (2008) Time scales of magmatic processes from modelling the zoning patterns of crystals. Rev Mineral Geochem 69:545–594
Davis MG, Garcia MO, Wallace P (2003) Volatiles in glasses from Mauna Loa Volcano, Hawai'i: implications for magma degassing and contamination, and growth of Hawaiian volcanoes. Contrib. Mineral Petrol. 144:570–591
Frost RB, Barnes CG, Collins WJ, Arculus RJ, Ellis DJ, Frost CR (2001) A geochemical classification of granitic rocks. J Petrol 42:2033–2048
Hammarstrom JM, Zen E (1986) Aluminum in hornblende, an empirical igneous geobarometer. Am Mineral 71:1297–1313
Holland T, Blundy J (1994) Non-ideal interactions in calcic amphiboles and their bearing on amphibole-plagioclase thermometry. Contrib Mineral Petrol 116:433–447
Ginibre C, Kronz A, Worner G (2002) High-resolution quantitative imaging of plagioclase composition using accumulated backscattered electron images: new constraints on oscillatory zoning. Contrib Mineral Petrol 142:436–448
Ginibre C, Worner G, Kronz A (2007) Crystal zoning as an archive for magmatic evolution. Elements 3:261–266
Goldstein JI, Newbury DE, Echlin P, Joy DC, Fiori C, Lifshin E (1981) Scanning electron microscopy and X-ray microanalysis, Plenum Press, 673 p
Ishihara S (1977) The magnetite-series and ilmenite-series granitic rocks. Mining Geol. 27:293–305
Jerram DA, Davidson JP (2007) Frontiers in textural and microgeochemical analysis. Elements 3:235–238

Kanaya K, Okayama S (1972) Penetration and electron-loss theory in solid targets. J Phys D Appl Phys 5:43–58
Nielsen RL, Michael PJ, Sours-Page R (1998) Chemical and physical indicators of compromised melt inclusions. Geochim Cosmochim Acta 62:831–839
Pant NC, Kundu Amitava, Joshi Sonalika, Dey Aloka, Bhandari Anubha, Joshi Anil (2009) Chemical dating of monazite—Testing of an analytical protocol against independently dated standards. Ind J Geosci 63(3):311–318
Parrish RR (1990) U-Pb dating of monazite and its application to geological problems. Can J Earth Sci 27:1431–1450
Pattison DRM, Chacko T, Farquhar J, McFarlane CRM (2003) Temperatures of granulite facies metamorphism: constraints from experimental phase equilibria and thermobarometry corrected for retrograde exchange. J Petrol 44:867–900
Potts PJ (1987) A handbook of silicate rock analysis. Blackie, 622 p
Potts PJ, Bowles JFW, Reed SJB, Cave MR (1995) Microprobe techniques in the earth sciences. Chapman & Hall, 419 p
Putirka KD (2008) Thermometers and barometers for volcanic systems. Rev Mineral Geochem 69:61–120
Slater L, McKenzie D, Gronvold K, Shimizu N (2001) Melt generation and movement beneath Theistareykir, NE Iceland. J Petrol 42:321–354
Sobolev AV (1996) Melt inclusions as a source of primary petrographic information. Petrology 4:209–220
Steiger RH, Jäger E (1977) Subcommission on geochronology: convention on the use of decay constants in geo- and cosmochronology. Earth Planet Sci Lett 36:359–362
Straub SM, Layne GD (2003) The systematics of chlorine, fluorine, and water in Izu arc front volcanic rocks: implications for volatile recycling in subduction zones. Geochim Cosmochim Acta 67:4179–4203
Streck MJ (2008) Mineral texture and zoning as evidence for open system processes. Rev Mineral Geochem 69:595–622
Suzuki K, Kato T (2008) CHIME dating of monazite, xenotime, zircon and polycrase: protocol, pitfalls and chemical criterion of possibly discordant age data. Gondwana Res 14:569–586
Wark DA, Watson EB (2006) Titaniq: a titanium-in-quartz geothermometer. Contrib Mineral Petrol 152:743–754
Watson EB, Wark DA, Thomas JB (2006) Crystallization thermometers for zircon and rutile. Contrib Mineral Petrol 151:413–433
Wells PRA (1977) Pyroxene thermometry in simple and complex systems. Contrib Mineral Petrol 62:129–139
Woldseth R (1973) X-ray energy spectrometry. Kevex Corp., Burlingame, CA 220p
Wu CM, Zhang J, Ren L-D (2004a) Empirical garnet-biotite-plagioclase-quartz (GBPq) geobarometry in medium- to high-grade metapelites. J Petrol 45:1907–1921
Wu CM, Zhang J, Ren L-D (2004b) Empirical garnet-muscovite-plagioclase-quartz geobarometry in medium to high-grade metapelites. Lithos 78:319–332
Xirouchakis D, Lindsley DH, Frost BR (2001) Assemblages with titanite (CaTiOSiO4), Ca-Fe-Mg olivine and pyroxenes, Fe-Mg-Ti oxides, and quartz: Part II. Application. Am Mineral 86:254–264
Zen E (1989) Plumbing the depths of batholiths. Am J Sci 289:1137–1157
Zen E, Hammarstrom JM (1984) Magmatic epidote and its petrologic significance. Geology 12:515–518

Hydrothermal Fluids of Magmatic Origin

Rajesh Sharma and Pankaj K. Srivastava

Abstract Hydrothermal fluids are natural heated water solutions wherein variety of elements, compounds and gases may be dissolved. They are generated by diverse crustal and mantle geological processes including basinal fluid interaction, magmatic differentiation and mantle degassing. Mixing of the fluids of two different origins is often possible, and their solidified product may be economic. As the magmatic hydrothermal fluid is formed during the course of magmatic evolution from low volatiles and CO_2-CH_4 rich at primary basaltic magmatic phase to the saline water rich during granite/pegmatite formation, large number of constituents including the common S, Cl, F, Na, K, N_2, metals and even REE may be found in it. The enrichment of halogen in the magmatic hydrothermal fluids promotes the partitioning of economically useful elements like Cu, Pb, Zn, W, Mn, Li, Rb, Sr and Ba from melt to the fluid phase. Salinity of this hydrothermal fluid, which varies from near 0 to >50 wt.% eq. NaCl, is a function of pressure. A wide range of immiscibility in the magmatic fluids results compositional modification of the gradually developing phases from nearly anhydrous melt to last residual low temperature water solution. Wet magma with higher mol% of water flows easily and exsolves water at low pressure regime. Influx of water in the heated rock suit can lower the liquidus temperature triggering melting at a lower temperature than the anhydrous melting. Fluid inclusions have been widely used to understand behaviour of ore forming fluids and the magmatic immiscibility such as silicate melt, H_2O-CO_2, hydrosaline melt, dense CH_4 and sulphide-metal melt. Ore deposition is generally linked with the late stage of magmatic hydrothermal fluid, and its repeated pulses may lead to the formation of large ore bodies. In addition to saline aqueous fluid, the volatiles like H_2S, CO_2, SO_2, SO_4, HCl, B and F, are

R. Sharma (✉)
Wadia Institute of Himalayan Geology, Dehra Dun 248001, India
e-mail: sharmarajesh@wihg.res.in

P. K. Srivastava
Department of Geology, University of Jammu, Jammu 180006, India
e-mail: pankajsrivastava.ju@gmail.com

S. Kumar and R. N. Singh (eds.), *Modelling of Magmatic and Allied Processes*, Society of Earth Scientists Series, DOI: 10.1007/978-3-319-06471-0_9,

found as significant ore-depositing agents in magmatic-hydrothermal fluids. A hydrothermal fluid may dissolve economically useful elements or simply act as carrier for them. PVTX conditions obtained from fluid inclusions are vital for defining hydrothermal system and resulting ore mineralization, though such interpretation is largely based on the knowledge of their phase equilibria. The fluid process related to the epithermal Au deposits, porphyry type deposits, Malanjkhand Cu deposit and Balda-Tosham tungsten province, India, have also been discussed briefly.

1 Introduction

Hydrothermal fluids were considered to be essentially derived from the cooling magmatic bodies, commonly ascending under low viscosity and negative density. This view has been extensively modified by the perception that hydrothermal fluids are fundamentally 'heated water' naturally generated below the surface. A fluid phase and heat source are required to generate the hydrothermal fluid. This fluid phase may be derived from variety of sources like magmatic, juvenile, metamorphic reactions, meteoric, connate or sea water, whereas heat can be obtained from magmatic system, geothermal gradient, radiogenic decay or metamorphic reactions. In view, the concepts of hydrothermal fluids have grown following the extensive studies on the hydrothermal to active geothermal systems. Much of these refinements focus on the timing of hydrothermal fluid evolution, variability in temperature, composition and salinity, the fluid source and the heat source (Hutchison 1983). The exsolution of hydrothermal fluids from the cooling magmatic body is a widespread phenomenon in a magmatic system. Numerous fossil hydrothermal systems linked to magmatic processes have been characterized and classified, and a dynamic magmatic hydrothermal fluid can be observed at the mid-oceanic ridge as well as in the active volcanoes on the continents. The hydrothermal fluids linked to the magmatic bodies have been dealt with wider details. Though the ascending hydrothermal fluid may or may not be having magmatic components (Fig. 1). Burnham (1979), Norton (1984) and Pirajno (2009) presented the classical review of the magmatic hydrothermal fluids.

The heating of pore fluids in the sedimentary sequence may be a result of burial, but before the initiation of metamorphism, thereby low temperature hydrothermal fluid circulation can occur in the basins. Hence, a hydrothermal fluid without magmatic component may be evolved. A comprehensive account of the hydrothermal fluids of sedimentary genesis has been given by Hanor (1979). Furthermore, a large number of metamorphic devolatization reactions take place in a rock pile depending upon their mineralogy and P-T regime. These reactions may produce metamorphic hydrothermal fluids which are commonly rich in H_2O and CO_2. One can find the products of the fluid amalgamation as there is severe possibility of mixing of two different fluids during the evolution of the crust. We present here an

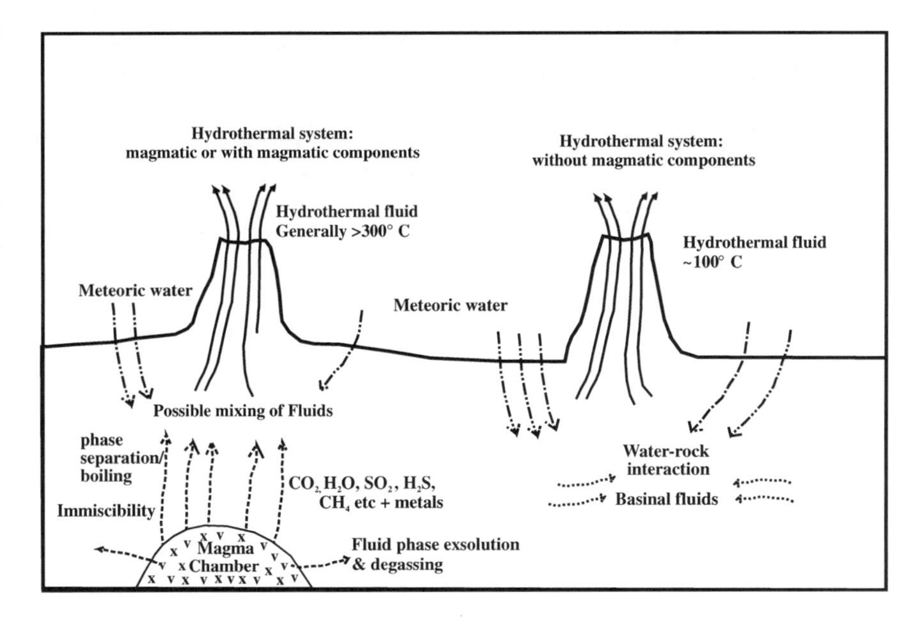

Fig. 1 Illustration of the hydrothermal fluid with or without magmatic components

introductory review of some of the issues of the magmatic-hydrothermal system, with an emphasis on the magmatic ore forming fluids in inclusions as observed in some studies including ours.

2 Composition of Magmatic-Hydrothermal Fluids

The processes of differentiation and crystal fractionation within a magmatic body result in the chemical variations of solid and fluid phases. This change occurs from the alleged fluid-absent melting to water-saturated conditions, wherein the mantle derived plumes govern the nature of early fluid phase which subsequently may be built up by the formation of magmatic series and the crustal contamination. The discontinuous nature of magma emplacement has significant consequences for change in the composition of hydrothermal fluids. The initial stage basaltic magma has low volatile contents with rich carbonic phase; late felsic rocks crystallize from a hydrous silicate melt and the quartz veins of residual magmatic affinity evolve in aqueous enriched fluid conditions. Water and its vapour is the most common and predominant component of the fluid that exsolve during cooling of a magmatic body, other significant constituents of such fluid include CO_2, N_2, SO_2, H_2S, CH_4, CO, HCl, HF, NH_3, O_2, rare argon, helium and neon. Bottrell and Yardley (1988) have analyzed fluid inclusions in the high temperature hydrothermal topaz-quartz-tourmaline rock samples of granite association, and found that the primary granite derived fluids were chloride brines, acidic with significant presence of Na, Cl, B. Active hydrothermal plumes at sea floor consist of aqueous- CO_2-CH_4 rich fluid phases, with presence of other gases like H_2 and N_2. These gases are exsolved because of their high concentrations in magma at higher temperature- pressure conditions. Whereas, Cl, HF, HBr, oxides and various elements may be found as traces dissolved in the hydrothermal fluids. Deep intrusions under high pressure release CO_2 rich volatiles as compared to those exsolved from shallow, more oxidizing, acidic, Cl rich magmatic phase. Metals and REE may also be found dissolved in such hydrothermal fluids. The chalcophile elements like Zn have strong affinity with the saline brine than for the melt, hence are carried out by the saline hydrothermal fluids (Holland 1972).

The composition of the crystallized part, melt and the fluid phase tend to change continuously as the fluids constantly exsolves from the melt, and the volatile phases are also partitioned between silicate phase and the fluid phase. Common volatiles dissolved in melt like H_2O, CO_2, HCl, HF, H_2S affect the position of the liquidus and solidus of the crystallizing phase. In addition to H_2O, increase of F, B and Li lead to lowering of the solidus temperature of the granitic magma. The distribution of metals in the melt or fluid phase of the system largely depends on the activity of Cl and F in it. The Cl preferentially get partitioned into the fluid phase, while F has tendency to partition in the felsic crystalline and fluid components (Burnham 1979; Eugster 1984; Shinohara 1994; Shinohara et al. 1989; Webster and Holloway 1990). The Cl and F partitioning suggested by various experimental works and the theoretical calculations (Kilinc and Burnham 1972; Manning and Pichavant 1984; Sinohara et al. 1989; Candela 1992; Webster and Holloway 1990) shows that the fluid/melt partition coefficients of Cl and F are of the expression: D_{Cl} (commonly >1) and D_F (commonly <1). The D_{Cl} increases with the increase in temperature, pressure and water in the fluid phase but decreases if F increases in the melts. While the D_F decreases if temperature or water in the fluid phase rises. The increase in F affects the system by (i) reducing the melt viscosity (ii) lowering down the solidus and liquidus temperatures (iii) increase the thermal stability of hydrous phase and (iv) increase in the solubility of silicate melt in the fluid phase. The increased concentration of Cl in the magmatic system favours the partitioning of Cu, Pb, Zn, W, REE, Mn, Li, Rb, Sr and Ba from melt to the fluid phase. This infers that the metals are removed from the melt phase to the magmatic hydrothermal fluid phase as the halogen concentration increases. Consumption of boron during growth of tourmaline crystals decreases solubility of water resulting in the release of magmatic water and formation of B, Li and F-rich greisens. The partitioning of sulphur is always dependent on the oxygen fugacity. Since H_2S is more soluble in hydrous silicate melt than SO_2, partitioning of SO_2 occurs into the magmatic aqueous phase. As the granites derived from the pelitic source have low oxygen fugacity, the sulphur contents of their exsolved aqueous phase are also low. This leads to the association of most Sn-W mineralization with such granites, and their deprived nature in sulphide minerals.

3 Salinity Variation

The concentrations of salts in the magmatic hydrothermal systems may greatly vary from near 0 wt.% in simple water solution to as high as 60–70 wt.% in hydrothermal brines. This variation is largely governed by the physicochemical conditions of melts during exsolution of hydrothermal fluid. Therefore, variation in salinity of the magmatic hydrothermal systems also help in identifying conditions during separation of hydrothermal fluid from a crystallizing granitic melt. Since NaCl salt is the most common constituent of such hydrothermal fluid, it can be used to understand the separation of hydrothermal fluid phase. Experimental and theoretical studies on the simple model system albite-NaCl-H_2O provide salinity of aqueous phase that exsolves from an evolving crystallizing melt. A systematic variation in NaCl concentration can be defined as a function of pressure (Bodnar 1992). The salinity of the initial melt at a higher pressure (cf. >1.3 kbar) is high but gradually decreases as the melt crystallizes. Final hydrothermal fluid exsolved from this melt would be near pure water. On contrary at low pressure salinity of initial melt is also low and is subsequently increases during the course of crystallization. The last fluid phase exsolved from such magmatic system is highly saline even with >50 wt.% eq NaCl. The salinity of initial melt at 1.3 kbar pressure is low to moderate, about 10–15 wt.%, and remains unchanged during crystallization history at constant pressure. Extensive experimental studies have been carried out on feldspar water system to understand behaviour of melt, crystallization and participation of fluid phase (e.g. Wen and Nekvasil 1994; Zeng and Nekvasil 1996; Hellmann 1994 etc.). Experimental studies show that an albite-H_2O system at water saturated liquidus at 832 °C: 2 kbar would release water after 57 % of the melt has crystallized. The salinity of this hydrothermal fluid would be 27.4 wt.% eq NaCl, which further decreases up to <1 wt.% at the last stage of hydrothermal phase separation. Thus, the late hydrothermal fluids at shallow depth would be high saline as compared to such fluids at great depths. Low water contents in the melt at the initial crystallization stage causes the water saturation at relatively late stage, therefore the early exsolving fluid becomes salt saturated at the room temperature. However, the most common cause of the salt saturation in the hydrothermal fluid at room temperature is aqueous fluid immiscibility. At the PTX conditions within the limit of liquid + vapour two phase system, immiscibility in the magmatic-hydrothermal fluid results in the high saline liquid and low saline vapour. This can be explained by the simple H_2O-albite solidus wherein at pressure <1.6 kbar, the exsolving fluid with salinity between 20 and 44 wt.% NaCl eq. would split into a vapour having 20 wt.% NaCl eq. and liquid having 44 wt.% NaCl eq. This continues until 90 % of the melt crystallizes and fluid enters to one phase field. The salts in the hydrothermal fluid are vital for the formation of ore deposits as more saline fluids can dissolve metals whereby salt-metal complexes may be formed. These salt-metal complexes can transport metals (Romberger 1982; Sharma et al. 2003), and subsequently leads to deposition and enrichment of metals. However, the formation of such salt-metal complexes also depends on the compatibility of these economically useful elements with the salts. Cu forms a compound with Cl and is transported in more saline medium, whereas, Mo concentration is not achieved with higher salinity because Mo does not have any affinity with Cl. The fluid inclusions can provide salt concentration of the fossil hydrothermal fluids which are expressed as wt.% NaCl eq. The eutectic temperatures of the saline aqueous fluids provide information about composition of salts dissolved in the fluid, whereas the final melting of the frozen aqueous fluid in inclusions suggests the salinity. More details of estimating salinity from fluid inclusions studies are available in number of studies (e.g. Crawford 1981; Roedder 1984).

4 Magmatic Fluid Immiscibility

Wide range of immiscibility exists in the magmatic fluids resulting in compositional modification of the gradually developing phase (Thompson et al. 2007). During magmatic evolution from nearly anhydrous melt to hydrothermal solution, the change in the composition of residual liquid may be continuous but release of volatiles from the magma in the forms of globules occur at decreasing temperature. As the cooling magma ascends towards a lower pressure regime, immiscibility of the lower density fluid phase in it results in exhalation of the hydrothermal fluids. A basaltic magma rising from the mantle releases dense CO_2 during ascent because of its reduced solubility and consequent exsolution during crystallization. At this stage, potassium and rare earth elements may move to the fluid phase along with CO_2. In addition, sulphide-melt globules may also be enriched in hydrothermal fluid at this stage. The participation from the mantle derived magmatic material is also possible, and with the persistence of silicate-sulphide immiscibility, metals from the main magma get separated into the sulphide phase. Subsequent to the separation of dense CO_2 from the rising basaltic magma at the reduced pressure conditions, water may joins to the large volume of CO_2 to form violent eruption at near surface conditions such as in volcanoes. As the CO_2 is released from the magma, it can hold molten state even at lower PT conditions and progressively becomes enriched in H_2O. The presence of NaCl in the melts promotes the unmixing of CO_2 in the vapour phase and more saline aqueous fluid to the liquid phase.

The concentration of CO_2, H_2O and Cl in crystallizing magma increases as the crystal formation takes place. Splitting of high saline water-'hydrosaline melt' occurs with the lowering of the P-T conditions during magmatic differentiation. This hydrosaline melt may further split into a low density CO_2 phase, low saline H_2O and a more saline melt. Fluorine dissolved in hydrothermal phase has also been observed in some granite melt, particularly during transition from granite to pegmatite (Webster and Holloway 1990). At the final stage of crystallization, the fluid saturated granitic magma exsolves aqueous fluid phase which is Cl rich and acidic. A dense H_2O phase may evolve from this magma at any stage from liquid line of descent, and the salinity of this hydrothermal phase depends on the initial halide contents of the magma. Hydrothermal fluid phase dominated by H_2O and

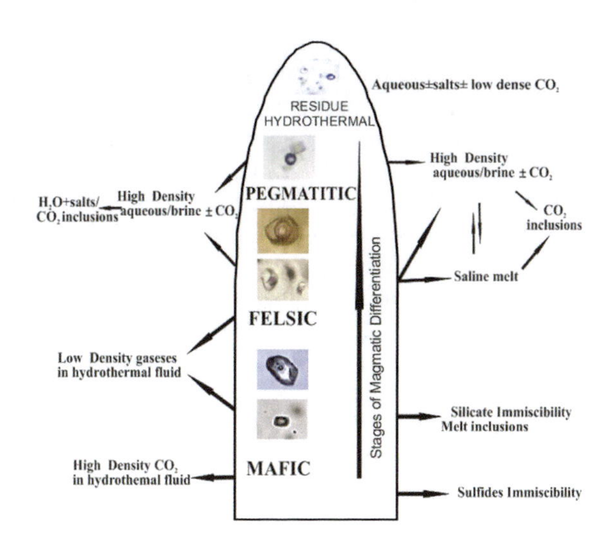

Fig. 2 Diagram representing fluid separation through immiscibility at various stages of magmatic differentiation along with some varieties of fluid inclusions observed in the representative samples (adapted after Roedder 1992)

enriched with Cl, CO_2, S and sometimes with metals are formed by the crystal fractionation of granitic magma whereby anhydrous alkali alumino-silicates are crystallized. Such hydrothermal phase is also formed during the fractional crystallization of pegmatitic bodies. The fluids evolved during formation of rare metal bearing pegmatites are low temperature aqueous phase with/without low density CO_2, whereas hydrothermal fluids released from the simple pegmatite are generally complex. Hydrothermal fluids evolving from a cooling granite body do not consist of Mg concentration because these are the end member in magmatic evolution. A general diagrammatic representation of fluid exsolved at various stages of the magmatic immiscibility (adapted after Roedder 1992) is presented together with the common fluid inclusions at various stages in Fig. 2.

The last residual liquid phase is water solution that can produce quartz veins even at 100–200 °C. In between formation of pegmatite and low temperature vein, a hydrothermal phase potential for ore deposition may also exist. Although separation of the fluid phase from the evolving magma is ubiquitous, but the existence of ore forming fluid is uncommon. Oxygen isotope systematics and fluid inclusion microthermometry can reveal the temperatures of successive pulses of hydrothermal fluids. These studies are also helpful in confirming whether hydrothermal fluid is formed from only magmatic derived water or also have some flux and mixing of meteoric component.

5 Evolution of Magmatic-Hydrothermal Systems

Melting behaviour of the granitic magma largely depend on its water contents which generally enrich gradually after crystal fractionation. But in the actively tectonised regions such as in subducting plate margins, the dehydration of hydrous minerals present in the subducting slab, may result in the release of water. The upwelling flux of this water can cause melting of the rocks and formation of wet granites. Such influx of water in a heated rock suit shifts the liquidus to a low temperature, causing their melting at much lower temperature than the anhydrous melting (Burnham 1975, 1967). Touret and Thompson (1993) demonstrated various aspects of fluids interaction with the rocks leading to their melting, fluid consumption and evolution. The maximum melting temperature of a simple albite system is 1100 °C at 1 atmospheric pressure which decreases with the addition of water to this system. The solubility of H_2O in silicate magmas is determined mainly by pressure and to a lesser extent temperature. The results of various experimental determinations regarding water content in melts of different compositions suggest that the water content is strongly dependent on pressure (Fig. 2). The magmas at the base of the crust are able to dissolve 10 to 15 % H_2O. For any given pressure, felsic melts can dissolve more water than mafic ones. Water dissolves in magma essentially as hydroxyl (OH) groups, although at higher pressure and with higher water contents discrete molecular water (H_2O) can also exists (Stolper 1982). Wet magma with higher mol% of water can flow easily but exsolves water as soon as it reaches to a low pressure regime. Formation of intrusion related ore deposits is linked with the irreversible magmatic transfer of heat and mass from the Earth's interior to its surface (Candela 1997). Several element-fractionating processes are superimposed in large magmatic-hydrothermal systems causing high degree of selective metals enrichment. These include early partitioning of elements during partial melting in the mantle, the separation of minerals and volatile phases during magma ascent and crystallization, and finally to the selective precipitation of ore minerals (e.g. Eugster 1984; Hedenquist and Lowenstern 1994; Shinohara and Hedenquist 1997). The fluid phase separation is also considered to be important mechanism to segregate metals at higher temperatures (Harris et al. 2003). Commonly, the rise of magma at epizonal levels leads to either magmatic volatile phase saturation if the system is initially volatile phase undersaturated, or further exsolution of volatiles if the magma is already saturated with one or more volatile phases (Candela 1997). This magmatic volatile phase is critical agent in the ore formation because of their high fluidity and buoyancy as well as preference of ore minerals to concentrate in aqueous/volatile phase. The occurrence of greisens, miarolitic cavities, presence of silicate melt together with aqueous fluid in inclusions in the mineralized granites are some of the evidences suggesting that silicic magmas exsolve large volume of magmatic-aqueous fluids (Candela and Blevin 1995; Lowenstern and Sinclair 1996).

Several models have been proposed for the evolution of intrusion-related magmatic-hydrothermal systems. The pioneer work of Burnham (1967, 1979) on the formation of zones of H_2O saturated magma and their localization towards the roof of a granite intrusion has encouraged experimental and theoretical work towards the granite related hydrothermal ore deposits (e.g. Whitney 1975, 1989; Candela 1991, 1997; Shinohara 1994 and many others). Physical models for the exsolution of volatiles from a convecting magma body, suggested by Shinohara and Kazahaya (1995) and Shinohara and Hedenquist (1997), show that at the low degrees of crystallization, vapour in the magma buoyantly rise and get saturated at the top of the magma body causing sudden and rapid fracturing in already crystallised magma and the adjacent wall rock. In case of crystallization of granitic magma, the liquidus assemblage is dominated by anhydrous minerals and the concentration of incompatible constituents including H_2O and other volatiles species increases by processes analogous to Rayleigh fractionation (Robb 2005). The granitic magma becomes water-saturated resulting in the exsolution of an aqueous fluid to form a chemically distinct phase in the silicate melt at some stage. This process is called H_2O-saturation, but it is also referred to as either "boiling" or "vapour-saturation". At the low pressure regime water released from magma may often boil because it shifts from one phase supercritical fluid to two phase (liquid-vapour) boiling fluid. Boiling of the hydrothermal fluid is an indication of the existence of hydrothermal fluid in magmatic conditions, and plays significant role in mineral deposition as the solubility of solutes changes with phase separation in the evolving hydrothermal fluid system (Drummond and Ohmoto 1985). It is possible to calculate the iso-density line (isochor) of a hydrothermal fluid on the P-T trajectory and to define the conditions of fossil hydrothermal fluid whether critical or not. Boiling occurs when the sum of the vapour pressures of the components of a liquid, is equal to the load pressure. It can result from a decrease in the load pressure or an increase in vapour pressure (referred as 'First Boiling'), or it is also possible to achieve vapour saturation by progressive crystallization of dominantly anhydrous minerals under isobaric conditions (referred as 'Second Boiling'). The first boiling is particularly applicable to shallow magmatic systems while second boiling generally occurs in deep-seated magmatic systems and only after a relatively advanced stage of crystallization. The aqueous fluid will tend to rise and concentrate in the roof, or carapace of the magma chamber due to its lower density than that of the granitic magma. The amount of magmatic-hydrothermal water formed this way, even after some of the original OH^- in magma is utilized to form hydrous minerals. CO_2, boron and fluorine bearing fluid fluxes can also exist thereby topaz and tourmaline can be formed in granitic pegmatite or in the granites altered by hydrothermal fluids.

Burnham (1979, 1997) examined in detail the magmatic hydrothermal system generated during the cooling of a hypothetical granodiorite intrusive stock containing 3 wt.% water. He assumed that the cooling of the intrusive body occurred in a subvolcanic environment. Therefore, it is implied that during the initial stages of cooling the system was open, allowing the escape of volatiles through fractures above the pluton. At a later stage, the intrusive body becomes a closed system by developing a solidified shell. Burnham showed that these processes lead to saturation with H_2O of the remaining interstitial melt, while the rest of the stock was still largely molten and H_2O unsaturated. This molten portion was enveloped by a zone of H_2O saturated interstitial melt, which in turn was enclosed by carapace (solidified shell). Thermodynamic calculations by Burnham suggest that the water saturated zone can expand up to 30 % upon complete crystallisation at a depth of 3 km, but not more than 5 % at deeper level of 5 km. The decrease in thickness of this H_2O saturated zone with depth resulted in an increase in the vapour pressure within the magma at later stage (Fig. 3). Boiling of fluid in the upper portion of the

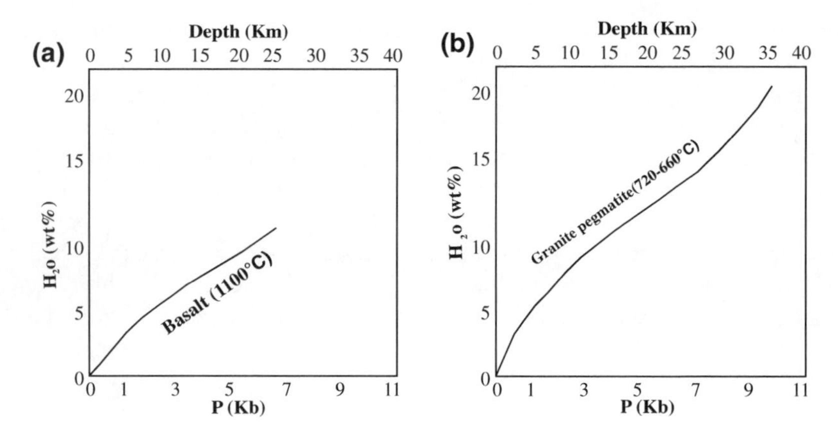

Fig. 3 The results of experimental determinations of solubility of water content in silicate melts: **a** solubility of water in basaltic melt with pressure **b** H_2O solubility in granitic or pegmatitic melt with pressure (after Burnham 1979)

H_2O saturated zone occurred at this stage which leads to the development of an abundant aqueous phase. As the confining vapour pressure increases more than the lithostatic/tensile strength of the rock, already solidified rock is fractured and brecciated. This reduced the fluid pressure causing evolution of more aqueous fluid phase wherein boiling of the fluid can also occur. The fractures, hence generated during the process, act as a major channel way for ore-bearing fluids and heat from the underlying and continuously cooling igneous body. The metal complexes are concentrated in the fluid phases and transported into the network of fractures, and are deposited as ore minerals. Ore mineralisation is usually associated with late pulses of magmatic hydrothermal fluid, and repeated pulses lead to the formation of large ore bodies. Most of these hydrothermal systems are restricted to the upper parts of the Earth's crust; being developed in epizonal setting wherein the mixing of meteoric waters can also occur with the magmatic hydrothermal fluids (Fig. 4).

The other model is suggested by Williams-Jones and Heinrich (2005) based on experimental studies on the stability of metallic species in aqueous vapours and fluid inclusion studies. They proposed that the low-density supercritical aqueous fluids or vapours are important agents for the transport of metallic elements in the hydrothermal systems. They suggested that the vapour phase in a hydrothermal system becomes increasingly dense with increasing temperature and pressure, while the co-existing aqueous liquid expands and at the critical point of 374 °C and 225 bars the two phases become indistinguishable with water being a supercritical fluid. According to this model, major mass-transfer occurs with vapour as metal transporting agent in magmatic hydrothermal systems, and is particularly efficient for metal deposition in fumaroles, porphyry-Cu, Cu-Mo, Cu-Au and Au-Cu epithermal systems (Heinrich et al. 1999; Williams-Jones et al. 2002; Williams-Jones and Heinrich 2005; Simon et al. 2007; Pirajno 2009; Seo et al. 2009; Landtwing et al. 2010; Mavrogenes et al. 2010; Zezin et al. 2011). Further, this model also explains that the depth at which fluids exsolve from the hydrous magma is a key factor in fluid

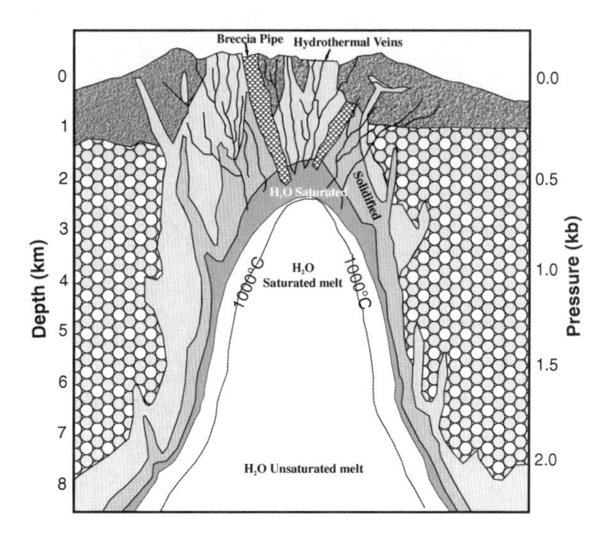

Fig. 4 Sketch showing model for the evolution of a magmatic-hydrothermal system during cooling of a granitic intrusion (reproduced from Burnham 1979)

behaviour. Hanson and Glazner (1995) suggested that the magmatic–hydrothermal system is a complex function of the rate of magma generation, upper-crustal strain rate, and the time-integrated physical properties of the magma. It is independent of the system produced either from single pulse or multiple pulses of magma, and whether the single pluton or multiple intrusive phases are formed.

6 Magmatic-Hydrothermal Fluids in Inclusions

Fluid inclusions are the microscopic droplets entrapped as closed geochemical and geothermobaric systems in minerals. They represent the parent fluid regime existed during growth history of mineral formation or during its evolution. The intrusive rocks have widest range of fluid inclusions. The best studied inclusions are immiscible silicate melt, H_2O-CO_2 inclusions, hydrosaline melt, dense CH_4 gas and occasionally sulphide-metal melt. Such diversity is a result of either entrapment of phases of immiscibility in magma or because of wide range of P-T conditions during crystallization from early magmatic differentiation to late low temperature hydrothermal actively. The immiscible fluid phase in silicate melt is found splitted into low density boiling aqueous rich phase and other carbonate, sulphide, hydrocarbon or metal bearing phases of varying density (Roedder 1992). The solid crystals, volatiles, aqueous fluid and melt vary drastically in the inclusions trapped in granitic, pegmatitic and alkalic rocks, and therefore are crucial to understand the evolution of magma (Sobolev and Kostyuk 1975). Studies on the alkaline rocks have indicated T_h values commonly above 1100 °C and a temperature between 1240 and 1260 °C have been reported for early minerals like plagioclase and pyroxenes. As the origin of carbonatites is controversial, the studies on these rocks also unravel inclusions with a wide range of temperature and

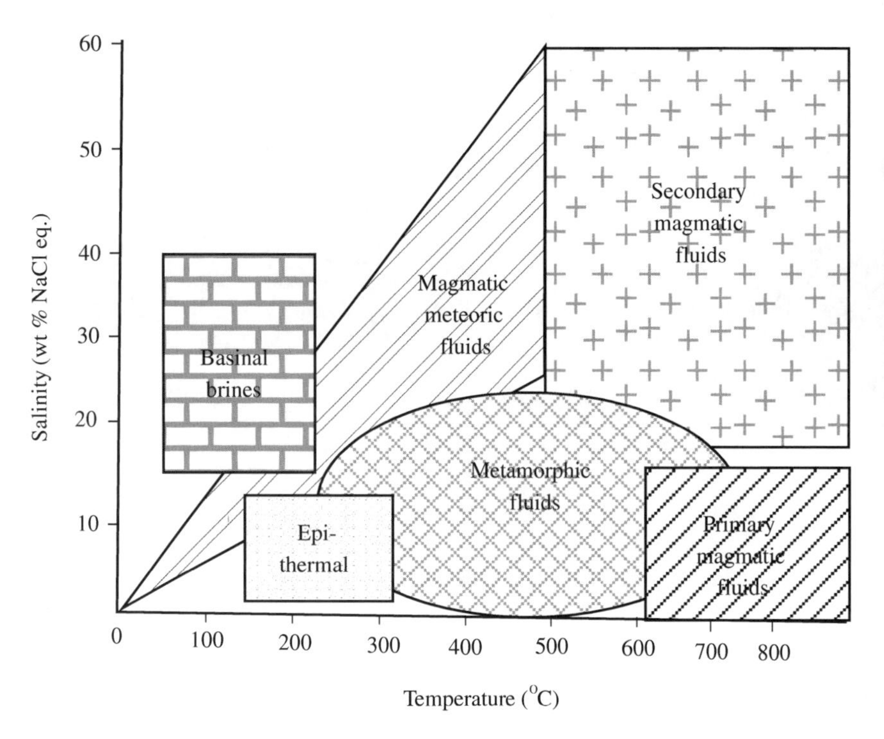

Fig. 5 Diagram showing salinity-temperature range for various hydrothermal fluids (summarised after Beane 1983; Roedder 1984; Bodnar et al. 1985; Lattanzi 1991; Wilkinson 2001 and our own observations.). The boundaries of the various fluids are general and not fixed

composition including carbonate-silicate melts, brines, CO_2 and H_2O etc., which shows that several processes and the mental fluid interaction in the genesis of these rocks (Frezzotti et al. 2002). Systematic studies of the fluid inclusions are capable of providing extremely useful information on the fluid phases associated with the evolution of magma. The variety of processes can be identified with the help of fluid inclusions data. Various information such as salinity of the fluid, temperature, density of the fluid etc., are significant for understanding process involved and the hydrothermal fluid evolution, and categorizing discrete fluids (Fig. 5).

7 Products of the Hydrothermal Fluids

Hydrothermal fluids discharged from crystallizing water-saturated magma may cause hydraulic fracturing in the country rocks. The ore forming hydrothermal fluids can propagate fractures culminating into the stock work mineralization. Although the tectonic setting and the composition of magma has critical control on the source material for mineralization, however, the later differentiation and fluid phase generation is vital for extracting the metals, their transport and deposition. This fluid may dissolve, or simply act as carrier for the economically useful elements. Magmatic hydrothermal ore deposits include porphyry deposits, skarn deposits, vein deposits and the exhalative deposits of volcano-sedimenatry linkage. Porphyry deposits are subduction zone related deposits of mainly Cu, Mo, Sn, W, Au which are associated with wet granites. Skarn is also a porphyry system formed when wet magma intrudes in the limestone rocks and fluid released from this magma interacts with carbonate rocks. Hydrothermal alteration zones with Cu, Mo, Sn, W, Fe and Au may be developed in the skarn deposits. Minerals in the skarn associated with least fractionated to highly fractionated granites are in order of F-Au-Cu-Zn-W-Mo-Sn. Vein deposits such as load gold deposits are classical mineralization deposited by the hydrothermal fluid flow in open space fractures. The diffusion of the volatiles in the wall rocks may deliver the mineralization even in the alteration zone. The economically useful elements may be exsolved together with the exsolution of fluid from the magma or these may be extracted from the country rocks that interact with circulating fluid.

The ore forming magmatic-hydrothermal fluids released from melts also contain volatiles like H_2S, CO_2, SO_2, SO_4, HCl, B and F in addition to aqueous fluid, and are powerful ore-depositing agents. Presence and abundance of these volatiles are generally related to the source region and composition of magma. Volcanic sublimates found around fumaroles provide direct evidence for the role of vapours in transporting significant amounts of metals. Large ore deposits can be formed due to different sets of transitional conditions during the circulation of ore forming fluids. According to Pirajno (2009), characteristics of exsolved aqueous fluids from magma are determined by the nature and redox conditions of an intrusion. Nb, REE and U deposits are generally found related with the peralkaline intrusions while base metals, Au, Cu and Mo deposits are associated with oxidised, I-type alkaline and calc-alkaline magmas, and the tin-tungsten mineralization is found deposited by hydrothermal fluids evolved from S-type granitic magmas. It is well demonstrated that subvolcanic, ore generative, hydrothermal systems are usually found associated with high-level granites (i.e. those emplaced at depths of $\sim$8 km or less), These high level granites represent the crystallized remains of magma (Candela 1997). Therefore, studies aimed at understanding magmatic evolution and volatile phase separation of shallow granites are important. The formation of granite-related magmatic-hydrothermal ore deposits is dependent on numerous physical and chemical factors. Some of these are composition and evolution of the parent magma, depth, geometry and temperature of magma emplacement, the rate of magma ascent etc. Dilles and Einaudi (1992), however, suggested that magma is not the only source of metals in some porphyry, skarn or high-temperature vein type ore deposits. The magmatic hydrothermal system associated with intermediate to acid magmatism produce a wide range of ore deposits with variable element ratios, from Cu, Mo or Au rich porphyries and skarns (e.g., John and Ballantyne 1998), to Sn, W or rare earth element (REE) rich veins and greisens replacements (Eadington 1983), to more distal lower temperature deposits rich in Au, Ag, As, and Hg (Rye 1993). These fluids are found invariably trapped as inclusions in various minerals in the ore deposits. The PTX trapping conditions of these inclusions can be estimated, which are vital for defining the processes leading to generation of the hydrothermal system and ore mineralization.

8 Fluid Inclusion Characteristics of Magmatic-Hydrothermal Ore Deposits

The ore forming fluids that exsolve from a silicate melt phase are invariably trapped as fluid inclusions in phenocrysts and other rock/ore forming minerals (Roedder 1984; Lattanzi 1991; Bodnar 1995 etc.). Studies of these fluid inclusions provide direct evidence for the PTX characteristics of the system during the time of fluid entrapment in inclusions (Ross et al. 2002; Seedorff and Einaudi 2004; Rusk et al. 2008; Mollai et al. 2009). However, the interpretation of inclusions largely relies on our current knowledge of the phase equilibria of geologically relevant fluids. As a consequence of the partitioning behavior of chlorine between melt and magmatic fluid, the aqueous fluids of magmatic origin observed in inclusions are moderately to highly saline (Kilinc and Burnham 1972; Shinohara et al. 1989; Cline and Bodnar 1991). The composition of magmatic fluids can be adequately modeled by the behavior of simple water salt solutions, such as H_2O-NaCl. Many experimental work (Keevil 1942; Sourirajan and Kennedy 1962; Haas 1976; Potter et al. 1977; Pitzer and Pabalan 1986; Sterner et al. 1988; Bischoff and Pitzer 1989; Bischoff 1991; Bodnar et al. 1985; Bodnar 1994) and theoretical concepts (Anderko and Pitzer 1993; Duan et al. 2003) of the H_2O-NaCl system provide a basis to interpret the PTX properties of most fluid inclusions. However, there are some conditions such as homogenization along the halite liquidus wherein further experimental data would improve our ability to interpret (Lemmlein and Klevtsov 1961; Roedder and Bodnar 1980; Bodnar 1994). In addition to the phase equilibrium studies, a number of previous hydrothermal fluid-flow models also infer the temporal and spatial thermal evolution of the magmatic-hydrothermal systems (Cathles 1981; Norton and Knight 1977; Norton 1984; Hayba and Ingebritsen 1997). The data from geochemical modelling (Kilinc and Burnham 1972; Shinohara et al. 1989; Cline and Bodnar 1991) and from the natural fluid inclusions can constrain the compositions of magmatic-derived aqueous fluids. Therefore, experimental and theoretical phase equilibria permit to interpret natural fluid inclusions and develop models of the magmatic-hydrothermal fluid evolution.

Different studies have confirmed that the solubility of CO_2 in magmas is typically an order of magnitude lower than that of H_2O. It is dissolved as molecular CO_2 in normal felsic or mafic melts, but in alkaline magmas it also exists as carbonate ionic complexes in solution (Lowenstern 2001). The presence of CO_2 in an evolving aqueous fluid within crystallizing granite will promote immiscibility between vapor and saline liquid phases of the solution. Such processes can be important during ore-formation since they promote the precipitation of metals from solution

(Lowenstern 2001). Effervescence of CO_2 from the fluid will also promote certain types of alteration in the host rocks and increases pH in the remaining fluid thereby influencing ore-forming process. Hence, the CO_2 may not be directly involved in the transport and concentration of metals, but it has a role in the distribution and precipitation of metals through the vapor phase. Since the liquid brine and lower density vapor can coexist at the solidus conditions of mid-crustal plutons to the near surface conditions (Bodnar et al. 1985), the low density gas rich fluids have a significant metal transporting capacity (e.g. Lowenstern et al. 1991; Wahrenberger et al. 2002). Heinrich et al. (1999) demonstrated that the chemical segregation of ore forming elements through separation of brine and vapor is a ubiquitous process which affects the composition of ore deposits across wide pressure- temperature range. The fluid inclusions reported from the magmatic-hydrothermal systems vary from the high temperature, high saline type to the low saline aqueous inclusions. Many of the magmatic-hydrothermal systems also contain aqueous-carbonic fluid inclusions (Pollard and Taylor 1986; Heinrich et al. 2004; Srivastava and Sharma 2008; Mollai et al. 2009). Magmatic fluids also show a wide range of salinities. The primary magmatic vapour exsolved directly from active volcanoes typically has salinities of only a few hundred to a few thousand ppm of TDS (Symonds 1992), whereas the primary magmatic vapour trapped at some depth below the top of the volcano might have salinity of several weight percent (Bodnar 1992, 1995). Many of the magmatic systems are also characterized by the presence of a vapour phase liberated as a result of aqueous fluid immiscibility ('secondary magmatic vapour' of Bodnar 1995). Thus the high salinity liquid produced in this manner is characteristics of magmatic fluid in many granite related ore deposits (e.g. Sharma et al. 1994; Beane and Bodnar 1995).

9 Models for the Formation of Porphyry-Type Cu, Mo and W Deposits

Considerable progress has been made in understanding the processes responsible for the formation of base metal deposits in crystallizing granitic intrusions (Candela and Holland 1984, 1986; Candela 1991, 1992; Candela and Picoli 1995; Sinclair 2007). Broadly, the porphyry Cu-Mo deposits are associated with arc related calc-alkaline or I type magmas, while Sn-W deposits are mostly associated with S-type granites derived from partial melting of continental crust. The magmatic-hydrothermal model is one of the most applicable genetic models for porphyry deposits. This model defines that the ore metals were derived from temporally and genetically related intrusions with large polyphase hydrothermal systems developed within them, and they commonly interact with meteoric fluids or possibly seawater at their top and peripheries (Sinclair 2007). During the waning stages of hydrothermal activity, the magmatic-hydrothermal systems collapsed inward and were replaced by waters of dominantly meteoric origin (Brimhall 1980; Brimhall and Ghiorso 1983). Burnham (1967, 1979), Philips (1973), and Whitney (1984) have suggested the orthomagmatic model, with some variations of the magmatic-hydrothermal model for the porphyry deposits. They envisaged that felsic and intermediate magmas were emplaced at high levels in the crust and underwent border zone crystallization along the walls and roof of the magma chamber. As a consequence, super saturation of volatile phases occurred within the magma resulting in separation of volatiles due to resurgent boiling. Ore metals and many other components were strongly partitioned into these volatile phases, which became concentrated in the carapace of the magma chamber (Christiansen et al. 1983; Manning and Pichavant 1984; Candela and Holland 1986; Cline and Bodnar 1991; Heinrich et al. 1992). A fundamental control on ore deposition was from the pronounced adiabatic cooling of the ore fluids due to their sudden expansion into the fracture and/or breccia systems. Thus, the structural control on ore deposition is important in porphyry deposits. Aplitic and micrographic textures in granitic rocks associated with porphyry deposits are the result of pressure-quench crystallization related to the rapid escape of the ore fluids (Shannon et al. 1982; Kirkham and Sinclair 1988).

Modification of the above orthomagmatic model is required for at least some, if not most, porphyry deposits, in view of studies by Shannon et al. (1982), Carten et al. (1988), Kirkham and Sinclair (1988), Shinohara et al. (1995) and Cloos (2002). Their studies pointed that the small porphyritic stocks associated with porphyry deposits were largely liquid until ore formation was essentially complete, and were conduits for enormous volumes of ore-forming fluids produced by degassing from large magma bodies. Shinohara et al. (1995) examined various mechanisms of fluid transport for porphyry Mo deposits, including gravitational ascent of bubbles in a static magma. They concluded that convection of the magma is the most favourable model for the transport of fluids to the site of ore formation. According to this model, fluid-charged and non-degassed magma rose by convection from a deep magma reservoir, and the fluid separation from the magma occurred at a shallow, subvolcanic depth. The volatiles were focused near the top of the magma column producing degassed magma that descended through the non-degassed magma because of its lower volatile contents and consequently greater density. Areas where ore-forming fluids accumulated in cupolas of intrusions associated with porphyry Mo and W-Mo deposits are characterized by abundant comb-quartz layers (Shannon et al. 1982; Carten et al. 1988; Kirkham and Sinclair 1988; Lowenstern and Sinclair 1996). The occurrence of comb-quartz layers in intrusions associated with some porphyry Cu, Cu-Au, and porphyry Mo deposits (Kirkham and Sinclair 1988; Atkinson and Hunter 2002) suggested that the model of Shinohara et al. (1995) may be applicable to porphyry deposits in general.

10 The Malanjkhand Copper Deposit, Central India

The Malanjkhand Cu–Mo–Au deposit is located in Balaghat district of Madhya Pradesh (Central India). It is a strategic and significant porphyry-style deposit and is the largest open pit copper mine in India with an estimated reserve of

789 million tons of copper ores with a grade of 0.83 % Cu. The Malanjkhand deposit is hosted by a granitoid complex of early-Proterozoic age that occurs in the northern part of Bhandara Craton, central India. This granitoid complex is surrounded by older granitic gneiss; metasedimentary rocks of the Saussar Group; volcano-sedimentary sequences of the Dongargarh Group and sedimentary rock units of Chhatisgarh Group. The age of the major phase of the Malanjkhand granitoid complex hosting the copper deposit has been constrained by U–Pb dating on zircons ($\sim$2476 $\pm$ 8 Ma; Panigrahi et al. 2004). The Re-Os dating of six molybdenite samples, both from the quartz reef and granite from the Malanjkhand deposit yielded a 'contemporaneous' age of 2490 $\pm$ 8 Ma (Stein et al. 2004). Furthermore, from high Re concentration (400–650 ppm), these authors argued in favor of a subduction related porphyry model for the Malanjkhand Cu-Mo deposit. It is also suggested that the Malanjkhand granite and the mineralization have undergone 50 Ma of protracted deformation. Mineralization in form of chalcopyrite-pyrite ores occurs as an arc shaped, 1.8 km-long and N-S oriented quartz reef. The granitoid enveloping the mineralized reef is less mineralized with sulfide-rich quartz and quartzo-feldspathic veins. Copper and subordinate molybdenum mineralization at Malanjkhand occurs within a fracture-controlled quartz-reef enclosed in a pink granitoid body surrounded by grey-granitoids constituting the regional matrix. Sulfide bearing stringers, pegmatites with only quartz, microcline and sulfide disseminations, all within the pink-granitoid, represent other minor modes of ore occurrences. Despite this diversity in mode of occurrence, the mineralogy of ores is quite consistent and confirm to a common paragenetic sequence comprising an early 'ferrous' stage of precipitation of magnetite (I) and pyrite (I) and, the main-stage chalcopyrite mineralization with minor sphalerite, pyrite (II), magnetite (II), molybdenite and hematite. Although the Malanjkhand deposit has been studied for more than four decades, the physicochemical environment of the hydrothermal fluid and its evolution is variedly discussed. The origin and evolution of the Malanjkhand copper- molybdenum deposit is also under debate with the issue of single versus multiple stages of mineralization (Sarkar et al. 1996; Panigrahi and Mookherjee 1997; Stein et al. 2004; Panigrahi et al. 2008). A porphyry-type affinity has been suggested for the Malanjkhand deposit (Sikka and Nehru 2002; Stein et al. 2004).

The fluid inclusion studies have been carried out by Jairath and Sharma (1986), Ramanathan et al. (1990) and Panigrahi and Mookherjee (1997). Jairath and Sharma (1986) inferred a CO_2-bearing hydrothermal fluid, with temperatures ranging from 210 to 470 °C and containing appreciable amount of Na and K. Boiling at 370 °C likely caused precipitation of the ore, with a later separation between a CO_2-rich fluid of low salinity and a more saline fluid having low CO_2 (Jairath and Sharma 1986). Based on the occurrence of K-feldspar, the absence of pyrrhotite and primary bornite, and considering pyrite-magnetite-chalcopyrite and hematite in the ore assemblages, they predicted log fO_2 of -30.2 to -34.4, log fS_2 of -8.8 to 11.6, pH of 4.5–6.5 and pressures from 225 to 440 bars during ore deposition. Panigrahi and Mookherjee (1997) inferred two-stage evolution of the ore fluid: initially a mixing of the two fluid components leading to the deposition of ore minerals, and a later separation of the carbonic component (preserved as pure CO_2 inclusions) resulting in the observed increase in salinity of the aqueous component. Using temperature-salinity plots, they concluded that the mineralization was brought about by a mixture of two fluids: high temperature ($\sim$375 °C) and low-saline (4–8 wt.% NaCl equivalent) fluid, containing appreciable amounts of CO_2 and sulfur, other a low temperature (180–200 °C), moderately saline (20–24 wt.% NaCl equivalent) metal-rich fluid. They also suggested that later was the dominant fluid component and was derived from the main phase of the Malanjkhand granitoid (Panigrahi et al. 1993, 2004). Pressure estimates vary from 550 to 1790 bars in the stringer ores.

11 Tungsten Province of Balda-Tosham Belt, Northwest India

The western Indian craton hosts number of the tungsten prospects which are associated with 850–750 Ma magmatic events (Srivastava and Sinha 1997; Sharma et al. 2003). A 500 km long W $\pm$ Sn belt, running from Balda in Rajasthan to Tosham in Haryana, has been recognized by Bhattacharjee et al. (1993). Characteristics of the individual deposits of the belt have been discussed by a number of workers (Kochher, 1973; Chattopadyaya et al. 1982, 1994; Bhattacharjee et al, 1993; Sharma et al. 1994, 2003; Banarjee and Pandit 1995; Srivastava 2004; Srivastava and Sukhchain 2005, 2007).

Almost continuous record of basic and acid magmatism dating from 3500 to 750 Ma, from Archean to late Proterozoic (Choudhary et al. 1984), has been observed in this region. The 850–750 Ma time span marks the most widespread acid magmatic activity represented by the anorogenic granite intrusions mainly in the western fringe of the Delhi Fold belt. Some of the granitic plutons of this age, popularly known as Balda granite, Degana granite, Sewariya granite and Tosham granite, are responsible for the W $\pm$ Sn mineralization (Srivastava and Sinha 1997). The culmination of this magmatism is characterized by a widespread, dominantly acidic and locally basic volcanism and the acid plutonic activities named as Malani Igneous Suits (745 $\pm$ 10 Ma, Bhushan 2000) which are present to the west of Delhi Fold Belt. The Degana granite and Tosham granite associated with W, Sn mineralization have been suggested to be part of Malani Igneous Suites (MIS) by Kochher (1973) and Chattopadhyay et al. (1994). However, Pareek (1981); Bhattacharjee et al. (1993); Bhushan (1995) argued that these granites are associated with older pre-Malani thermal events. The W-mineralization in Balda-Tosham belt is tectonically controlled mineralization which is found in shears and fractures. It occurs as vein type mineralization associated with quartz veins/sheets commonly intruded along the shear zones and/or joints. The wolframite is the main tungsten mineral distributed as disseminations/thin bladed crystals in the mineralized quartz veins or as fine to coarse dissemination in the greisenised granites.

The ore assemblage is nearly similar at all the locations with wolframite being the dominant mineral and subordinate minerals are scheelite, ilmenite, pyrite, arsenopyrite, chalcopyrite ± cassiterite ± molybdenite. Quartz is the main gangue mineral.

The fluid inclusion studies carried out on such deposits have demonstrated that saline aqueous ± carbonic hydrothermal fluids were responsible for the ore mobilization and deposition processes (Pollard et al. 1991; Linnen 1998; Sharma et al. 1994 etc.). Fluid inclusions in the vein quartz of the mineralized veins from the tungsten belt of northwestern India represent the syn- to post- tungsten mineralizing events and also unravel the fluid flow through the shears. Sharma et al. (1994) identified three main stages of the ore fluid evolution in the fluid inclusions study of Balda tungsten deposit. The three discrete groups of fluids in inclusions suggested from the tungsten belt of Rajasthan are: (a) high saline and high temperature fluids (b) high temperature low to moderately saline carbonic fluids and (c) the moderate to low saline and low temperature fluids. The presence of miarolitic cavities in the host granite, and similar fluid inclusions from both the granites and vein quartz further support the progressive hydrothermal evolution from the magma responsible for granite crystallization for the Degana tungsten deposit. Presence of high fluorine in mica associated with the tungsten mineralization from Degana is pointed by Srivastava and Sukhchain (2005), suggesting high HF activities during the evolution of the hydrothermal fluid. Throughout the belt, the spatial association of tungsten mineralized quartz veins within the host granite, their close association with the shear zones within the granites, pervasive greisenization in the mineralized zone and the evidence for boiling of fluid suggest a genetic linkage between the granite and tungsten mineralization.

Fluid inclusion data from the tungsten deposits of the belt further indicate a complex evolutionary history of the hydrothermal fluid. Owing to the existence of high temperature and high salinity fluids with the crystallizing host granite, wide spread greisenization took place in this stage. The wolframite mineralization occurred at this stage and is present as disseminations in greisenized granite. In the second stage of the fluid evolution complex carbonic fluid evolved which is represented by the entrapment of aqueous-carbonic fluids with variable CO_2-H_2O ratios. The boiling of the fluid is also envisaged in this stage (Sharma et al. 1994, 2003). Srivastava and Sharma (2008) proposed that the CO_2 rich inclusions are the results of continued boiling and volatile segregation. As the emplacement of the host granites in the belt is in epizonal condition (Chattopadhyay et al. 1994; Srivastava and Sinha 1997; Srivastava 2004; Srivastava and Sukhchain 2005), the gas saturation was very high and caused boiling of the fluid during the late stage of magma crystallization. Pandit and Sharma (1999) presented a fluid evolution path for the lithium mineralization of Sewariya, closely related to tungsten province of Rajasthan, India. A boiling hydrothermal fluid wherein gradual increase in salinity occurred was identified, which finally cooled down and culminated to a low to moderate salinity –temperature hydrothermal fluid system. A near similar fluid evolution was also proposed for W-mineralization at Sewariya (Sharma et al. 2003) (Fig. 6).

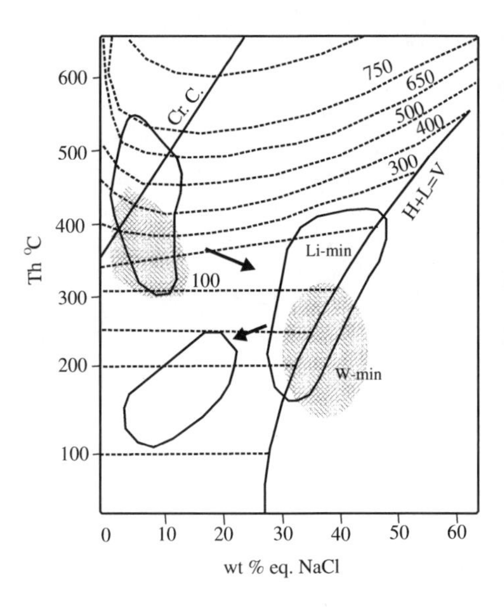

Fig. 6 Fluid evolution path on temperature-composition field for Li and W-mineralization in South Delhi Fold Belt (modified after Pandit and Sharma 1999 and Sharma et al. 2003)

In case of Degana and Tosham deposits, the excess vapour pressure of the fluid in the crystallizing granite caused the brecciation of the granite. The formation of breccia in the Degana and Tosham was a result of sudden decrease of the pressure. This caused the boiling of the fluid and emplacement of the quartz sheets/veins. Major wolframite mineralization took place during this period. A number of integrated studies of ore forming magmatic hydrothermal systems associated with acid rocks composition, volcanic—plutonic rock association have shown that salinity variation and temperature and pressure decrease are indeed major factors contributing to the formation of ore deposits (Heinrich 2006). At the late stage of hydrothermal fluid evolution, the cool meteoric water was mixed with it resulting decrease in its density and salinity. The crystallization of wolframite was ceased as a result of CO_2 depletion in the fluids. Deposition of the minor fluorite and sulphides took place at this stage.

12 Near-Surface Magmatic-Hydrothermal Processes: The Epithermal Gold Deposits

Epithermal gold deposits represent lode deposits that consist of economic concentrations of gold, silver and in some cases copper, lead and zinc. Gold is the principal commodity of epithermal deposits, and can be found as native gold, or alloyed with silver. A lode deposit is epithermal deposits characterized as having minerals either disseminated through the ore-body, or contained in a network of veins. The epithermal gold deposits are the world class gold deposits associated with either active or geologically recent volcanic environments. The term 'epithermal' is derived from Lindgren's (1933) classification of ore deposits and refers

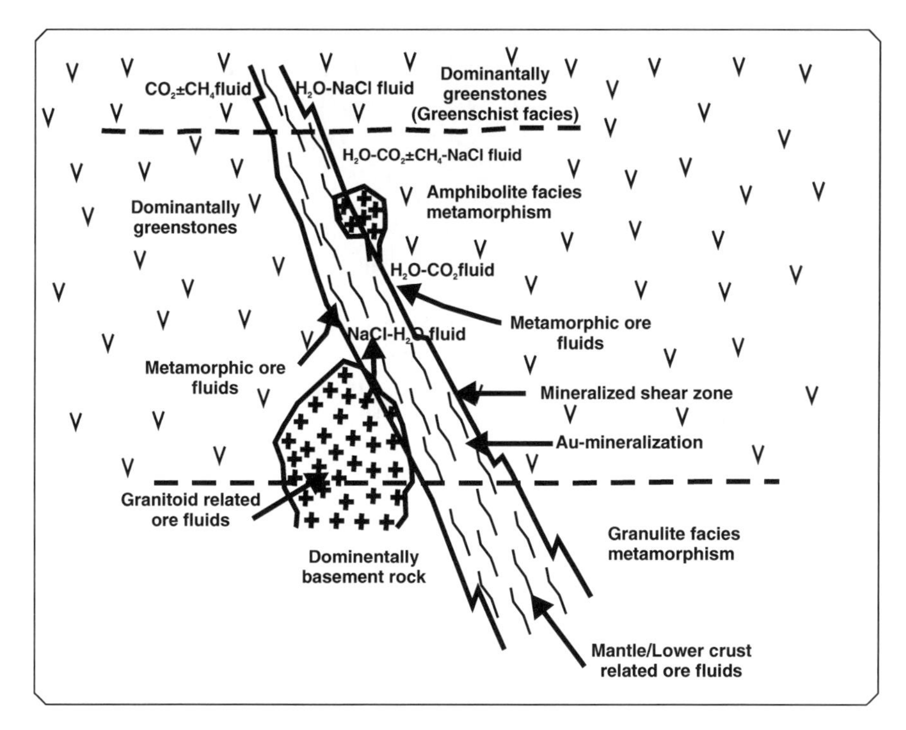

Fig. 7 Magmatic-metamorphic model of the orogenic gold as applied to Indian gold deposits, shown with general nature of the ore forming fluids. Model reproduced from Sarkar (2010), who adapted from Groves (1993), and the broad nature of ore fluids are from various fluid inclusion studies mentioned in the text

to those that formed at shallow crustal levels (i.e. epizone). Many studies of this ore-forming environment and their comparisons with active, modern analogues such as the Taupo volcanic zone on the north island of New Zealand, have shown that epithermal deposits typically form at temperatures between 160 and 270 °C and pressures equivalent to depths between 50 and 1000 m (Cooke and Simmons 2000; Hedenquist et al. 2000). There are two contrasting styles of mineralization that are now recognized in epithermal deposits: high-sulfidation and low-sulfidation types. These terms refer specifically to the oxidation state of sulfur in the ore fluid. Their chemistry and pH also relates to the nature of alteration associated with each type. It is now widely acknowledged that the gold in ore-forming quantities in magmatic hydrothermal systems may be transported to the site of deposition by low-density supercritical aqueous fluids or vapours (Heinrich et al. 1999; Williams-Jones et al. 2002; Williams-Jones and Heinrich 2005; Simon et al. 2007; Pirajno 2009; Seo et al. 2009; Landtwing et al. 2010; Mavrogenes et al. 2010). Broadly, all these records of the hydrothermal systems are found preserved as fluid inclusions in ore and gangue minerals, and show that the portions of these systems are invariably vapour-dominated (Heinrich et al. 2004; Landtwing et al. 2010). Furthermore, analyses of the fluid inclusions indicate that commonly the concentration of gold is substantially higher in the vapour than in the coexisting brine (Ulrich et al. 1999; Seo et al. 2009; Monecke et al. 2011). This provides very strong evidence that the associated gold deposits have been formed in large part by magmatic vapour. On the other hand, models of the orogenic gold (Groves 1993; Groves et al. 2003; Sarkar 2010) show involvement of the fluids from diverse origin in the mineralization. Various workers have studied fluid inclusions to investigate the ore forming fluids responsible for the gold mineralization in south Indian craton (Misra 2010; Misra and Panigrahi 1999; Pandalai et al. 2003; Nevin et al. 2010; Krishnamurthi et al. 2010). The diversity in their opinions is regarding the involvement of metamorphogenic ore fluid with or without mixing of the granite derived hydrothermal fluid and mantle/lower crust related ore fluids migrating in the shear conduits. A general nature of the ore forming fluids investigated by these workers is shown in the model of orogenic gold mineralization adapted by Sarkar (2010) from Groves (1993) for the Indian gold deposits (Fig. 7).

Overall, the hydrothermal fluids through the flow of elements and energy contribute significantly in controlling the chemical budget of the lithosphere and in the formation of economically useful mineral deposits.

Acknowledgements RS thanks Director, Wadia Institute of Himalayan Geology for the encouragement and permission. Authors are thankful to Prof. Santosh Kumar for giving this opportunity to contribute. PKS thanks his students for the help in preparation of some figures. Reviewers: Prof. M. Obeid, Fayoum University, Egypt and Dr. R. Krishnamurthy IIT, Roorkee, India are thanked for their comments. Authors thank John Wiley and Narosa publications for permission to reproduce some figures. This lecture note article has been greatly benefited by many published work including those of Burnham, Candela, Heinrich, Roedder, Bodnar, Holloway, Shinohara, Thomson and many others.

References

Anderko K, Pitzer KS (1993) Equation-of-state representation of phase equilibria and volumetric properties of the system NaCl-H_2O above 573 K. Geochim Cosmochim Acta 57:1657–1680

Atkinson WW, Hunter W (2002) Comb quartz layers in the porphyry copper deposit at Yerington, Nevada. Geol Soc Am Abstracts with Programs 34(5):15

Banerji S, Pandit MK (1995) Lithium and tungsten mineralization in Sewariya pluton, South Delhi fold belt, Rajasthan : evidences for preferential host rock affinity. Curr Sci 69:252–256

Beane RE (1983) The magmatic-meteoric transition. Geothermal resources council, Special Report No. 13, pp 245–253

Beane RE, Bodnar RJ (1995) Hydrothermal fluids and hydrothermal alteration in porphyry copper deposits. In: Pierce FW, Bohm JG (eds) Porphyry copper deposits of the American Cordillera, Arizona Geol. Soc. Digest, vol 20, pp 83–93

Bhattacharjee J, Fareeduddin, Jain SS (1993) Tectonic setting, petrochemistry and tungsten metallogeny of Sewariya granite in south Delhi fold belt, Rajasthan. J Geol Soc Ind 42:3–16

Bhushan SK (1995) Late Proterozoic continental growth; implications from geochemistry of acid magmatic events of western Indian Craton Rajasthan. Memorial Geol Soc Ind 34:339–355

Bhushan SK (2000) Malani rhyolite-a review. Gondwana Res 3:65–77

Bischoff JL, Pitzer KS (1989) Liquid-vapor relations for the system NaCl-H_2O: summary of the P-T-x surface from 300 to 500 °C. Am J Sci 289:217–248
Bischoff JL (1991) Densities of liquids and vapors in boiling NaCl-H_2O solutions: a PVTX summary from 300 to 500 °C. Am J Sci 291:309–338
Bodnar RJ (1992) Can we recognize magmatic fluid inclusions in fossil system based on room temperature phase relations and microthermometric behaviour? Geol Surv Jpn Rep 279:26–30
Bodnar RJ (1994) Synthetic fluid inclusions. XII. experimental determination of the liquidus and isochors for a 40 wt % H_2O-NaCl solution. Geochim Cosmochim Acta 58:1053–1063
Bodnar RJ (1995) Fluid inclusion evidence for a magmatic source for metals in porphyry copper deposits. In: Thompson JFH (ed) Mineralog. Assoc. Canada Short Course, vol 23, pp 139–152
Bodnar RJ, Reynolds TJ, Kuehn CA (1985) Fluid-inclusion systematics in epithermal systems. In: Berger BR, Bethke PM (ed) Geology and Geochemistry of Epithermal Systems. Reviews in Economic Geology, Soc. Econ. Geologists, vol 2, pp 73–97
Bottrell SH, Yardley BWD (1988) The composition of a primary granite derived ore fluid from S.W. England, determined by fluid inclusion analysis. Geochim Cosmochim Acta 52:585–588
Brimhall GH Jr (1980) Deep hypogene oxidation of porphyry copper potassium-silicate proto-ore at Butte, Montana: a theoretical evaluation of the copper remobilization hypothesis. Econ Geol 75:384–409
Brimhall GH Jr, Ghiorso MS (1983) Origin and ore-forming consequences of the advanced argillic alteration process in hypogene environments by magmatic gas contamination of meteoric fluids. Econ Geol 78:73–90
Burnham CW (1967) Hydrothermal fluids at the magmatic stage. In: Barnes HL (ed) Geochemistry of hydrothermal ore deposits, holt. Rinehart and Winston Inc, New York, pp 34–76
Burnham CW (1975) Water and magmas: a mixing model. Geochim Cosmochim Acta 39:1077–1084
Burnham CW (1979) Magma and hydrothermal fluids. In: Barnes HL (ed) Geochemistry of hydrothermal ore deposits, 2nd edn. Wiley Interscience, New York, pp 71–136
Burnham CW (1997) Magmas and hydrothermal fluids. In: Barnes HL (ed) Geochemistry of hydrothermal ore deposits, 3rd edn. Wiley and Sons, New York, pp 63–123
Candela PA (1991) Physics of aqueous phase exsolution in plutonic environments. Am Mineral 76:1081–1091
Candela PA (1992) Controls on ore metal ratios in granite-related ore systems: an experimental and computational approach. Trans R Soc Edinburgh: Earth Sci 83:317–326
Candela PA (1997) A review of shallow, ore-related granites: textures, volatiles, and ore metals. J Petrol 38:1619–1633
Candela PA, and Blevin PL (1995) Physical and chemical magmatic controls on the size of magmatic–hydrothermal ore deposits. In: Clark AH (ed) Giant ore deposits- II. Kingston Queens University, pp 2–37
Candela PA, Holland HD (1984) The Partitioning of copper and molybdenum between silicate melts and aqueous fluids. Geochim Cosmochim Acta 48:373–380
Candela PA, Holland HD (1986) A mass transfer model for copper and molybdenum in magmatic hydrothermal systems: origin of porphyry-type ore deposits. Econ Geol 81:1–19
Candel PA, Piccoli PM (1995) Model ore-metal partitioning from melts into vapor and vapor–brine mixtures. In: Thompson JFH (ed) Magmas fluids and ore deposits, Mineralalog. Assoc. Canada, Short-Course, vol 23, pp 101–127
Carten RB, Geraghty EP, Walker BM (1988) Cyclic development of igneous features and their relationship to high-temperature hydrothermal features in the Henderson porphyry molybdenum deposit. Colorado Econ Geol 83:266–296
Cathles LM (1981) Fluid flow and genesis of hydrothermal ore deposits. Econ Geol 75th Anniversary 424–457
Chattopadhyay B, Chattopadhyay S, Bapna VS (1994) Geology and geochemistry of Degana Pluton a Proterozoic Rapakivi granite in Rajasthan, India. Mineral Petro 50:69–82

Choudhary AK, Gopalan K, Sastry CA (1984) Present status of the geochronology of the precambrian rocks of Rajasthan. Tectonophysics 105:131–140
Christiansen EH, Burt DM, Sheridan MF, Wilson RT (1983) The petrogenesis of topaz rhyolites from the western United States. Contrib Miner Petrol 83:16–30
Cline JS, Bodnar RJ (1991) Can economic porphyry copper mineralization be generated by a typical calc-alkaline melt? J Geophys Res 96:8113–8126
Cloos M (2002) Bubbling magma chambers, cupolas and porphyry copper deposits. In: Ernst G (ed) Frontiers in geochemistry, organic, solution, and ore deposit geochemistry, international book series, vol 6, pp 191–217
Cooke DR, Simmons SF (2000) Characteristics and genesis of epithermal gold deposits. Rev Econ Geol 13:221–244
Crawford ML (1981) Phase equilibria in aqueous fluid inclusions. In: Hollister LS, Crawford ML (eds) Short course in fluid inclusions, application to petrology, Mineralog. Ass. Canada, vol 6, pp 75–100
Dilles JH, Einaudi MT (1992) Wall-rock alteration and hydrothermal flow paths about the Ann-Mason porphyry copper deposit, Nevada–a 6-km vertical reconstruction. Econ Geol 87:963–2001
Drummond SE, Ohmoto H (1985) Chemical evolution and mineral deposition in boiling hydrothermal systems. Econ Geol 80:126–147
Duan Z, Moller N, Weare JH (2003) Equations of state for the NaCl–H_2O–CH_4 system and the NaCl- H_2O–CO_2–CH_4 system: phase equilibria and volumetric properties above 573 k. Geochim Cosmochim Acta 67(4):671–680
Eadington PJ (1983) A fluid inclusion investigation of ore formation in a tin-mineralized granite, New England, New South Wales. Econ Geol 78:204–1221
Eugster HP (1984) Granites and hydrothermal ore deposits: a geochemical framework. Miner Mag 49:7–23
Frezzotti ML, Andersen T, Neumann ER, Simonsen SL (2002) Carbonatite melt–CO_2 fluid inclusions in mantle xenoliths from Tenerife, Canary Islands: a story of trapping, immiscibility and fluid–rock interaction in the upper mantle. Lithos 64:77–96
Groves DI (1993) The crustal continuum model for late-Archaean lode-gold deposits of the Yilgarn Block, Western Australia. Miner Deposita 28:366–374
Groves DI, Goldfarb RJ, Robert F, Hart CJR (2003) Gold deposits in metamorphic belts: overview of current understanding, outstanding problems, future research and exploration significance. Econ Geol 98:1–29
Haas JLJ (1976) Physical properties of the coexisting phases and thermochemical properties of the H_2O component in boiling NaCl solutions: Geol Surv Bull 1421-A:73
Hanor JS (1979) The sedimentary genesis of hydrothermal fluids. In: Barnes HL (ed) Geochemistry of hydrothermal ore deposits, 2nd edn. John Wiley and Sons, New York, pp 137–168
Hanson RB, Glazner AF (1995) Thermal requirements for extensional emplacement of granitoids. Geology 23:213–216
Harris AC, Kamenetsky VS, White NC, Van-Achterbergh E, Ryan CG (2003) Silicate-melt inclusions in quartz veins: linking magmas and porphyry Cu deposits. Science 302:2109–2111
Hayba DO, Ingebritsen SE (1997) Multiphase ground water flow near cooling plutons. J Geophys Res 102:12235–12252
Hedenquist JW, Lowenstern JB (1994) The role of magmas in the formation of hydrothermal ore deposits. Nature 370:519–527
Hedenquist JW, Arribas RA, Gonzalez UE. (2000) Exploration for epithermal gold deposits. In: Hagemann S, Brown PE (eds) Gold in 2000. Soc. Econ. Geologists, Rev. Econ. Geol., vol 13, pp 245–77
Heinrich CA (2006) From fluid inclusion microanalysis to large scale hydrothermal mass transfer in the Earth's interior. J Mineral Petro Sci 101:110–117
Heinrich CA, Ryan CG, Mernagh TP, Eadington PJ (1992) Segregation of ore metals between magmatic brine and vapor: a fluid inclusion study using PIXE microanalysis. Econ Geol 87:1566–1583

Heinrich CA, Gunther D, Audetat A, Ulrich T, Frischknecht R (1999) Metal fractionation between magmatic brine and vapour, determined by microanalysis of fluid inclusions. Geology 27:755–758

Heinrich CA, Driesner T, Stefansson A, Seward TM (2004) Magmatic vapour contraction and the transport of gold from the porphyry environment to epithermal ore deposits. Geology 32:761

Hellmann R (1994) The albite-water system: Part I. The kinetics of dissolution as a function of pH at 100, 200 and 300°C. Geochim Cosmochim Acta 58:595–611

Holland HD (1972) Granites, solutions and base metal deposits. Econ Geol 67:281–301

Hutchison CS (1983) Economic deposits and their tectonic setting. Macmillan Press, London, p 365

Jairath S, Sharma M (1986) Physico-chemical conditions of ore deposition in Malanjkhand copper sulphide deposit. Proc Ind Acad Sci Earth Planet Sci 95:209–221

John DA Ballantyne GH (1998) Geology and ore deposits of the Oquirrh and Wasatch mountains, Utah: society of economic geologists guide book series, vol 29, p 256

Keevil NB (1942) Vapor pressures of aqueous solutions at high temperatures. Am Chem Soc J 64:841–850

Kilinc IA, Burnham CW (1972) Partitioning of chloride between a silicate melt and coexisting aqueous phase from 2 to 8 kilobars. Econ Geol 67:231–235

Kirkham RV Sinclair WD (1988) Comb quartz layers in felsic intrusions and their relationship to the origin of porphyry deposits. In: Taylor RP, Strong DF (eds) Recent advances in the geology of granite-related mineral deposits. The Canadian institute of mining and metallurgy, vol 39, pp 50–71

Kochher N (1973) The occurrence of ring dyke in the Tosham igneous complex, Hissar, Haryana. J Geol Soc Ind 14:190–193

Krishnamurthi R, Sen AK, Pradeepkumar T, Sharma R (2010) Gold mineralization in the Southern granulite terrane of Peninsular India. In: Deb M, Goldfarb RJ (eds) Gold metallogeny in India and beyond. Narosa Pub, House Delhi, pp 222–233

Landtwing M, Furrer C, Redmond P, Pettke T, Guillong M, Heinrich CA (2010) The Bingham canyon porphyry Cu–Mo– Au deposit. III. Zoned copper–gold ore deposition by magmatic vapour expansion. Econ Geol 105:91–118

Lattanzi P (1991) Applications of fluid inclusions in the study and exploration of mineral deposits. Eur J Miner 3:689–701

Lemmlein GG, Klevtsov PV (1961) Relations among the principle thermodynamic parameters in a part of the system H_2O-NaCl. Geochemistry 2:148–158

Lindgren W (1933) Mineral deposits, 4th edn. McGraw-Hill, New York, p 930

Linnen RL (1998) Depth of emplacement, fluid provenance and metallogeny in granitic Terrains: a comparison of western Thailand with other tin belts. Mineral Deposita 33:461–476

Lowenstern JB, Sinclair WD (1996) Exsolved magmatic fluid and its role in the formation of combined layered quartz at the Cretaceous Logtung W-Ma deposit, Yukon Territory, Canada. Trans R Soc Edinburgh: Earth Sci 87:291–304

Lowenstern JB (2001) Carbon dioxide in magmas and implications for hydrothermal systems. Mineral Deposita 36:490–502

Lowenstern JB, Mahood GA, Rivers ML, Sutton SR (1991) Evidence for extreme partitioning of copper into a magmatic vapor phase. Science 252:1405–1408

Manning D, Pichavant M (1984) Experimental studies of the role of fluorine and boron in the formation of late-stage granitic rocks and associated mineralization. Int Geol Cong 27:386–387

Mavrogenes JA, Henley RW, Reyes AG, Berger B (2010) Sulphosalt melts: evidence of high-temperature vapour transport of metals in the formation of high-sulphidation lode gold. Econ Geol 105:257–262

Misra B (2010) Metamorphism and hydrothermal fluid evolution in relation to gold metallogeny, Dharwar Craton, southern India. In: Deb M, Goldfarb RJ (eds) Gold metallogeny India and Beyond. Narosa publishing house Pvt. Ltd, New Delhi, pp 154–167

Misra B, Panigrahi MK (1999) Fluid evolution in the Kolar gold field: evidence from fluid inclusion studies. Miner Deposita 34:173–181
Mollai H, Sharma R, Pe-Piper G (2009) Copper mineralization around the Ahar Batholith, north of Ahar, (NW Iran): evidence for fluid evolution and the origin of the skarn ore deposit. Ore Geol Rev 35:401–414
Monecke T, Kempe Ulf, Trinkler M, Thomas R, Dulski P, Wagner T (2011) Unusual rare earth element fractionation in a tin-bearing magmatic hydrothermal system. Geology 39:294–298
Nevin GC, Malli VM, Pandalai HS (2010) Modeling of Hutti gold deposits: Challenges and constraints. In: Deb M, Goldfarb RJ (eds) Gold metallogeny India and Beyond. Narosa publishing house Pvt. Ltd, New Delhi, pp 168–190
Norton DL (1984) Theory of hydrothermal systems. Ann Rev Earth Planet Sci 12:155–177
Norton D, Knight J (1977) Transport phenomena in hydrothermal systems: cooling plutons. Am J Sci 277:937–981
Pandalai HS, Jadav GN, Mathew B, Panchapakesan V, Raju KK, Patil ML (2003) Dissolution channels in quartz and the role of pressure changes in gold and sulphide deposition in Archaean, greenstone-hosted, Hutti gold deposits, Karnataka, India. Miner Deposita 38:597–624
Pandit MK, Sharma R (1999) Lithium mineralization in Proterozoic leucogranite, South Delhi Fold Belt, Western India : role of fluids in ore mobilization and deposition. Anais da Academia Brasileira de Ciencias 71:67–88
Panigrahi MK, Mookherjee A, Pantulu GVC, Gopalan K (1993) Granitoids around the Malanjkhand, copper deposit: Types, age relationship. Proc Ind Acad Sci Earth Planet Sci 102:399–413
Panigrahi MK, Mookherjee A (1997) The Malanjkhand copper (+Molybdenum) deposit: mineralization from a low temperature ore fluid of granitoid affiliation. Mineral Deposita 32:133–148
Panigrahi MK, Mishra KC, Bream B, Naik RK (2004) Age of granitic activity associated with copper molybdenum mineralization at Malanjkhand, Central India. Mineral Deposita 39:670–677
Panigrahi MK, Naik RK, Pandit D, Mishra KC (2008) Reconstructing physico-chemical parameters of hydrothermal mineralization of copper at the Malanjkhand deposit, India, from mineral chemistry of biotite, chlorite and epidote. Geochem J 42:443–460
Pareek HS (1981) Petrochemistry and petrogenesis of Malani igneous suit, India. Geol Soc Am Bull 92:206–273
Philips WJ (1973) Mechanical effects of retrograde boiling and its probable importance in the formation of some porphyry ore deposits. Inst Min Metall Sect 82:90–97
Pirajno F (2009) Hydrothermal processes and mineral systems. Springer science. Springer, Berlin, p 1250
Pitzer KS, Pabalan RT (1986) Thermodynamics of NaCl in steam. Geochim Cosmochim Acta 50:1445–1454
Pollard PJ, Taylor RG (1986) Progressive evolution of alteration and tin mineralization: controls by interstitial permeability and fracture-related tapping of magmatic fluid reservoirs in tin granites. Econ Geol 81:1795–1800
Pollard PJ, Andrew AS, Taylor RG (1991) Fluid inclusions and stable isotope evidence for interaction between granites and magmatic hydrothermal fluid during formation of disseminated and pipe type mineralization at Zariplaats Tin mine. Econ Geol 86:121–141
Potter RW II, Babcock RS, Brown DL (1977) A new method for determining the solubility of salts in aqueous solutions at elevated temperatures. U.S Geol Surv J Res 5:389–395
Ramanathan A, Bagchi J, Panchapakesan J, Sahu BK (1990) Sulphide mineralization at Malanjkhand—a study. Geol Surv India Spec Publ 28:585–598
Robb L (2005) Introduction to ore forming processes. Blackwell Science Ltd, Oxford, p 373
Roedder E (1984) Fluid inclusions: reviews in mineralogy. Rev Mineral Soc Am 12:644
Roedder E (1992) Fluid inclusion evidence for immiscibility in magmatic differentiation. Geochim Cosmochim Acta 56:5–20

Roedder E, Bodnar RJ (1980) Geologic pressure determinations from fluid inclusions studies. Ann Rev Earth Planet Sci 8:263–301
Romberger SB (1982) Transport and deposition of gold and the transport of gold in hydrothermal ore solutions. Geochim Cosmochim Acta 37:370–399
Ross PS, Jebrak M, Walker BM (2002) Discharge of hydrothermal fluids from a magma chamber and concomitant formation of a stratified breccia zone at the Questa porphyry molybdenum deposit, New Mexico. Econ Geol 97:1679–1699
Rusk BG, Reed MH, Dilles JH (2008) Fluid inclusion evidence for magmatic-hydrothermal Fluid Evolution in the Porphyry Copper—molybdenum deposit at Butte. Montana Econ Geol 103:307–334
Rye RO (1993) The evolution of magmatic fluids in the epithermal environment: the stable isotope perspective. Econ Geol 88:733–752
Sarkar SC (2010) Gold mineralization in India: an introduction. In: Deb M, Goldfarb RJ (eds) Gold metallogeny India and Beyond. Narosa publishing house Pvt. Ltd, New Delhi, pp 95–122
Sarkar SC, Kabiraj S, Bhattacharya S, Pal AB (1996) Nature, origin, evolution of granitoid-hosted early Proterozoic copper-molybdenum mineralization at Malanjkhand, Central India. Mineral Deposita 31:419–431
Seedorff E, Einaudi MT (2004) Henderson Porphyry Molybdenum system, Colorado: II. decoupling of introduction and deposition of metals during geochemical evolution of hydrothermal fluids. Econ Geol 99:39–72
Seo J, Guillong M, Heinrich CA (2009) The role of sulphur in the formation of magmatic-hydrothermal copper -gold deposits. Earth Planet Sci Lett 282:323–328
Shannon JR, Walker BM, Carten RB, Geraghty EP (1982) Unidirectional solidification textures and their significance in determining relative ages of intrusions at the Henderson mine, Colorado. Geology 19:293–297
Sharma R, Srivastava P, Naik MS (1994) Hydrothermal fluids of the tungsten mineralization near Balda, the district of Sirohi, Rajasthan, India. Petrology 2:589–596
Sharma R, Banerjee S, Pandit MK (2003) W- mineralization in Sewariya area, South Delhi fold belt, Northwesten India: fluid inclusion evidence for tungsten transport and conditions of ore formation. J Geol Soc Ind 61:37–50
Shinohara H (1994) Exsolution of immiscible vapor and liquid phases from a crystallizing silicate melt: implications for chlorine and metal transport. Geochim Cosmochim Acta 58:5215–5221
Shinohara H, Kazahaya K (1995) Degassing processes related to magma chamber crystallization. In: Thompson JFH (ed) Magmas fluids and ore deposits. Mineralog. Ass. Canada Short Course, vol 23, pp 47–70
Shinohara H, Kazahaya K, Lowenstern JB (1995) Volatile transport in a convecting magma column: Implications for porphyry Mo mineralization. Geology 23:1091–1094
Shinohara H, Hedenquist J (1997) Constraints on magma degassing beneath the far Southeast Porphyry Cu–Au deposit, Philippines. J Petrol 38:1741–1752
Shinohara H, Iiyama JT, Matsuo S (1989) Partition of chlorine compounds between silicate melt and hydrothermal solutions; I partition of NaCl-KCl. Geochim Cosmochim Acta 53:2617–2630
Sikka DB, Nehru CE (2002) Malanjkhand copper deposit, India: is it not a porphyry type? J Geol Soc India 59:339–362
Simon AC, Pettke T, Candela P, Piccoli P, Heinrich C (2007) The partitioning behaviour of As and Au in S-free and S-bearing magmatic assemblages. Geochim Cosmochim Acta 71:1764–1782
Sinclair WD (2007) Porphyry deposits. In: Goodfellow WD (ed) Mineral deposits of Canada: a synthesis of major deposit-types, district metallogeny, the evolution of geological provinces, and exploration methods, Geol. Ass. Canada, mineral deposits division, no 5, pp 223–243
Sobolev VS, Kostyuk VP (1975) Magmatic crystallization based on a study of melt inclusions. Fluid Incl Res 9:182–235
Sourirajan S, Kennedy GC (1962) The system H_2O-NaCl at elevated temperatures and pressures. Am J Sci 260:115–141

Srivastava PK (2004) Geochemistry and tungsten potential of Balda granite, Balda, Sirohi district, Northwestern India. J Econ Geol Res Manage 1:205–216
Srivastava PK, Sharma R (2008) Hydrothermal fluids linked with tungsten province of Balda-Tosham belt, Northwest India. Mem Geol Soc Ind 72:125–144
Srivastava PK, Sinha AK (1997) Geochemical Characterization of tungsten-bearing granites from Rajasthan, India. J Geochem Exp 60:173–182
Srivastava PK, Sukhchain (2005) Petrographic characteristics and alteration geochemistry of granite-hosted tungsten mineralization at Degana, NW India. Res Geol 55:373–384
Srivastava PK, Sukhchain (2007) Geochemistry of the trioctahedral micas from Degana Granite. J Geol Soc Ind 69:1203–1208
Stein HJ, Hannah JL, Zimmerman A, Markey RJ, Sarkar SC, Pal AB (2004) A 2.5 Ga porphyry Cu–Mo–Au deposit at Malanjkhand, Central India: implications for late Archean continental assembly. Precamb Res 134:189–226
Sterner SM, Hall DL, Bodnar RJ (1988) Synthetic fluid inclusions. V. Solubility relations in the system NaCl-KCl-H_2O under vapor-saturated conditions. Geochim Cosmochim Acta 52:989–1006
Stolper E (1982) The speciation of water in silicate melts. Geochim Cosmochim Acta 46:2609–2620
Symonds RB, Reed MH, Rose WI (1992) Origin, speciation and fluxes of trace element gases at Augustine volcano, Alaska: insights into magma degassing and fumarolic processes. Geochim Cosmochim Acta 56:633–657
Thompson AB, Aerts M, Hack AC (2007) Liquid immiscibility in silicate melts and related systems. In: LiebscherA, Heinrich C (eds) Fluid equilibria in the crust, Mineralog. Soc. Am., Rev. Mineralogy Geochemistry, vol 65, pp 99–128
Touret JLR, Thompson AB (1993) Fluid-rock interaction in the deeper continental lithosphere. Chem Geol 108:230
Ulrich T, Gunther D, Heinrich C (1999) Gold concentrations of magmatic brines and the metal budget of porphyry copper deposits. Nature 399:676–679
Wahrenberger C. Seward TM, Dietrich V (2002) Volatile trace element transport in high temperature gases from Kudriavy volcano (Iturup, Kurile Islands, Russia). In: Hellmann R, Wood SA (eds) Water-rock Interaction, Ore deposits and environmental geochemisty: atribute to David A. Crerar, Geochem. Soc. no 7, pp 307–327
Webster JD, Holloway JR (1990) Partitioning of F and Cl between magmatic hydrothermal fluids and highly evolved granitic magmas. Geol Soc Am Spec Pap 246:21–34
Wen S, Nekvasil H (1994) Ideal associated solutions: application to the system albite-quartz-H_2O. Am Mineralogist 79:316–331
Whitney JA (1975) Vapour generation in a quartz monzonite magma: a synthetic model with application to porphyry copper deposits. Econ Geol 70:346–358
Whitney JA (1984) Volatiles in magmatic systems. In: Whitney JA, Henley RW, Truesdell AH, Barton PB (eds) Fluid-mineral equilibria in hydrothermal systems. Soc. Econ. Geologists, Rev. Econ. Geol., vol 1, pp 155–175
Whitney J (1989) Origin and evolution of silicic magmas. Rev Econ Geol 4:83–201
Wilkinson JJ (2001) Fluid inclusions in hydrothermal ore deposits. Lithos 55:229–272
Williams-Jones AE, Heinrich CA (2005) Vapour transport of metals and the formation of magmatic-hydrothermal ore deposits. Econ Geol 100:1287–1312
Williams-Jones AE, Migdisov AA, Archibald S, Xiao Z (2002) Vapour-transport of ore metals. In: Hellmann R, Wood S (eds) Water–rock interaction, ore deposits and environmental geochemistry: atribute to David A. Crerar. The Geochemical Society, no. 7, pp 279–306
Zeng Q, Nekvasil H (1996) An associated solution model for albite-water melts. Geochim Cosmochim Acta 60:59–73
Zezin DYu, Migdisov AA, Williams-Jones AE (2011) PVTx properties of H_2O–H_2S fluid mixtures at elevated temperature and pressure based on new experimental data. Geochim Cosmochim Acta 75:5483–5495

Oxidized Granitic Magmas and Porphyry Copper Mineralization

Shunso Ishihara and Akira Imai

Abstract Absence or presence of modal magnetite and whole-rock Fe_2O_3/FeO ratio of granitoids have been potentially used to recognize the bimodal series of granioids in Japan viz. magnetite-series and ilmenite-series granitoids corresponding to oxidized and reduced granitoids respectively. Oxidized granitoids are mostly responsible for the origin of metallic (copper and gold) sulfide deposits particularly of porphyry nature. Oxidation status of some copper-hosting Precambrian and Phanerozoic granitoids have been assessed in order to understand intrinsic properties of source regions and prevailed oxygenated environment evolving through time. Chemistry of hydrous mafic silicates (amphibole and biotite) and apatite has also been discussed specifically to characterize the mineralized, oxidized granitoids and explaining the key role of F, Cl and SO_3 as potential carrier of ore metals in hydrothermal system emanated from granitic magmas respectively.

1 Introduction

Based on variation in modal content of opaque minerals two series of granitoids were recognized in the Inner Zone of Southwest Japan, which was equivocally reflected in wet-chemically analysed ferric/ferrous ratio of whole rock. Thus, the difference in Fe_2O_3/FeO ratio indeed reflect the presence or absence of magnetite, which is essentially controlled by prevailing high and low oxygen fugacity of the host granitic magmas respectively (Ishihara 1971a). Later, these two-series were referred as

S. Ishihara (✉)
National Institute of Advanced Industrial Science and Technology (AIST),
Tsukuba, Japan
e-mail: s-ishihara@aist.go.jp

A. Imai
Akita University, Akita, Japan

S. Kumar and R. N. Singh (eds.), *Modelling of Magmatic and Allied Processes*, Society of Earth Scientists Series, DOI: 10.1007/978-3-319-06471-0_10,

magnetite-series and ilmenite-series, implying magnetite-bearing and magnetite-free nature of granitoids respectively (Ishihara 1977). Here, the magnetite-series and ilmenite-series are called as oxidized and reduced granitoids, respectively.

The oxidized granitoids are commonly associated with important metallic mineral deposits where the ore minerals are present as sulfides. In Japanese island, for example, 98 % of molybdenum occurs in oxidized granitoids; 97 and 96 % of lead and zinc occur as veins and skarns commonly hosted in sedimentary and igneous rocks of the oxidized igneous terrains respectively, while copper occurs ca 86 % in the same igneous terrain. Gold–silver quartz veins are seen mostly (97 %) in the oxidized series volcanic terrains (Fig. 1).

Porphyry copper deposit is a typical product of the oxidized magmas, and the most discovered ores are strongly biased to Cenozoic age (Fig. 2). It is interesting to understand the oxidation status of granitic rocks throughout the earth history, particularly of older granitic terrains (early Earth) containing porphyry copper deposits. Whole-rock major oxides viz. $Fe_2O_3 + TiO_2$–$FeO + MnO$–MgO commonly hosted in opaque and ferromagnesian minerals act as potential chemical parameters to assess the redox nature of granitic rocks (Ishihara 1971a). This clearly shows the presence or absence of magnetite among mafic minerals commonly found in granitoids, which essentially indicate oxidizing or reducing nature of magmas. Here, we examined the oxidation status of some Precambrian granitoids, which are related to copper and gold deposits in India, South Africa and Western Australia, and Phanerozoic porphyry copper deposits of Chile and the Circum-Pacific regions. Compositions of hydrous ferro-magnesian silicates and apatite from some mineralized granitoids are also utilized to decipher the redox nature of ore-bearing hydrothermal fluids and potentiality to carry ore elements, which were emanated from granitic magmas.

2 Oxidation Status of Some Precambrian Granitoids

Major chemistry of some Precambrian granitoids is plotted in terms of Fe_2O_3 + -TiO_2–$FeO + MnO$–MgO (Fig. 3a–d). Most of reduced series granitoids plot between the $FeO + MnO$ and MgO corners (Fig. 3a), depending upon FeO/MgO ratio of biotite and hornblende. Oxidized granitoids of the Mo– province plot between the biotite-hornblende field and the $Fe_2O_3 + TiO_2$ apex which are influenced by the magnetite contents.

The oldest gneissic granitoids, 3,750–3,600 Ma, occurring in the Central Dome of western Greenland appear to be magnetite-free belonging to reduced-series (Ishihara et al. 2006). In the Barberton Mountains in South Africa, reduced-series TTG (tonalite-trondhjemite-granodiorite) predominates in the Theespruit (3,450 Ma) and Kaap Valley (3,230 Ma) plutons. Nelspruit granitic batholith (3,105 Ma) is composed of oxidized-series biotite granites, which may be considered the first oxidized granite formed in the early Earth crust (Ishihara et al. 2006).

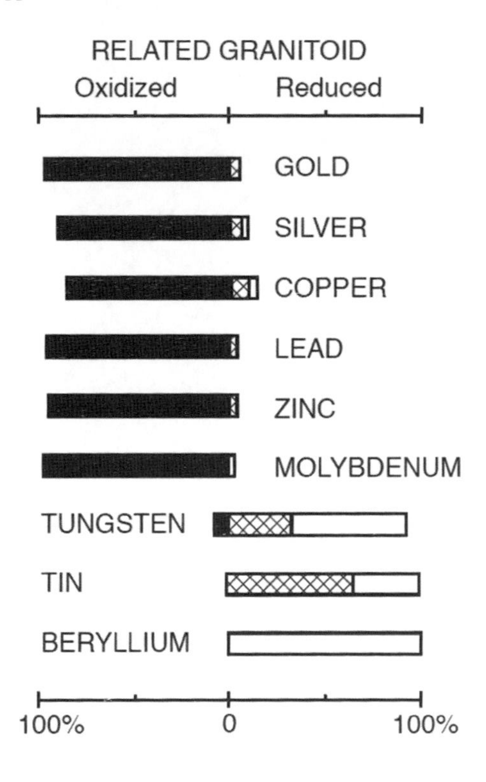

Fig. 1 Some metals in veins and skarns of the oxidized and reduced granitic terrains observed in Japanese islands. Simplified from Ishihara (1981)

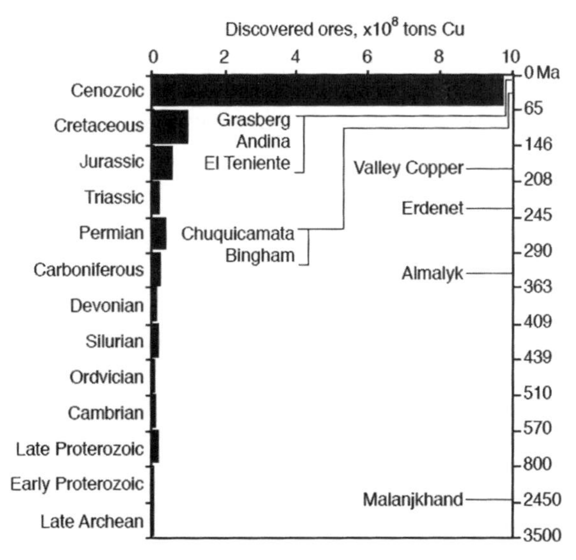

Fig. 2 Tonnage of discovered porphyry copper ores *versus* geological time. Modified from Ishihara et al. (2006)

Toward the Late Archean to Proterozoic in the Yilgarn Craton of Western Australia, oxidized-series granitoids become more common in mineralized Eastern Goldfield region. Here, the calc-alkaline series of 2,690–2,660 Ma granitoids

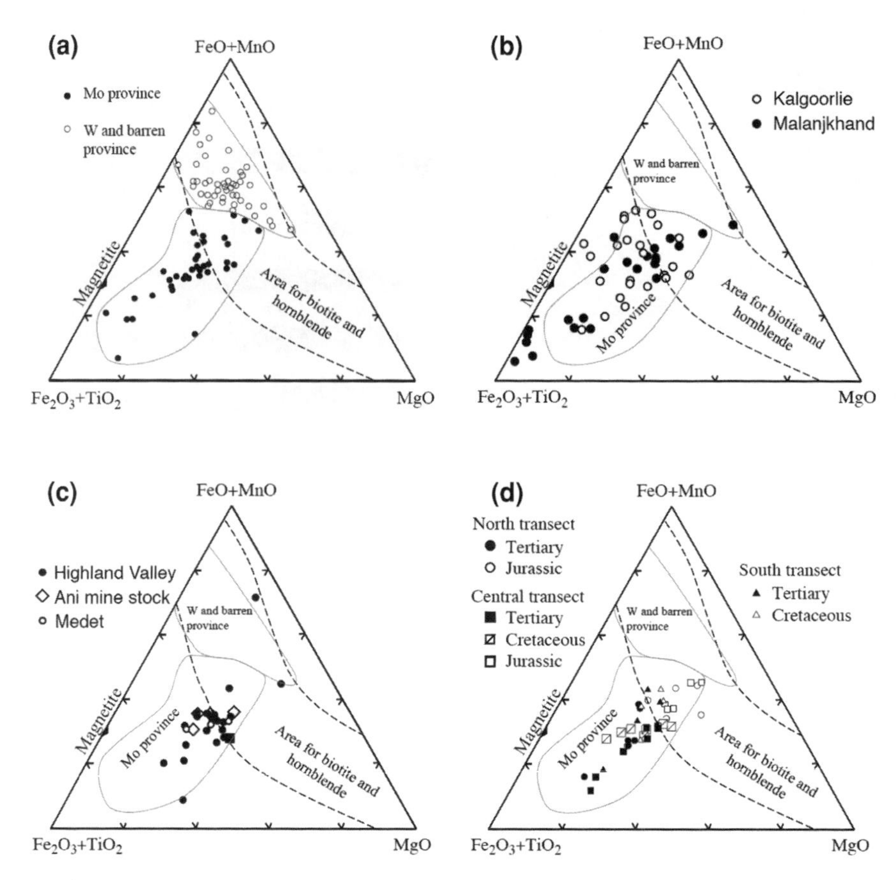

Fig. 3 Mafic components diagram plotted for the two series of granitoids of various regions. **a** Type area of Southwest Japan, **b** Kalgoorlies region of Western Australia, original data taken from Witt and Davy (1997), and Malanjkhand, India original data taken from Sarkar et al. (1996). **c** Erdenet, valley copper, Medet, and Ani mine. **d** North, Central and South transects of Chile (Ishihara and Chappell 2010)

(Smithies and Witt 1997) are all composed of magnetite-bearing rocks, and the felsic alkaline granitoids of 2,650–2,630 Ma (Smithies and Champion 1999) are even more enriched in the Fe_2O_3 component (Ishihara et al. 2006). Archean granitoids of the Kalgoorlie region of the Yilgarn Craton are predominantly biotite granodiorite and monzogranite with some hornblende-bearing varieties (Witt and Davy 1997).

Although copper occurs commonly in volcanogenic massive sulfide deposits (Mariko 2008), intrusion-related porphyry copper deposits are actually none in the Precambrian terrains (Fig. 3). However, a large copper deposit of Palaeoproterozoic age occurs at Malanjikhand, central India, which contains about 1.2×10^6 tons of copper. The bulk mineralization occurs in sheeted quartz-sulfide veins hosted in calc-alkaline tonalite-granodiorite of ~2,490 Ma intimately associated with

K-alteration defined by alkali feldspar + dusty hematite in feldspar ± biotite ± sericite (Sarker et al. 1996). The Malanjkhand granitoids plot not only in the field of oxidized-series granitoids of the Mo province but also have proximity more close to the Fe_2O_3–TiO_2 apex (Fig. 3b). This implies severe hematitization of the rock-forming magnetite present in the granitoids, which may partly be due to the post-magmatic hydrothermal activities. Interestingly copper deposits at Malanjkhand have been advocated to be of porphyry origin based on Re–Os isotopic investigations (Stein et al. 2004, 2006) and redox series evaluation of granitoids (Kumar and Rino 2007). If the copper deposits at Malanjkhand are porphyry type in strict sense then it may represent unique occurrence in Precambrian terrain.

3 Porphyry Copper-Related Granitoids

Phanerozoic granitoids related to porphyry copper mineralization have been identified as oxidized series using various methods in British Columbia, Canada (Ishihara 1971b) and northern Chile (Ishihara et al. 1984). The available data are plotted in the $Fe_2O_3 + TiO_2$–$FeO + MnO$–MgO diagram (Fig. 3a–d). The porphyry copper deposits in the Triassic (210 Ma) granitoids of the Guichon Creek batholith occur predominantly as net-veining system rather than dissemination with potassic alteration (Casselman et al. 1995). The published (Ishihara 1971b) and unpublished (W. J. Davis, pers comm) data of unaltered and least-altered granitoids plot within the field of Mo– province of Japan and shift slightly towards MgO apex, implying MgO-rich character of the granitoids (Fig. 3c). Cretaceous granitoids sampled from Medet Mine pit and Miocene granitoids of Ani Mine stock veined with chalcopyrite-quartz and some chlorites (Ishihara and Chappell 2010) plot equivocally in the region of the Guicheon Creek batholiths which suggest their similar characters.

In the northern part of Chile, basaltic andesite and quartz dioritic magmatism initiated in Jurassic along the Pacific coast and progressively advanced into the interior. Associated with the high-level intrusions, the world leading porphyry copper deposits are known to occur in three N–S trending belts (Fig. 4). The western-most Paleocene-Early Eocene porphyry copper belt contains all the three major deposits of Toquepara, Cuajone and Cerro Verde in Peru, and Cerro Colorado and Chimborazo in Chile. Late Eocene-Early Oligocene belt, next to the east, is composed of gigantic deposits containing more than 10 million tons of copper metal at Collahuasi (33.2 MT), Chuquicamata (66.4 MT), Escondida (14.1 MT), El Sarvador (11.3 MT), Escondida Norte (14.1 MT), Collahuasi (33.2 MT) and Radomiro Tornic (19.9 MT). Another gigantic deposit is distributed in the easternmost part of the Middle Miocene to Early Pliocene belt, such as Los Pelambres (26.8 MT), Rio Blando/Los Broncos (56.7 MT) and El Teniente (94.4 MT) (Camus 2003).

Regional and ore-related granitoids were studied in Northern, Central and Southern transects of Chile belt as shown in Fig. 4, which are plotted in

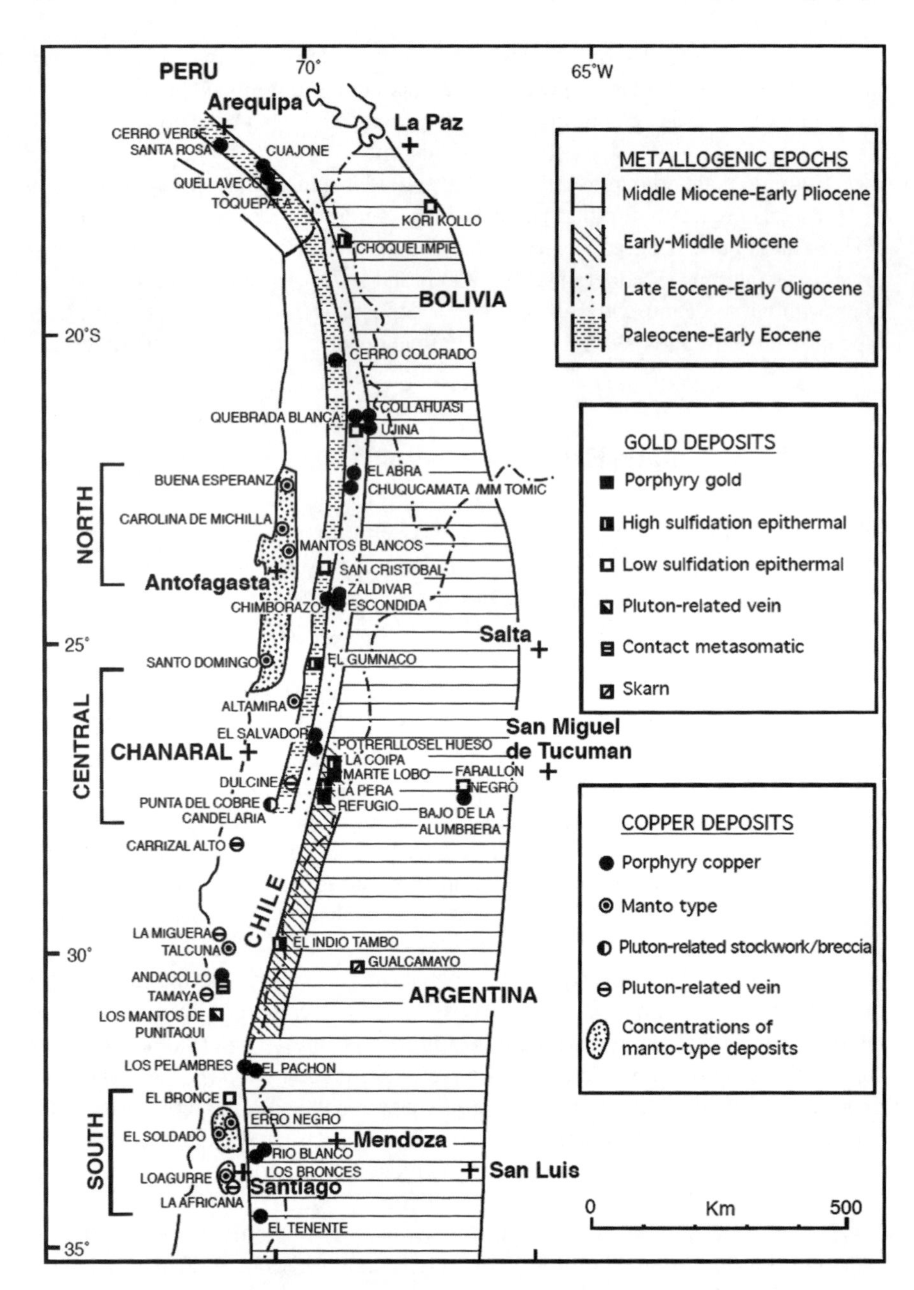

Fig. 4 Distribution of metallic ore deposits of igneous origin in the northern Chile-Peru and locations of the studied transects (from Ishihara and Chappell 2010)

the Fe_2O_3 + TiO_2–FeO + MnO–MgO diagram (Fig. 3d). All the samples are distributed in region of the Mo– province, and the older Jurassic and Cretaceous granitoids plot more closely to the biotite-hornblende region. Tertiary granitoids plot closer to the Fe_2O_3 + TiO_2 apex, implying that they are largely oxidized in nature.

4 Mafic Silicates and Apatite Compositions of Chilean Rocks

All the intrusives hosting the porphyry copper are composed of hornblende and biotite bearing granitic rocks and are absolutely devoid of two-mica (muscovite-biotite) granites. The electron-probe micro-analysis (EPMA) of hornblende, biotite and apatite have been carried out following the methods as described by Imai (2002), in order to understand nature of host granitic magmas and further to assess their potential to mineralization. The analytical results are shown in Tables 1, 2 and 3.

4.1 Hornblende and Biotite

As the iron is consumed to form magnetite in the early stage of crystallization of oxidized granitoids, both hornblende and biotite are expected to become Mg-rich with increasing silica content. On the other hand, absence of magnetite causes precipitation of Fe-rich mafic silicates in reduced granitoids. These have been demonstrated citing an example from the Japanese granitoids (Czamanske et al. 1981). Hornblende and biotite of the Chilean granitoids were analyzed and listed in Tables 1 and 2, and are plotted in Fig. 5.

In the Chuquicamata area, the Andina granodiorite is not related to the copper mineralizations in a strict sense, but Fortuna granodiorite and West porphyry, the granitoids from El Sarvador, Rio Branco and El Teniente deposits, are all ore-related. Amphibole and biotite of the Chilean granitoids are lower in the Fe/(Fe + Mg) ratio than those of the typical oxidized series of the Mo– province in Japan. The MgO-rich character of the Chilean rocks implies the difference in source rocks and also suggests higher oxygen fugacity of the granitic melts.

4.2 Apatite

Primary (euhedral) apatite crystallizing in melts contains fluorine, chlorine and sulfur, which are critical carriers of ore components. In the type locality of Japan, fluorine contents of apatite in the oxidized granodiorite and granitic series of the Mo-mineralized Sanin province vary from 2.3 to 3.6 wt% whereas fluorine

Table 1 Electron-probe micro-analysis (EPMA) of amphiboles from ore-related granodiorites in the Chuquicamata, El Salvador and Rio Blanco mines

Sample ID	79CHU06 (63.1 %)		79CHU03 (64.3 %)		79CHU05 (65.1 %)		791104-2 4(61.7 %)		79031401 (66.6 %)	
	Andina Gd		West Porphyry		Fortune Gd		El Salvador		Rio Blanco	
Number of analysis	24		31		26		24		40	
	av	($\pm 1\sigma$)	av	($\pm 1\sigma$)	Av	($\pm 1\sigma$)	av	($\pm 1\sigma$)	av	($\pm 1\sigma$)
SiO_2 (wt%)	50.65	(1.48)	51.35	(0.94)	53.12	(1.04)	51.46	(1.16)	51.54	(1.82)
Al_2O_3	4.25	(1.01)	4.15	(0.65)	2.84	(0.72)	3.88	(0.86)	4.11	(1.31)
TiO_2	0.81	(0.25)	0.80	(0.20)	0.40	(0.23)	0.84	(0.18)	0.93	(0.42)
FeO[a]	10.71	(0.74)	10.28	(0.56)	9.31	(0.52)	10.18	(0.53)	10.66	(0.94)
MnO	0.46	(0.04)	0.58	(0.04)	0.46	(0.06)	0.27	(0.04)	0.38	(0.05)
MgO	16.65	(0.80)	16.90	(0.55)	18.00	(0.43)	17.07	(0.60)	16.61	(0.97)
CaO	11.72	(0.44)	11.79	(0.14)	12.08	(0.22)	11.64	(0.17)	11.70	(0.37)
Na_2O	0.85	(0.20)	0.76	(0.14)	0.50	(0.17)	1.13	(0.22)	0.86	(0.34)
K_2O	0.48	(0.31)	0.34	(0.06)	0.20	(0.09)	0.36	(0.07)	0.34	(0.13)
Cl	0.12	(0.03)	0.06	(0.01)	0.05	(0.04)	0.06	(0.01)	0.09	(0.04)
=O	−0.03		−0.01		−0.01		−0.01		−0.02	
F	0.26	(0.09)	0.13	(0.08)	0.14	(0.09)	0.22	(0.09)	0.11	(0.07)
=O	−0.11		−0.05		−0.06		−0.09		−0.05	
Total	96.82		97.07		97.03		97.00		97.26	
Cation (O = 23)										
Si	7.384	(0.151)	7.418	(0.104)	7.611	(0.117)	7.418	(0.104)	7.435	(0.200)
Al	0.732	(0.178)	0.708	(0.114)	0.480	(0.124)	0.708	(0.114)	0.701	(0.228)
Ti	0.090	(0.028)	0.087	(0.022)	0.043	(0.025)	0.087	(0.022)	0.102	(0.047)
Fe	1.307	(0.101)	1.243	(0.072)	1.115	(0.066)	1.243	(0.072)	1.287	(0.123)
Mn	0.057	(0.006)	0.071	(0.004)	0.056	(0.007)	0.071	(0.004)	0.046	(0.007)
Mg	3.619	(0.141)	3.638	(0.107)	3.845	(0.078)	3.638	(0.107)	3.571	(0.179)
Ca	1.831	(0.065)	1.825	(0.021)	1.854	(0.029)	1.825	(0.021)	1.808	(0.045)

(continued)

Table 1 (continued)

Sample ID	79CHU06 (63.1 %)		79CHU03 (64.3 %)		79CHU05 (65.1 %)		791104-2 4(61.7 %)		79031401 (66.6 %)	
	Andina Gd		West Porphyry		Fortune Gd		El Salvador		Rio Blanco	
Number of analysis	24		31		26		24		40	
	av	($\pm 1\sigma$)	av	($\pm 1\sigma$)	Av	($\pm 1\sigma$)	av	($\pm 1\sigma$)	av	($\pm 1\sigma$)
Na	0.240	(0.059)	0.214	(0.039)	0.140	(0.047)	0.214	(0.039)	0.241	(0.096)
K	0.090	(0.058)	0.062	(0.012)	0.036	(0.016)	0.062	(0.012)	0.063	(0.024)
Cl	0.029	(0.007)	0.015	(0.003)	0.012	(0.010)	0.015	(0.003)	0.022	(0.009)
F	0.120	(0.041)	0.059	(0.037)	0.065	(0.040)	0.059	(0.037)	0.051	(0.033)
X_{Mg}	0.735	(0.022)	0.745	(0.016)	0.775	(0.014)	0.745	(0.016)	0.735	(0.028)

Total Fe content is expressed as FeO[a] . X_{Mg} denotes atomic Mg/(Mg + Fe). The values written in parenthesis with sample IDs are whole-rock silica content.
Gd Granodiorite

Table 2 Electron-probe micro-analysis (EPMA) of biotites from ore-related granodiorites in the Chuquicamata, El Teniente and Rio Blanco mines

Sample ID	79CHU06 (63.1 %)		79CHU03 (64.3 %)		79CHU05 (65.1 %)		79PTL (59.0 %)		79031401(66.6 %)	
	Andina Gd		West Porphyry		Fortune Gd		El Teniente		Rio Blanco	
Number of analysis	42		29		30		79		52	
	av	($\pm 1\sigma$)	av	($\pm 1\sigma$)	Av	($\pm 1\sigma$)	av	($\pm 1\sigma$)	av	($\pm 1\sigma$)
SiO_2 (wt%)	37.98	(0.37)	37.62	(0.23)	37.83	(0.36)	38.23	(0.50)	37.63	(0.28)
Al_2O_3	12.95	(0.16)	13.47	(0.16)	13.50	(0.25)	13.43	(0.42)	13.21	(0.19)
TiO_2	4.18	(0.17)	4.08	(0.35)	3.73	(0.37)	4.26	(0.25)	4.10	(0.19)
FeO^a	14.63	(0.30)	14.77	(0.31)	14.77	(0.26)	14.13	(0.49)	15.94	(0.38)
MnO	0.26	(0.04)	0.30	(0.04)	0.28	(0.04)	0.08	(0.03)	0.21	(0.03)
MgO	15.13	(0.23)	14.91	(0.36)	15.01	(0.30)	15.33	(0.48)	14.05	(0.24)
CaO	0.02	(0.03)	0.05	(0.04)	0.04	(0.10)	0.01	(0.02)	0.01	(0.02)
Na_2O	0.11	(0.02)	0.10	(0.03)	0.10	(0.02)	0.18	(0.03)	0.08	(0.02)
K_2O	9.12	(0.48)	9.05	(0.23)	9.12	(0.46)	8.89	(0.14)	9.28	(0.19)
Cl	0.33	(0.02)	0.18	(0.03)	0.22	(0.02)	0.19	(0.02)	0.23	(0.02)
=O	−0.07		−0.04		−0.05		−0.04		−0.05	
F	0.44	(0.11)	0.18	(0.09)	0.18	(0.11)	0.10	(0.09)	0.14	(0.11)
=O	−0.19		−0.08		−0.08		−0.04		−0.06	
Total	94.88		94.60		94.67		94.74		94.77	
Cation (O=22)										
Si	5.749	(0.034)	5.680	(0.023)	5.707	(0.053)	5.719	(0.052)	5.707	(0.033)
Al	2.310	(0.025)	2.397	(0.027)	2.401	(0.044)	2.368	(0.076)	2.361	(0.032)
Ti	0.476	(0.019)	0.463	(0.039)	0.423	(0.041)	0.479	(0.029)	0.467	(0.022)
Fe	1.852	(0.043)	1.865	(0.041)	1.863	(0.032)	1.767	(0.065)	2.023	(0.046)
Mn	0.033	(0.005)	0.039	(0.006)	0.037	(0.004)	0.010	(0.004)	0.027	(0.004)
Mg	3.413	(0.053)	3.356	(0.079)	3.376	(0.064)	3.418	(0.099)	3.177	(0.058)
Ca	0.004	(0.004)	0.008	(0.006)	0.007	(0.017)	0.002	(0.002)	0.002	(0.003)

(continued)

Table 2 (continued)

Sample ID	79CHU06 (63.1 %)		79CHU03 (64.3 %)		79CHU05 (65.1 %)		79PTL (59.0 %)		79031401(66.6 %)	
	Andina Gd		West Porphyry		Fortune Gd		El Teniente		Rio Blanco	
Number of analysis	42		29		30		79		52	
	av	($\pm 1\sigma$)	av	($\pm 1\sigma$)	Av	($\pm 1\sigma$)	av	($\pm 1\sigma$)	av	($\pm 1\sigma$)
Na	0.031	(0.007)	0.030	(0.007)	0.029	(0.007)	0.051	(0.008)	0.024	(0.006)
K	1.762	(0.088)	1.743	(0.040)	1.756	(0.086)	1.697	(0.028)	1.796	(0.036)
Cl	0.085	(0.006)	0.046	(0.007)	0.057	(0.005)	0.049	(0.004)	0.058	(0.005)
F	0.212	(0.052)	0.087	(0.041)	0.088	(0.053)	0.047	(0.040)	0.066	(0.052)
X_{Mg}	0.648	(0.006)	0.643	(0.008)	0.644	(0.007)	0.659	(0.013)	0.611	(0.008)

Total Fe content is expressed as FeO[a] . X_{Mg} denotes atomic Mg/(Mg + Fe). The values written in parenthesis with sample IDs are whole-rock silica content.
Gd Granodiorite

Table 3 Cl, F and SO_3 contents of apatite from ore-related granodiorites in the Chuquicamata, El Salvador, El Teniente and Rio Blanco mines

Sample ID	79CHU06		79CHU03		79CHU05		791104-24		79PTL		79031401	
	Andina Gd		West Porphyry		Fortune Gd		El Salvador		El Teniente		Rio Blanco	
Number of analysis	21		40		31		21		28		42	
	av	($\pm 1\sigma$)	av	($\pm 1\sigma$)	av	($\pm 1\sigma$)	av	($\pm 1\sigma$)	av	($\pm 1\sigma$)	av	($\pm 1\sigma$)
Cl (wt%)	1.18	(0.19)	0.91	(0.27)	0.94	(0.16)	0.44	(0.04)	1.94	(0.55)	0.43	(0.13)
F	2.99	(0.31)	2.72	(0.31)	2.87	(0.37)	2.91	(0.24)	1.97	(0.33)	3.14	(0.40)
SO_3	0.07	(0.03)	0.21	(0.14)	0.24	(0.13)	0.17	(0.09)	0.20	(0.12)	0.06	(0.07)
F/Cl(atom)	4.73		5.56		5.67		12.38		1.89		13.63	

Gd Granodiorite

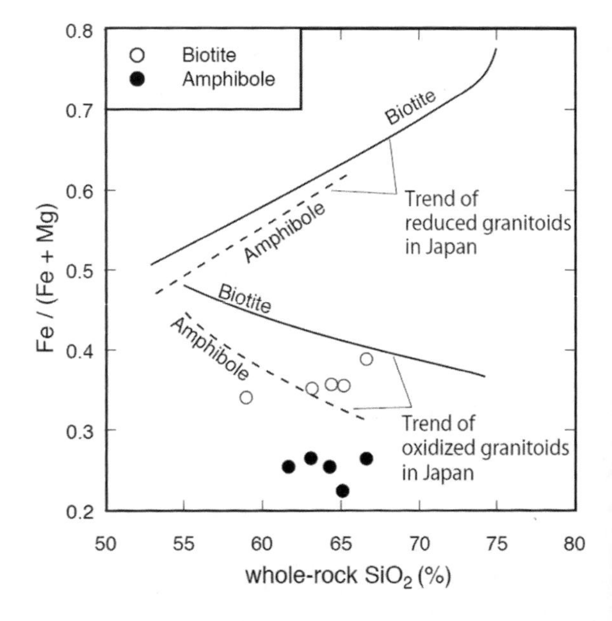

Fig. 5 Fe/Mg ratios of amphibole and biotite versus whole-rock SiO_2 content (wt%) of typical host granitoid series of Chilean Belt. Oxidizing and reducing trends of Japanese granitoids are shown for comparison after Czamanske et al. (1981)

measures still higher content of 3.3–4.2 wt% in the reduced series granitoids of the W–Sn mineralized Sanyo province.

Chlorine content, on the other hand, is higher up to 0.69 wt% in apatite of mafic rocks belonging to the oxidized series, but ranges from 0.33 to 0.02 wt% in oxidized-series granodiorite and granite. In the reduced series, chlorine content in apatites of mafic rocks is also higher than those of felsic rocks, but in the apatite of reduced-series granodiorite-granite has very low content of chlorine, generally below the detection limit of 0.03 wt% (Czamanske et al. 1981).

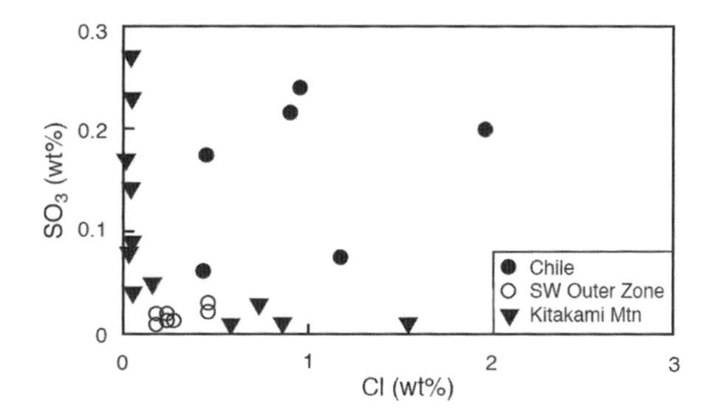

Fig. 6 SO_3 and Cl contents of apatite from typical host granitoid series in Japan and Chilean granitoids. Data of SW Outer Zone (reduced series) and Kitakami mountain (oxidized series) are from Imai (2004)

Chlorine, fluorine and sulphate (SO_3) contents of apatites from granitoids are listed in the Table 3. The fluorine contents of apatites in selected Chilean granitoids are as low as 1.97–3.14 wt%, but chlorine contents are much higher 0.43–1.94 wt% than those of the Japanese granitoids. Therefore, the Chilean granitods appear to have had an advantage to transport and concentrate base metals during the magmatic fractionations in the ore locales of the Chilean belt.

Large amount of sulfur must have been expelled out from the magma chambers to the site of the porphyry copper deposits. The sulfur contents of the magmas could be recorded in terms of SO_3 contents of the rock-forming apatites. The SO_3 contents of microphenocrystic apatites were extensively studied by Imai (2002, 2004) in silicic volcanic and subvolcanic rocks of the Western Pacific rim. Among plutonic rocks, apatites in Cretaceous oxidized-series granitoids at Kitakami Mountains, which are chalcopyrite-magnetite mineralized, show variable amounts of SO_3 and Cl, but apatites in reduced-series granitoids associated with tin mineralization in the Outer Zone of Southwest Japan are very low in both the elements (Fig. 6).

Apatites in the studied Chilean granitoids contain 0.06–0.24 wt% of SO_3 (Table 3) and roughly show a positive correlation with the Cl content (Fig. 6). The high SO_3 contents of apatite implies relatively abundant sulfate sulfur dissolved in the oxidized magmas, which might have been concentrated in the latest magmatic hydrothermal fluid phase together with ore-element copper.

In the reduced granitoids like those of the Outer Zone of Southwest Japan, sulfur contained therein is sometimes abundant, because the interacted sediments are enriched in sedimentary sulfur and carbon. This reduced sulfur crystallizes as pyrrhotite within the granitoids (Ishihara et al. 1999), and *neither* accumulates as oxidized sulfur in apatite *nor* concentrated in the post magmatic hydrothermal fluids. Therefore, no large copper deposits have been formed related to reduced-series granitoids.

5 Conclusions

(1) Oxygen fugacity of granitic rocks was generally very low in the earliest Precambrian time, which increased gradually with time. It was affected strongly by the availability of free oxygen and organic carbon in the source regions and on the earth surface including the hydrosphere. Porphyry copper metal was therefore progressively increased toward the present.
(2) The oxidized granitoids are characterized by high Mg/Fe ratio of hornblende and biotite of the host granitoids, in addition to the presence of magnetite (±hematite).
(3) Apatite in the granitoids serves as good indicator of ore metal carrier such as fluorine, chlorine and sulfate sulfur. Intrusive rocks of the porphyry copper mineralized regions have generally high contents of these elements.

Acknowledgments The authors wish to acknowledge Joseph B. Whalen and W. J. Davis for providing unpublished data. Santosh Kumar is thanked for generous comments on the earlier version and meticulous editorial work.

References

Camus F. (2003) Geologia de los sistemas porfiricos en los Andes de Chile. SERNAGEOMIN, Santiago, 267 p
Casselman MJ, McMillan WJ, Newman KM (1995) Highland Valley porphyry copper deposits near Kamloops, British Columbia: a review and update with emphasis on the valley deposit. In: Porphyry deposits of the Northwestern Cordillera of North America, CIM Special Volume (vol 46). The Canadian Institute of Mining, Metallurgy and Petroleum, pp 161–191
Czamanske GK, Ishihara S, Atkin SA (1981) Chemistry of rock-forming minerals of the Cretaceous-Paleogene batholiths in southwestern Japan and implications for magma genesis. J Geophy Res 86:10431–10469
Imai A (2002) Metallogenesis of porphyry Cu deposits of the Western Luzon Arc, Phillipines: K-Ar ages, SO_3 contents of microphenocrystic apatite and significance of intrusive rocks. Resour Geol 52:147–161
Imai A (2004) Variation of Cl and SO_3 contents of microphenocrystic apatite in intermediate to silicic igneous rocks of cenozoic Japanese Island arcs: implications for porphyry Cu metallogenesis in the Western Pacific arcs. Resour Geol 54:357–372
Ishihara S (1971a) Modal and chemical composition of the granitic rocks related to the major molybdenum and tungsten deposits in the inner zone of Southwest Japan. J Geol Soc Jpn 77:441–452
Ishihara S (1971b) Some chemical characteristics of the intrusive rocks of the bethlehem porphyry copper deposits, B.C Canada. Bull Geol Surv Jpn 22:535–546
Ishihara S (1977) The magnetite-series and ilmenite-series granitic rocks. Mining Geol 27:293–305
Ishihara S (1981) The granitoid series and mineralization. Econ Geol, 75th Anniversary volume, 458–484

Ishihara S, Chappell BW (2010) Petrochemistry of I-type magnetite-series granitoids of the northern Chile, highland valley, Southern B. C., Canada, Erdenet mine, Mongolia, Dexing mine, China, Medet mine, Bulgaria, and Ani mine. Jpn Bull Geol Surv Jpn 61:383–415

Ishihara S, Ulriksen CE, Sato K, Terashima S, Sato T, Endo Y (1984) Plutonic rocks of North–Central Chile. Geol Surv Jpn 35:503–536

Ishihara S, Yamamoto M, Sasaki A (1999) Sulfur and carbon contents and δ^{34}S ratio of Miocne ilmenite-series granitoids: Osumi and Shibisan plutons SW Japan. Bull Geol Surv Jpn 50:671–682

Ishihara S, Ohmoto H, Anhaeusser CR, Imai A, Robb LJ (2006) Discovery of the oldest oxic granitoids in the Kaapvaal Craton and its implications for the redox evolution of early Earth. In: Kesler SE, Ohmoto H (eds) Evolution of early Earth's atmosphere, hydrosphere, and biosphere—constraints from ore deposits, vol 198. Geological Society of America Memoir, pp 67–80

Kumar S, Rino V (2007) Redox series evaluation of Cu (±Mo ±Au) hosting Palaeoproterozoic Malanjkhand granitoids and enclaves, central India: evidence from magnetic susceptibility, phase petrology and geochemistry. J Econ Geol Geores Manage 4:105–127

Mariko T (2008) Geology of Ore Deposits. Kokin-Shoin, Tokyo, 580 p (in Japanese)

Sarkar SC, Kabiraj S, Bhattacharya S, Pal AB (1996) Nature, origin and evolution of the granitoids-hosted early Proterozoic copper-molybdenum mineralization at Malanjkhand, central India. Miner Deposita 31:419–431

Smithies RH, Champion DC (1999) Late Archean felsic alkaline igneous rocks in the Eastern Goldfields, Yilgarn Craton, Western Australia: a result of lower crustal delamination? J Geol Soc London 156:561–576

Smithies RH, Witt WK (1997) Distinct basement terranes iodentified from granite geochemistry in late Archean granite-greenstones, Yilgarun Craton, Western Australia. Precam Res 83:185–201

Stein HJ, Hannah JL, Zimmerman A, Markey RJ, Sarkar SC, Pal AB (2004) A 2.5 porphyry Cu-Mo-Au deposit at Malanjkhand, central India: implication for Late Archean continental assembly. Precam Res 134:189–226

Stein HJ, Hannah JL, Zimmerman A, Markey RJ (2006) Mineralization and deformation of the Malanjkhand terrain (2490–2440 Ma) along the Southern margin of Central Indian Tectonic Zone. Miner Deposita 40:755–765

Witt WK, Davy R (1997) Geology and geochemistry of Archean granites in the Kalgoorie region o the Eastern Goldfields, Western Australia: a syn-collisional tectonic setting? Precam Res 83:133–183

Mass Balance Modelling of Magmatic Processes in *GCDkit*

Vojtěch Janoušek and Jean-François Moyen

Abstract The freeware Geochemical Data Toolkit 3.0 (www.gcdkit.org) or, in short, *GCDkit*, offers a flexible environment for handling, recalculation and plotting of whole-rock geochemical data from igneous and metamorphic rocks. The current contribution demonstrates that the system can be easily expanded by the plugin modules, short yet powerful chunks of user-defined code, which can be easily and freely re-distributed. It describes their internal architecture, as well as the way how they may communicate with the core of the system and can be integrated into its Graphical User Interface (GUI). The plugins in *GCDkit* provide an appropriate platform for development of modules for numerical modelling of igneous processes. The presented simple plugins for direct and reverse modelling of the major-element mass balance in course of the fractional crystallization provide a sound, and potentially useful, proof of this concept.

1 Introduction

The Geochemical Data Toolkit (*GCDkit*; http://www.gcdkit.org) is an open-source software package for handling, recalculation and plotting of geochemical data from (meta-) igneous rocks (Janoušek et al. 2003, 2006). It is written in the

V. Janoušek (✉)
Czech Geological Survey, Klárov 3, 118 21 Prague 1, Czech Republic
e-mail: vojtech.janousek@geology.cz

V. Janoušek
Institute of Petrology and Structural Geology, Charles University in Prague,
Albertov 6, 128 43 Prague 2, Czech Republic

J.-F. Moyen
UMR 6524, Université Jean-Monnet, 23 rue du Docteur Michelon,
42023 Saint-Etienne, France
e-mail: jean.francois.moyen@univ-st-etienne.fr

S. Kumar and R. N. Singh (eds.), *Modelling of Magmatic and Allied Processes*,
Society of Earth Scientists Series, DOI: 10.1007/978-3-319-06471-0_11,

open-source R language (http://www.r-project.org), version for Windows, which represents a rich environment for data analysis, graphics and software development. The *GCDkit* not only offers a Graphical User Interface (GUI) to some of the powerful statistical and graphical functions built in R, but also provides a number of specialized tools designed specifically for igneous geochemists. Core routines for easy import, modification, searching, subsetting, classification, recalculation, plotting and output of the geochemical data are available. In particular, *GCDkit* comes with a wealth of built-in publication quality plots, which are mostly defined as templates in the internal format *Figaro*. Such diagrams can be easily edited, used as a basis for classification, or can form collections, termed *plates*.

In the new *GCDkit,* version 3.0 (Janoušek et al. 2011), most functions are not only accessible via GUI, but can run fully in interactive or even in a batch mode, without any potentially pestering dialogues. This means efficiency for repetitive recalculations/plotting tasks and a large degree of automation. As a long-term target, the system aims to facilitate routine and tedious operations involving large whole-rock geochemical datasets, including those coming from online databases.

2 Rationale

The current contribution examines one of the prominent features of the *GCDkit,* namely its modular structure, which allows expansion by means of the so-called plugin modules. The basic concepts needed for development of such a new plugin are explained, including the meaning of relevant internal variables of *GCDkit*, functions serving for building custom elements of the GUI and structure of a typical plugin module.

Furthermore, the chapter aims to address a major shortcoming of the system, i.e. the current lack of tools for forward and reverse numerical modelling of the main magmatic processes, such as fractional crystallization, partial melting, binary mixing or assimilation and fractional crystallization (AFC, DePaolo 1981). Principles of mass balance calculations are explained, and on this basis, simple plugins for direct and reverse modelling of fractional crystallization developed based on the major-element data.

As an example to test the code, we use the dataset for Sázava suite of the Variscan Central Bohemian Plutonic Complex (Janoušek et al. 2000). The compositions of the selected whole-rock samples (file `sazava.data`) and the typical constituting minerals (`sazava_mins.data`) are given in the subdirectory '`Test_data`' of the *GCDkit* distribution. This working directory can be set by a command: `setwd(paste(gcdx.dir, ''Test_data'', sep=''/''))`.

On the other hand, this chapter does not aim to be a tutorial to R programming, as the fundamental skills needed can be readily acquired with aid of several publications, or even the online PDF document "An Introduction to R" available from the R help system (Venables et al. 2012). For further details regarding the *GCDkit,* reader can consult its independent help system (menu *GCDkit|Help*).

3 Stepping Stones

3.1 User's Guide to Internal Variables of GCDkit

Besides the native format (`*.data` files, i.e. tab-delimited plain text tables with samples in rows), data can be loaded into *GCDkit* in several ways. Thanks to the *RODBC* package (Ripley and Lapsley 2013), it is possible to import from Microsoft Excel, Microsoft Access and DBF files, as well as the data formats used by the popular geochemical packages such as *NewPet* (Clarke et al. 1994), *IgPet* (Carr 1995), *MinPet* (Richard 1995) and *PetroGraph* (Petrelli et al. 2005). Moreover, data can be loaded, via text (`*.csv`) files, from web-based databases such as GEOROC (http://georoc.mpch-mainz.gwdg.de/georoc) and PETDB (http://www.petdb.org).

Regardless the method used, the data are always split between two objects. Numeric matrix **`WR`** stores all the chemical analyses. It has row names (sample names, found in the first column of the file), and column names which correspond to the variables (individual elements or oxides in arbitrary order). The rest of the imported information is saved into dataframe **`labels`**. Its rownames are, again, sample names (so `WR` and `labels` can be stitched together). The columns contain all the textual information, potentially useful for characterization, subsetting and grouping of the numeric data, as well as plotting symbol attributes "`Symbol`", "`Colour`" and "`Size`" (either as defined by user or default values automatically assigned by the system).

Both variables can be accessed and manipulated just like any other R object, for instance `WR[,''SiO2'']` will return a named vector containing the SiO_2 contents of all your samples. In fact, most of the *GCDkit* functionality involves manipulating `WR`.

The most recent values calculated (e.g., CIPW normative compositions, or normalized values used for the freshly plotted spiderplot) are stored in the variable **`results`**. It can be a vector, matrix or list depending on the character of the input data and function which has manipulated them. The results (i.e., the namesake variable) can be copied to clipboard (and then pasted into any Windows application), appended to the data currently in memory (i.e. to the data matrix `WR`) for later manipulation/plotting or saved into a variety of formats. The two most relevant functions, `r2clipboard()` and `addResults()`, respectively, can be accessed from a menu, which appears after right-clicking the R Console window (and are also included in the menu: *GCDkit|Copy results to clipboard* and *GCDkit|Append results to data*).

3.2 Graphical User Interface in MS Windows

The R supplies a series of simple and convenient functions allowing user interaction through GUI. As these functions are poorly documented outside the R-help system, we provide here a simple overview (Table 1). Unfortunately, most of them are MS

Table 1 An overview of GUI-related functions in R

Syntax	Description	Windows specific?
`winDialog(type, message)`	Opens Windows-like dialog box showing the `message`, returns the name of the button pressed (the available choices are defined by `type` as `''okcancel''`, `''ok''`, `''yesno''` or `''yesnocancel''`)	Yes
`winDialogString(message, default)`	Opens Windows-like dialog box showing the `message`, returns the text string entered by the user (with optional `default`)	Yes
`select.list(choices, multiple = FALSE, title = NULL)`	Selects a text item (or items if `multiple = TRUE`) from a scrollable modal dialog box window (with `title`) containing a list of several possibilities (character vector `choices`)	No
`winMenuAdd(menuname)`	Adds a new menu named `menuname`	Yes
`winMenuAddItem(menuname, itemname, action)`	Changes the action attached to a given item (`itemname`) of the menu (`menuname`); `action` is a character string describing what should happen when that menu item is invoked (most typically the name of the function to be called, or `''enable''` or `''disable''` toggling the menu item availability)	Yes
`winMenuNames()`	Lists names of all user-defined menus	Yes
`winMenuItems(menuname)`	Lists names of all items within a user-defined menu specified by the `menuname`	Yes
`winMenuDel(menuname)`	Deletes the menu specified by the `menuname`	Yes
`winMenuDelItem(menuname, itemname)`	Deletes the given item (`itemname`) from a menu (`menuname`)	Yes
`choose.files(caption = ''Select files'')`	Opens the file chooser dialog box with title `caption`, and returns a string with the full path to the file selected	No
`data.entry(object_name)`	A spreadsheet-style editor to edit tabular data (vector, matrix or data frame specified by the `object_name`)	No
`edit(object_name)`	Offers a default text editor for the same purpose	No

Windows-specific and thus any code using them is not portable. For instance, the function `winDialog` invokes a Windows-like dialogue box and returns the name of the button pressed, while `winDialogString` makes possible to type in a short text. The function `select.list` serves for selecting one (or more) from predefined list of multiple choices. The function `choose.files` opens the file chooser dialogue box. Another set of functions appends/removes menus or their items to/from the *R Console* window: `winMenuAdd`, `winMenuAddItem`, `winMenuDel` and `winMenuDelItem`. The functions `winMenuNames` and `winMenuItems` provide information on the current structure of the user-defined menus.

3.3 Architecture of a Plugin Module

The functionality of the *GCDkit* is easy to expand by means of the so-called plugins. This mechanism provides a simple method of adding effortlessly new items to *GCDkit* menus. Upon successful loading a datafile, the system will execute any R code (plain text files recognized by a suffix of `.r` or `.R`) found in the `...library\GCDkit\Plugin` directory of the R installation. By convention, functions specified by plugins are to be attached to the *Plugins* menu of the *GCDkit* system. Thus a minimal *GCDkit* plugin could look something like Code box 1.

Code box 1

```
# Minimal plugin skeleton
   hello<-function(){                        # define function
    cat("Hello world!\n")
   }
  winMenuAdd("Plugins/Test plugin")         # create a menu entry
  winMenuAddItem("Plugins/Test plugin","Say hello", "hello()")
```

After invoking the relevant menu item, *Plugins|Test plugin*, we get:

```
hello( )
Hello world!
```

4 Mass Balance Calculations: A Key to Major-Element Modelling of Igneous Processes

4.1 Mixing is Everywhere!

Central to numerical modelling of whole-rock geochemical data, and major-element analyses in particular, are mass balance calculations. These utilize a set of mixing equations, of which the easiest, for binary mixing, is:

$$C_M = fC_A + (1-f)C_B \tag{1}$$

where C_M is the concentration of the given element in the mixture, C_A, C_B those in end-members A and B, and f the proportion of A.

There is, therefore, one such equation for each element considered. If more components are involved (multicomponent mixing), and several elements are considered, Eq. (1) can be written for each element, and expanded to accommodate several components, resulting in the following system:

$$\begin{aligned} C_M^{SiO_2} &= \left[f_1 C_1^{SiO_2} + f_2 C_2^{SiO_2} + \cdots + f_n C_n^{SiO_2}\right] \\ C_M^{TiO_2} &= \left[f_1 C_1^{TiO_2} + f_2 C_2^{TiO_2} + \cdots + f_n C_n^{TiO_2}\right] \\ &\cdots \\ C_M^{P_2O_5} &= \left[f_1 C_1^{P_2O_5} + f_2 C_2^{P_2O_5} + \cdots + f_n C_n^{P_2O_5}\right] \end{aligned} \tag{2}$$

where f_i denotes the proportion of component i, and C_i^{ox} the concentration of oxide ox in end member i.

This system can be recast into a matrix form, using a matrix multiplication:

$$\vec{C}_M = \vec{f}\,\overline{\overline{C}} \tag{3}$$

where $\vec{C}_M$ is the vector of concentrations of individual elements in the mixture, $\vec{f}$ the vector of proportions of individual components (summing up to 1), and $\overline{\overline{C}}$ a matrix of concentrations of the elements of interest in the individual components (minerals in rows, elements/oxides in columns).

Code box 2

```
mins<-read.table("sazava_mins.data",sep="\t",fill=TRUE)    # import the data
mins[is.na(mins)]<-0         # replace all missing values by zeros
mins<-as.matrix(mins)        # convert to matrix (required for matrix multiplication)
f<-c(0.1,0.5,0.1,0.3)        # vector with mineral proportions
print(f%*%mins)              # print result of the matrix multiplication
```

The approach using matrix multiplication (in R implemented via the `%*%` operator) can be used to calculate the bulk chemistry of a rock, from wt.% proportions of its mineral constituents and their chemical compositions. For instance, the script whose listing is given in Code box 2 calculates the chemistry of a rock consisting of 0.1 Qtz, 0.5 Plg, 0.1 Bi and 0.3 Hb whose respective compositions are given in the file `sazava_mins.data` (note that the proportions must be supplied in exactly the same order as the mineral compositions appear in the mineral composition file):

```
       SiO2   TiO2   Al2O3   FeOt    MgO    CaO   Na2O    K2O
[1,] 53.842 0.628 19.112 7.972 3.851 9.212 2.859 1.347
```

Precisely the same task is performed by the *GCDkit* function `WRComp(mins, f)`.

As the name suggests, mixing equations are directly applicable to magma mixing problems. However, the approach can be generalized to many other processes as well, such as fractional crystallization, assimilation or partial melting, which we shall examine in some detail.

4.2 Forward Modelling of Fractional Crystallization

In the fractional crystallization scenario, it is possible to view the parental magma as a mixture of the (future) differentiated melt (l) and crystallized minerals (s). Thus the mass balance Eq. (1) can be expressed, for an element k, as:

$$C_0^k = fC_l^k + (1-f)C_S^k \tag{4}$$

Since the cumulate composition itself is a mixture of the minerals, each with a proportion m_1, m_2,…, m_n, respectively:

$$C_s^k = \sum_{i=1}^{n}(m_i C_i^k) \tag{5}$$

$$C_0^k = fC_l^k + (1-f)\sum_{i=1}^{n}(m_i C_i^k) \tag{6}$$

it is easy to express the composition of the differentiated liquid as:

$$C_l^k = \frac{\left(C_0^k - (1-f)\sum_{i=1}^{n}(m_i C_i^k)\right)}{f} \tag{7}$$

or, in a vector form:

$$\vec{C}_l = \frac{\left(\vec{C}_0 - (1-f) \times \vec{m}.\overline{\overline{C}}\right)}{f} \tag{8}$$

where, similar to Eq. (3), $\overline{\overline{C}}$ is a matrix with the composition of all minerals in the cumulate and $\vec{m}$ a vector with their modal proportions.

Let us design a simple plugin, which will calculate the effects of fractional crystallization, given the chemistry of the parental melt, compositions of all the mineral phases and their proportions in the crystallizing cumulate.

Code box 3

```
FC_direct<-function(){
      # select sample representing the parental melt
      pm.name<-select.list(rownames(WR),title="Parental melt")

      # composition of fractionating minerals
      mineral_file<-choose.files(caption="Select the mineral file")
      mins<-read.table(mineral_file,fill=TRUE)
      mins<-as.matrix(mins)              # conversion is needed to a matrix
      mins[is.na(mins)]<-0               # replace missing values in mins by zeros

      # parental magma composition
      pm<-WR[pm.name,colnames(mins)]

      # mineral proportions in the cumulate
      m<-winDialogString("Enter the proportions of crystallizing minerals",
          "0,0.35,0.05,0.6")
      m<-unlist(strsplit(m,","))         # split a single comma separated text string
                                         # into individual items of a vector
      m<-as.numeric(m)                   # and make it a numeric vector

      # fraction of the melt remaining
      f<-winDialogString("Enter the fraction of the remaining melt","0.4")
      f<-as.numeric(f)                   # input is a character, convert to number

      # Calculate!
      cs<-m%*%mins                       # composition of the cumulate (eq. 5)
      fm<-(pm-(1-f)*cs)/f                # and of fractionated melt (eq. 8)

      # Print the results
      z<-rbind(pm,mins,cs,fm)            # bind the results as rows of an new matrix
      rownames(z)<-c("Parental",paste(m,rownames(mins)),"Cumulate","Fractionated")
      cat("System after",(1-f)*100,"% fractional crystallization:\n")
      print(round(z,2))
      return(z)
  }
  winMenuAdd("Plugins/Modelling")
  winMenuAddItem("Plugins/Modelling","Forward FC","results<-FC_direct()")
```

The listing of the plugin is given in Code box 3. The plugin includes a rudimentary GUI, prompting the user for the composition of the parental melt (one of the currently loaded samples), as well as the mineral file and the phase proportions. Running it with the default parameters of 60 % fractionation (fraction of melt remaining is 0.4) of a cumulate consisting of 0.35 Plg, 0.05 Bi and 0.6 Hb from parental magma Sa-4 of the Sázava dataset gives the following output:

```
System after 60 % fractional crystallization:
                  SiO2 TiO2 Al2O3  FeOt  MgO   CaO Na2O  K2O
Parental         50.72 0.83 17.57  9.62 5.18  9.92 2.83 1.60
0 Qtz           100.00 0.00  0.00  0.00 0.00  0.00 0.00 0.00
0.35 Plg         53.41 0.00 29.48  0.09 0.00 11.27 5.05 0.12
0.05 Bi          35.32 2.11 15.31 23.56 9.05  0.01 0.10 9.81
0.6 Hb           45.35 1.39  9.47 18.57 9.82 11.92 1.08 1.02
Cumulate         47.67 0.94 16.77 12.35 6.34 11.10 2.42 1.14
Fractionated     55.30 0.67 18.78  5.52 3.43  8.15 3.44 2.28
```

Please note that the default values (the ones that appear initially in the dialog boxes) are hard-wired in the code in order to correspond to the parameters in this exercise. For ordinary use, these can be omitted or replaced by some more sensible defaults. The oxides used for modelling are exactly those for which there are data available in the mineral data file; it is the user's responsibility to ensure that `WR` and the mineral file use the same oxide list (beware of missing oxides such as P_2O_5, or of alternate names such as `FeOt` versus `FeOT` or `FeO*`; note also that R is case sensitive). Likewise the user must ensure that the mineral proportions are supplied in the same order in which the minerals appear in the composition file: this plugin is rather minimal and could be improved to make it more robust and/or user-friendly.

4.3 Reverse Modelling of Fractional Crystallization

When determining the degree of fractional crystallization and proportions of the individual minerals in the cumulate, from the known (assumed) compositions of the parental and fractionated melt, as well as the chemistries of the fractionating mineral phases, the problem is effectively one of (reverse) mixing. In other words, we try to re-combine cumulus phases with the fractionated melt in order to best reproduce the known composition of the mixture (the parental melt).

Deriving a differentiated melt C_L from a parent C_0, with a cumulate C_S made of n minerals is equivalent to mixing $(n + 1)$ components (differentiated melt and n minerals) to form the parental liquid. In this case, if f is the proportion of melt (decreasing from 1 to 0 in course of the crystallization), and m the mineral proportions in the cumulate, the proportions in the mixture are f for the melt and $(1-f) \times m_i$ for each mineral i. The balance for all elements modelled can be expressed by the following system of equations:

$$\begin{aligned} C_0^{SiO_2} &= fC_L^{SiO_2} + (1-f)\left[m_1 C_1^{SiO_2} + m_2 C_2^{SiO_2} + \cdots + m_n C_n^{SiO_2}\right] \\ C_0^{TiO_2} &= fC_L^{TiO_2} + (1-f)\left[m_1 C_1^{TiO_2} + m_2 C_2^{TiO_2} + \cdots + m_n C_n^{TiO_2}\right] \\ &\cdots \\ C_0^{P_2O_5} &= fC_L^{P_2O_5} + (1-f)\left[m_1 C_1^{P_2O_5} + m_2 C_2^{P_2O_5} + \cdots + m_n C_n^{P_2O_5}\right] \end{aligned} \qquad (9)$$

which can be written in a matrix form (Albarède 1995) as:

$$\overrightarrow{C_0} = \overline{\overline{C_1}}\vec{X} \tag{10}$$

where:

$$\overrightarrow{C_0} = \begin{pmatrix} C_0^{SiO_2} \\ C_0^{TiO_2} \\ \vdots \\ C_0^{P_2O_5} \end{pmatrix} \tag{11}$$

$$\overline{\overline{C_1}} = \begin{pmatrix} C_L^{SiO_2} & C_1^{SiO_2} & C_2^{SiO_2} & \dots & C_n^{SiO_2} \\ C_L^{TiO_2} & C_1^{TiO_2} & C_2^{TiO_2} & \dots & C_n^{TiO_2} \\ \vdots & \vdots & \vdots & \ddots & \vdots \\ C_L^{P_2O_5} & C_1^{P_2O_5} & C_2^{P_2O_5} & \dots & C_n^{P_2O_5} \end{pmatrix} \tag{12}$$

$$\vec{X} = \begin{pmatrix} f \\ (1-f)m_1 \\ (1-f)m_2 \\ \vdots \\ (1-f)m_n \end{pmatrix} \tag{13}$$

Using the Eq. 10 as above, we can extract from the data the information required to define $\overrightarrow{C_0}$ (the composition of the rock we consider as the primitive liquid that did fractionate), and the matrix $\overline{\overline{C_1}}$ (Eq. 12) that stores the evolved melt chemistry (C_L) in the first column, followed by the compositions of the fractionating minerals. We are now trying to solve the Eq. (10) for vector $\vec{X}$ which should contain the degree of fractional crystallization and relative proportions of minerals in the cumulate as per Eq. 13.

If the system has more independent equations than variables (i.e. more elements than minerals), it is overdetermined. This would be the case for most of petrogenetic calculations, as fractionation is typically a multi-variant reaction having more components than phases (in phase relations terminology); therefore there is no exact solution. The 'best' can be obtained by several approaches, the most common being the least-squares or linear programming methods (Bryan et al. 1969; Wright and Doherty 1970; Stormer and Nicholls 1978; Banks 1979; Albarède 1995).

In R, the least-squares method is implemented by the function `lsfit(A, y, intercept = FALSE)`, where *A* is a matrix whose rows correspond to cases and columns to variables (our matrix $\overline{\overline{C_1}}$), *y* is the vector with the "responses" (in our case the expected $\overrightarrow{C_0}$ composition), and intercept is set to `FALSE` as the solution should pass through the origin. The outcome is a variable of the mode list (i.e., essentially a container for loosely connected several other variables, or components), of which the most interesting is the component `$coefficients`, corresponding the vector $\vec{X}$ as defined above, and a component `$residuals` with deviations between the calculated and observed magma compositions for each element. The sum of squares of residuals (R^2) then provides a useful measure for the goodness of fit, while residuals for individual elements give information on the most troublesome among them. As a word of caution, it is important to remember that the numerical "best" solution may, or may not be geologically relevant; the user must think in terms of petrology and decide for himself whether the mathematical result has a geological meaning. Also, the more phases are involved in the calculation, the better the fit tends to be.

In the following example, using the plugin defined in Code box 4, we shall assume Sa-3 as parental melt, Sa-1 as the fractionated melt and crystallizing minerals compositions given in the file `sazava_mins.data`.

Code box 4

```
FC_reverse<-function(){
     # select sample representing the parental melt
     pm.name<-select.list(rownames(WR),title="Parental melt")

     # select sample representing the fractionated melt
     fm.name<-select.list(rownames(WR),title="Fractionated melt")

     # composition of fractionating minerals
     mineral_file<-choose.files(caption="Select the mineral file")
     mins<-read.table(mineral_file,fill=TRUE)
     mins<-as.matrix(mins)
     mins[is.na(mins)]<-0

     # parental melt composition
     C0<-WR[pm.name,colnames(mins)]

     # fractionated melt composition
     CL<-WR[fm.name,colnames(mins)]

     #  Calculate!
     C1<-t(rbind(CL,mins))          # eq. 12, matrix needs to be transposed first
     model<-lsfit(C1,C0,intercept=FALSE)
     r.squared<-sum(model$res^2)
     f<-model$coeff[1]              # see eq.13, first item is fraction of melt
                                    # left, i.e. 1-degree of fract. crystallization
     cum.prop<-model$coeff[-1]/(1-f) # remaining items are proportions of minerals
                                    # in the cumulate

     # Print
     cat(round(100*(1-f),3),"% fc ","\n")
     print(round(cum.prop,3))
     cat("Quality of the model (squared residuals):",round(r.squared,2),"\n")
     return(results)
  }
 winMenuAdd("Plugins/Modelling")
 winMenuAddItem("Plugins/Modelling","Reverse FC"," results<-FC_reverse()")
```

The code yields the following result:

```
40.962 % fc
  Qtz    Plg     Bi     Hb
0.021 0.491 0.098 0.360
Quality of the model (squared residuals): 0.08
```

Again, the user must ensure that the oxides have the same names in the mineral file and in `WR`. We can test the code by reversing the direct modelling exercise (Code box 3) by selecting the sample Sa-4 as the primitive melt and replacing the fractionated melt composition by the direct modelling outcome. This means that the line:

```
CL <- WR[fm.name, colnames(mins)]
```

in Code box 4 is to be replaced by:

```
CL <- results["Fractionated",]
```

giving:

```
60 % fc
 Qtz  Plg   Bi   Hb
0.00 0.35 0.05 0.60
Quality of the model (squared residuals): 0
```

In this rather naïve example, the R^2 is equal to 0, which indicates a "perfect" model. This is hardly surprising since the assumed composition of the fractionated liquid has indeed been calculated previously.

4.4 Generalized Norm

A similar approach can be employed also for calculation of a generalized norm, given the composition of the bulk rock and all its mineral constituents (all in wt.%).

Code box 5

```
normative<-function(){
   # composition of minerals
   mineral_file<-choose.files(caption="Select the mineral file")
   mins<-read.table(mineral_file,fill=TRUE)
   mins<-as.matrix(mins)
   mins[is.na(mins)]<-0

   # select sample representing the rock
   rock.name<-select.list(rownames(WR),title="Rock")
   rock<-WR[rock.name,colnames(mins)]

   # Calculate!
   results<-lsfit(t(mins),rock,intercept=F)$coeff
   print(round(results*100,2))
  }
 winMenuAddItem("Plugins/Modelling","Normative comp."," results<-normative()")
```

The Code box 5 produces the following output for sample Sa-4 and mineral compositions given in the file `sazava_mins.data`:

```
 Qtz   Plg    Bi    Hb
6.12 42.05  7.90 42.67
```

Please note that the individual percentages do not sum up to 100 %, and thus the more sophisticated constrained regression algorithm (assuming a certain sum of the vector with the solution) would be superior (Albarède 1995). Unconstrained "modal" contents of minerals, both raw and recast to 100 %, are calculated by the *GCDkit* function `Mode`; constrained solution is available via the function `ModeC`. Results of both calculations are returned by the function `ModeMain(WR)`, which is attached to the menu *Calculations|Norms...|Mode*. Further details can be found on the relevant help page.

5 Conclusions

GCDkit 3.0 offers a rich and versatile environment for interpretation of whole-rock geochemical data, with numerical modelling of igneous processes, both direct and reverse, being no exception. In particular, the plugins mechanism enables writing short yet powerful chunks of code, which are independent of the *GCDkit* core, can be easily and freely re-distributed but at the same time integrated seamlessly into the current GUI. The presented simple plugins for direct and reverse modelling of the major-element mass balance in course of the fractional crystallization provide a sound, and potentially useful, proof of the concept.

Acknowledgments The authors are indebted to a number of people without whose work or feedback the *GCDkit* system would not be working, in particular Colin M. Farrow (Glasgow), Vojtěch Erban (Prague) and Jakub Trubač (Prague) as well as the whole R development team which have designed such a powerful and flexible environment. Moreover, we are thankful to Vojtěch Erban, who has provided a stimulating and insightful review of this paper. Lastly, we are grateful to a number of users, who provided invaluable feedback and motivation to our work. In particular we are thankful to organizers of *GCDkit/R* workshops at National Geophysical Research Institute (NGRI) in Hyderabad, India (Vysetti Balaram 2013), University of Stellenbosch, South Africa (Gary Stevens 2012) and University of Helsinki, Finland (Tapani Ramö 2011). Special thanks go to Santosh Kumar (Kumaun University, Nainital) for being the mastermind behind the Indian workshop, for having invited us to this volume and for companionship during our trip to India. This contribution has been financed by the French-Czech Program Mobility 7AMB13FR026 and Czech Science Foundation (GAČR) project P210/11/2358.

References

Albarède F (1995) Introduction to geochemical modeling. Cambridge University Press, Cambridge, pp 1–543

Banks R (1979) The use of linear programming in the analysis of petrological mixing problems. Contrib Mineral Petrol 70:237–244

Bryan WB, Finger LW, Chayes F (1969) Estimating proportions in petrographic mixing equations by least-squares approximation. Science 163:926–927

Carr M (1995) IgPet for Windows. Terra Softa, Somerset

Clarke D, Mengel F, Coish RA, Kosinowski MHF (1994) NewPet for DOS, version 94.01.07. Department of Earth Sciences, Memorial University of Newfoundland, Canada

DePaolo DJ (1981) Trace element and isotopic effects of combined wallrock assimilation and fractional crystallization. Earth Planet Sci Lett 53:189–202

Janoušek V, Bowes DR, Rogers G, Farrow CM, Jelínek E (2000) Modelling diverse processes in the petrogenesis of a composite batholith: the Central Bohemian Pluton, Central European Hercynides. J Petrol 41:511–543

Janoušek V, Farrow CM, Erban V (2003) GCDkit: new PC software for interpretation of whole-rock geochemical data from igneous rocks. Geochim Cosmochim Acta 67:186

Janoušek V, Farrow CM, Erban V (2006) Interpretation of whole-rock geochemical data in igneous geochemistry: introducing Geochemical Data Toolkit (GCDkit). J Petrol 47:1255–1259

Janoušek V, Farrow CM, Erban V, Trubač J (2011) Brand new Geochemical Data Toolkit (GCDkit 3.0): is it worth upgrading and browsing documentation? (Yes!). Geol výzk Mor Slez 18:26–30

Petrelli M, Poli G, Perugini D, Peccerillo A (2005) PetroGraph: a new software to visualize, model, and present geochemical data in igneous petrology. Geochem Geophys Geosyst 6. doi:10.1029/2005GC000932

Richard LR (1995) MinPet: mineralogical and petrological data processing system, version 2.02. MinPet Geological Software, Québec

Ripley B, Lapsley M (2013) RODBC, an ODBC database interface, version 1.3-2. http://cran.r-project.org/. Accessed 25 Feb 2013

Stormer JC, Nicholls J (1978) XLFRAC: a program for the interactive testing of magmatic differentiation models. Comput Geosci 4:143–159

Venables WN, Smith DM, R Development Core Team (2012) An Introduction to R. Notes on R: a programming environment for data analysis and graphics. Version 2.15.2 (2012–10–26). http://cran.r-project.org/doc/manuals/R-intro.pdf. Accessed 9 Dec 2012

Wright TL, Doherty PC (1970) A linear programming and least squares computer method for solving petrologic mixing problems. Geol Soc Am Bull 81:1995–2008

About the Editors

Dr. Santosh Kumar is Professor and Head, Department of Geology, Centre of Advanced Study in Geology, Kumaun University, Nainital, India. Earlier he served as Reader in Geology, Nagaland Central University, Kohima, Nagaland, India. He obtained his Ph.D. degree in Geology from Banaras Hindu University, Varanasi, India and also in Geochemistry from Comenius University, Bratislava. Prof. Kumar is extensively researching on enclaves and granitoids of Himalayan and Cratonic Belts of India in order to establish the magma chamber processes and crustal growth. He has over 27 years of research experience in the field of mineralogy, petrology and geochemistry of igneous rocks; guided six Ph.D. students and published 41 research papers in reputed international and national journals apart from editing a book on Magmatism, Tectonism and Mineralization. He has visited several institutions in Czech Republic, Slovak Republic, Switzerland, former USSR, Japan, USA, China, Korea and Spain to pursue collaborative research activities. Prof. Kumar is Fellow of the Geological Society of India, Mineralogical Society of India, Indian Society for Applied Geochemists, The Society of Earth Scientists and International Association of Gondwana Research. He is also a Member of Expert Panel, Program Advisory Committee for Young Scientists in the field of Earth and Atmospheric Sciences, Science and Engineering Research Board, New Delhi. He has been awarded prestigious Uttarakhand Ratna-2012 for his significant contribution in the field of research and extension education.

S. Kumar and R. N. Singh (eds.), *Modelling of Magmatic and Allied Processes*,
Society of Earth Scientists Series, DOI: 10.1007/978-3-319-06471-0,

Dr. Rishi Narain Singh, FNA is currently an INSA Senior Scientist with CSIR-National Geophysical Research Institute, Hyderabad, India. He obtained his Ph.D. degree in Geophysics from Banaras Hindu University, Varanasi, India in 1969. Dr. Singh has been mostly with CSIR-NGRI since 1964. During 1977–1978 he was an Associate Professor at the Indian Institute of Geomagnetism. During 1996–1999, he was the Scientist-in-charge of CSIR-Centre for Mathematical Modelling and Computer Simulation (CMMACS), Bangalore. He was Project Director for DOD's program on Indian Ocean Modelling and Dynamics (INDOMOD). During 1999–2003, Dr. Singh worked as the Director, National Environmental Engineering Research Institute, Nagpur. He was awarded with the prestigious Shanti Swaroop Bhatnagar Prize in 1985 and elected to the Fellowship of Indian Academy of Sciences, Indian National Science Academy, Indian Geophysical Union, Association of Exploration Geophysicists and Maharashtra Academy of Sciences. He has been leader of the Indian delegation to IUGG in 1999. The Indian Geophysical Union awarded him with the Decennial Award in 2004.

Dr. Singh has carried out extensive research in mathematical modelling of various geophysical and environmental processes. He has modelled thermal and mechanical processes in the Indian continental and oceanic lithosphere by constraining solutions of thermomechanical equations, including moving boundary problems, with geophysical observations. He constructed models for origins of geological features such as Cuddapah basin, south Indian high grade metamorphic terrains, Afanasi seamount, Carlsberg Ridge, and fracture zones in the Indian Ocean. Dr. Singh has over one hundred and fifty publications; coordinated over one hundred environmental impact and risk assessment reports. He has co-authored the book "Municipal Water and Wastewater Treatment". His current research interest is in the field of near surface geophysical and environmental modelling incorporating physical, chemical and biological aspects.

Printed by Printforce, the Netherlands